Natural Resource Management

Volume 47

Series Editors

David Zilberman, College of Natural Resources, University of California, Berkeley, CA, USA

Renan Goetz, Department of Economics, University of Girona, Girona, Spain

Alberto Garrido, ETS, Technical University of Madrid, Madrid, Spain

There is a growing awareness to the role that natural resources, such as water, land, forests and environmental amenities, play in our lives. There are many competing uses for natural resources, and society is challenged to manage them for improving social well-being. Furthermore, there may be dire consequences to natural resources mismanagement. Renewable resources, such as water, land and the environment are linked, and decisions made with regard to one may affect the others. Policy and management of natural resources now require interdisciplinary approaches including natural and social sciences to correctly address our society preferences. This series provides a collection of works containing most recent findings on economics, management and policy of renewable biological resources, such as water, land, crop protection, sustainable agriculture, technology, and environmental health. It incorporates modern thinking and techniques of economics and management. Books in this series will incorporate knowledge and models of natural phenomena with economics and managerial decision frameworks to assess alternative options for managing natural resources and environment.

More information about this series at http://link.springer.com/series/6360

Sara Valaguzza · Mark Alan Hughes
Editors

Interdisciplinary Approaches to Climate Change for Sustainable Growth

Editors
Sara Valaguzza
University of Milan
Milan, Italy

Mark Alan Hughes
Kleinman Center for Energy Policy
University of Pennsylvania
PA, USA

ISSN 0929-127X   ISSN 2511-8560 (electronic)
Natural Resource Management and Policy
ISBN 978-3-030-87566-4   ISBN 978-3-030-87564-0 (eBook)
https://doi.org/10.1007/978-3-030-87564-0

© The Editor(s) (if applicable) and The Author(s), under exclusive license to Springer Nature Switzerland AG 2022
This work is subject to copyright. All rights are solely and exclusively licensed by the Publisher, whether the whole or part of the material is concerned, specifically the rights of translation, reprinting, reuse of illustrations, recitation, broadcasting, reproduction on microfilms or in any other physical way, and transmission or information storage and retrieval, electronic adaptation, computer software, or by similar or dissimilar methodology now known or hereafter developed.
The use of general descriptive names, registered names, trademarks, service marks, etc. in this publication does not imply, even in the absence of a specific statement, that such names are exempt from the relevant protective laws and regulations and therefore free for general use.
The publisher, the authors and the editors are safe to assume that the advice and information in this book are believed to be true and accurate at the date of publication. Neither the publisher nor the authors or the editors give a warranty, expressed or implied, with respect to the material contained herein or for any errors or omissions that may have been made. The publisher remains neutral with regard to jurisdictional claims in published maps and institutional affiliations.

This Springer imprint is published by the registered company Springer Nature Switzerland AG
The registered company address is: Gewerbestrasse 11, 6330 Cham, Switzerland

Contents

1	**Introduction**.. Mark Alan Hughes and Sara Valaguzza	1
Part I	**Interpreting, Communicating and Composing a Complex Phenomenon**	
2	**The Ideological Trick of Climate Change and Sustainability**....... Gianluca Schinaia	11
3	**Words Count: The Role of Language in Overcoming Climate Inertia**.. Eleonora Ciscato and Marianna Usuelli	27
4	**Measuring Complex Socio-economic Phenomena. Conceptual and Methodological Issues**................................... Filomena Maggino and Leonardo Salvatore Alaimo	43
Part II	**The Scientific Debate**	
5	**Glaciers: Vanishing Elements of Our Mountains and Precious Witnesses of Climate Change**.............................. Guglielmina Diolaiuti, Maurizio Maugeri, Antonella Senese, Veronica Manara, Giacomo Traversa, and Davide Fugazza	63
6	**Rural Revival and Coastal Areas: Risks and Opportunities**........ Felice D'Alessandro	93
7	**Climate Change: From Science to Policies, Backward and Forward**.. Sara Valaguzza	107

Part III Ethics and Policies

8 The EU Perspective from Setbacks to Success: Tackling Climate Change from Copenhagen to the Green Deal and the Next-Generation EU 127
Agostino Inguscio

9 Carbon Pricing from the Origin to the European Green Deal 141
Isabella Alloisio and Marzio Galeotti

10 Technology Innovation in the Energy Sector and Climate Change: The Role of Governments and Policies 159
Francesco Ciaccia

11 How Emerging Technologies Are Finally Matching the Policy Leverage of Cities with Their Political Ambitions 181
Mark Alan Hughes, Angela Pachon, Oscar Serpell, and Cornelia Colijn

12 Sustainable Transportation 199
Ethan Elkind

13 From Green to Social Procurement 217
Laura Carpineti

14 Climate, Sustainability, and Waste: EU and US Regulatory Approaches Compared 245
Madeline June Kass

15 Construction Industry and Sustainability 261
Adriana Spassova

16 Impact of Climate Change in Agriculture: Estimation, Adaptation, and Mitigation Issues 289
Alessandro Olper and Daniele Curzi

Part IV People's Behaviors to Address Climate Change: Proactive and Conflicting Actions

17 Climate Change and Consumer Behavior 315
Elisa De Marchi, Alessia Cavaliere, and Alessandro Banterle

18 Climate Change Litigation: Losing the Political Dimension of Sustainable Development 333
Sara Valaguzza

19 The Judicial Review of Administrative Decisions with Environmental Consequences 349
Eduardo Parisi

20 Mediation in Environmental Disputes 365
Roberta Regazzoni

Index 375

Chapter 1
Introduction

Mark Alan Hughes and Sara Valaguzza

Abstract The chapter opens the discussion on the multi-faceted complexity of climate change and presents the structure of the volume, describing the interconnections between all the aspects discussed in the single chapters and guiding the readers towards the envisaged path. It underlines the elements that characterize the adopted approach, which critically observes the interactions among interdependent systems in a multi- and inter-disciplinary fashion.

Keywords Climate change · Complexity · Interdisciplinary perspectives · Adaptation · Mitigation · Book outline

Scholarship on climate change policies must confront both complexity, in which interacting parts generate emergent outcomes that are more than simply the sum of those parts, and uncertainty, in which the capacity for robust adaptation to many possible future conditions defines many policies. Together these challenges require scholarship that is multi- and inter-disciplinary in order to observe and understand the critical interactions among interdependent systems, both physical and social. This volume is an attempt to gather and integrate multiple scholarly perspectives on some of these interdependent systems that are producing policy responses to complex and uncertain climate change.

Interdisciplinary perspectives that attempt to see beyond well-defended disciplinary boundaries risk seeing only the simplest contours of the terrain. And

---

M. A. Hughes
Kleinman Center for Energy Policy, University of Pennsylvania, Philadelphia, PA, USA
e-mail: mahughes@upenn.edu

S. Valaguzza (✉)
University of Milan, Milan, Italy
e-mail: sara.valaguzza@unimi.it

© The Author(s), under exclusive license to Springer Nature Switzerland AG 2022
S. Valaguzza, M. A. Hughes (eds.), *Interdisciplinary Approaches to Climate Change for Sustainable Growth*, Natural Resource Management and Policy 47, https://doi.org/10.1007/978-3-030-87564-0_1

interdisciplinary dialogues that attempt to communicate between highly evolved disciplinary languages risk conveying only the simplest ideas of a literature. We accept those risks and indeed expect a high degree of simplification in this volume. Our gambit is that a higher-level system view from several disciplines is worth a degree of simplification. Indeed, simplification might be an antidote to the superficiality of much policy discourse on climate change.

Our intention in this volume is the reader will allow the complexity to emerge in successive chapters, which are structured and sequenced to make connections and provide a mutually reinforcing set of observations and tentative conclusions.

The volume is divided into four Parts. The first Part introduces the main concepts of the book: climate change and sustainability, wellbeing, and mitigation and adaptation. Each theme will be critically discussed and examined by the main experts of each sector who contributed to the drafting of the volume.

Part II presents the scientific understanding of climate change and explores some of the more pressing issues driving policy development, such as the melting of the glaciers and the impact on coastal areas.

Part III discusses significant experiences in the environmental policies both in the European Union and in the United States of America. The choice of the topics includes diversified but complementary aspects. They have been selected to permit a comprehension of both the general policies, such as the European Green Deal and carbon pricing, and the sectorial policies on energy, waste management, green and social procurement, transportation, energy, agriculture and construction.

The fourth Part closes the circle by explaining possible approaches about climate change, by exploring the legal and economic aspects of both adversarial and more lenient approaches towards a more sustainable world. It faces four main issues in the economic and juridical context: consumer behaviors, climate litigations, environmental litigations and the alternative forms of dispute resolution on environmental matters, with particular regard to environmental mediation.

The first Part of the book paves the way for the technical discussion, which is carried out in Parts II, III and IV. In the second chapter of Part I, entitled "The ideological trick of climate change and sustainability", the issues of climate change and sustainability are analyzed under different angles, by underlining the links with the economic, social and political factors to trace their causes and implications. The initiatives of the governments around the world, the "go green" wave, the impacts of the renewed attention of the entrepreneurial world towards the environment on people's behaviors are examined with particular attention to the recent events and democratic debates, from climate skepticism to the right-wing populism. The conclusions reached demonstrate that "ideological bias play a significant role in polarizing public debates on climate change policies" and point to a particularly problematic aspect, namely the misinformation on climate change issues. It is thus stated the necessity to launch a network system including experts of communication, sociologists and scientists to assess, in a preliminary phase, the truthfulness of the information related to climate change and sustainable topics.

The relevance of the mass media in the debate on climate and sustainable policies emerges in Chap. 3, entitled "Words count: the role of language in overcoming climate inertia." This chapter continues the reasoning began in the first chapter, by deepening the issue of the approach from the media and the communication systems to the environmental issues since the Nineties—when the debate started to have a significant relevance. The chapter presents research carried out on the main Italian newspapers and illustrates how the presence of news on climate change and sustainability has grown. The study demonstrates that climate change is described by a constant reference to metaphors that evoke situations such as war, diseases or travels. The conclusions push forward the necessity of a reframe which could put the debate on climate change in a non-adversarial environment: the use of new metaphors could be a powerful communication tool to promote new values and sustainable practices, in a proactive climate change narrative.

In order to deal with this innovative narrative, it must however be reminded that the apparent simplicity of the debate on climate change conceals a deep complexity which must be constantly instructed to avoid unrealistic reconstructions and ineffective solutions. Therefore, Chap. 4, "Measuring complex socio-economic phenomena. Conceptual and methodological issues" sets forth a definition of complexity employed in the second Part of the volume. The relationship between complexity and knowledge is here depicted and the concept of "system" is introduced as "an organic, global and organized entity, made up of many different parts, aimed ad performing a certain function." The chapter points out the necessity to promote effective measurement system of the mentioned complexity and favors the understanding that the scientific data (also pertaining to the climate science) can be read in different ways: "Each observation evaluated within a theoretical framework represents a datum. Consequently, we can obtain different types of data from the same empirical observation based on different theoretical frameworks."

After setting forth the basis to understand the complexity of the discussed issues, the volume gets to the heart of the scientific debate on climate change. Part II includes two scientific contributions which deal with some of the most problematic issues deriving from the rising of the temperature worldwide. Chapter 5, on "Glaciers: vanishing elements of our mountains and precious witnesses of Climate Change", contains the testimony of a group of researchers of the University of Milan working on the Italian Alps—precious and unique chain—that realized a monitoring protocol through which it is possible to measure the reduction of the glaciers basing on satellite detection systems. This survey makes the rapidity of the phenomenon extremely evident and indicates the urgency to intervene, also to contrast the less-known effects such as the negative effects for human health and the environment of the toxic substances released with ice melting. The irreversible character of the phenomenon indirectly clarifies the meaning, still pressing, of the mitigation policies.

Chapter 6, "Rural revival and coastal areas: risks and opportunities" is dedicated to describe another serious environmental issue: the dangers for the human and natural systems deriving from extreme atmospheric events. In particular, the problems regarding the coastal areas and the marine ecosystems are here considered and analyzed. After underlining the numerous and complex vulnerabilities detected by the IPCC reports (among the others) the chapter carries out a deep analysis of the adaptation policies aiming at minimizing the negative effects of climate change in a context of resilient actions. Therefore, Chaps. 5 and 6, read together, witness how science is an endless source of data, capable of pushing the adoption of environmental policies in regard to dangerous natural events (Chap. 6), as well as an instrument of the adaptation policies (Chap. 7).

Certainly, both in the field of mitigation and adaptation policies, the link between science and policy is crucial. In particular, it is necessary to answer three crucial questions: how the scientific evidence can make the adoption of public policies necessary, how can public policy lean against science and how science and the new technologies can be an instrument of resilience policies. Chap. 7, "Climate Change: from science to policies, backward and forward", is dedicated to the mentioned topic, which is analyzed by highlighting similarities and differences between the methods of politics and science. This chapter closes Part II but opens Part III, which intends to provide the readers with a framework of exemplifications of the main public policies adopted in terms of mitigation and adaptation.

Public policies are at the center of an aggressive debate, in which it is emphasized their capacity of protecting the planet and humanity from the negative consequences of the growing rise of the global temperature. Consequently, the analysis of some of them will allow to capture the areas needing more implementation and the success experiences that may eventually be replicated in the future. As for the rest of the volume, the interdisciplinary approach also characterizes this Part, so that the contributions expose economic, juridical, technical and political aspects, depending on the matter dealt with by the authors. Chapter 7serves as a bridge to Part III, where science and policy meet in the concrete implementation of sustainability strategies.

Chapter 8, "The EU perspective from setbacks to success: Tackling Climate Change from Copenhagen to the Green Deal and the Next Generation EU", starts with this finding: While the world will remember Twenty-Twenty as the year of the COVID-19 pandemic, it was also an exceptional year in at least one other aspect. Scientists at Copernicus and at NASA have confirmed that, globally, 2020 tied with 2016 as the hottest year on record. Then, the reflection shifts to European Green Deal policies, showing why we must look at the future with optimism.

This chapter shows the role EU is playing in relaunching a sustainable and inclusive growth and offers a deep analysis of successes and failures in shaping the global climate agenda, also stressing that, as a consequence of COVID-19, nowadays "when navigating in uncertain waters, the voice of experts is a valuable currency." This fact creates new and important laws also between science and climate policy, where "Science is of course at the basis of the European Green Deal and of the EU commitment to reach net zero by 2050. This target is not arbitrary—it is evidence-informed, based on a detailed impact assessment, and underpinned by scientific modelling."

Chapter 9, "Carbon pricing from the origin to the European Green Deal", is focused on the economic theories of climate change and examines the mechanisms of carbon trade tax and of the emission trading system in a critical and systematic way, considering the European Union context in which these policies have been firstly implemented. The meaning of the exposed theories and their ambits of application in the European countries are discussed also with punctual argumentations aiming at letting the costs and benefits of the considered instruments emerge.

Chapter 10, "Technology innovation in the energy sector and climate change—the role of Governments and policies", develops the driving force of technological innovation where adaptation policies are essentially based on the edginess and research, such as in the energy sector. There, the role of governments is particularly crucial, setting the overall legislative and regulatory framework for markets and finance in which the prerequisites and essential conditions for developing technologies are identified. Around the world, governments can leverage on different tools, which are described and commented on in the chapter. According to the findings of the analysis, energy transition and the strictly connected development of innovative technologies will expose the global energy system to face new crucial challenges, such as data protection and privacy, potential vulnerabilities and sabotage, but also inequality and social disruption.

Chapter 11, "How emerging technologies are finally matching the policy leverage of cities with their political ambitions", elaborates the hypothesis that in light of new technologies the energy policies of local and regional governments are sources of experimentation that may produce useful results.

The chapter deals with the complex issues of climate change governance and discusses the importance of appropriately aligning powers with efficacy and innovation.

The chapter concludes that technology innovation "is the driver of energy policy that makes it possible for cities to play the impactful role that they have sought for decades." The offered perspective allows to launch interesting reflections both in the United States and in Europe, by detecting network action which allow to integrate supranational and national policies with a bottom-up government that is inspired to the European concept of vertical subsidiarity.

For instance, the cities may avail themselves of energy building codes, set energy performance standards, offer weatherization services to residents who meet income qualifications, rationalize and reduce the energy consumption through municipally-owned-vertically-integrated eclectic utilities, which have control of their supply mix helping achieve climate targets.

The chapter also introduces a reasoning on the role of public undertakings, by offering data that demonstrate how, at a local level, they sometimes can be referred to as interesting examples of energy consumption reductions.

Chapter 12, which goes under the title "Sustainable transportation", provides a framing of the mitigation and adaptation policies that accompany the re-thinking of the transportation system, both public and private, to make it adequate to face the next challenges.

The selected approach focuses on transportation law and policies and is not limited to climate policies in a strict sense; differently, it demonstrates how the adaptation policies include evaluations and choices of social and economic character, in execution of the sustainable development principle. Moreover, in presenting a zero-emission transportation future, the author proposes an approach that is based on the efficiency of the actual transportation systems, on which the lifestyles of our communities rely. The analysis foresees a positive and continuous development: "The good news for climate advocates is that the industry is scaling and innovating rapidly, the result of careful and deliberate policy around the globe. Central to this progress has been declining prices for lithium ion batteries. Policy makers can continue this progress by focusing on transportation policies such as mandates, incentives, subsidies and permit streamlining and electricity rate reform."

The strategic role of public entities, also as market operators, is crucial and meaningful in the public procurement field, to which Chap. 13, "From green to social procurement", is dedicated.

The contribution presents a very topical subject, by pondering on the public policies through which the public entities can drive the competition towards quantitative parameters that valorize and promote the sustainability values, by operating on their tender procedures for the purchase of works, services and supplies.

The chapter illustrates good praxis and positive experiences but highlights the necessity to induce techniques that are able to make the application of social procurement criteria mandatory and systematic, especially in Europe. In this regard, adoption of command and control politics supported by an adequate monitoring system is suggested.

Chapter 14 regards "Climate, sustainability and waste—EU and USA regulatory approaches compared." The contribution explores the links between sustainability, waste regulation, and climate change, by comparing the different regulatory approaches towards waste management in the European Union and in the United States. Despite highlighting significant similarities between the two systems, including a decentralized administrative approach, a strict waste management hierarchy that favors recycling and composting, significant regulatory differences are observed. In the optic of the learning from each other, the chapter considers the European approach to be particularly interesting as it serves a wider purpose than just the protection of human health and the environment.

Chapter 15, "Construction industry and sustainability" deepens the difficult challenges that interest the construction sector. It investigates different areas, from the extraction and reuse of construction materials to energy and water consumption reduction, to circular economy.

Chapter 16, "Impact of climate change in agriculture: estimation, adaptation, and mitigation issues", deals with the relationship between climate change, agriculture and food security, with an overview of the main adaptation and mitigation strategies. It includes a summary of the emerging methods developed in the last two decades to empirically quantify the economic costs of climate change. This chapter will be focused on the agricultural sector, admittedly the most sensitive sector to climate change, although the econometric methods described could be applied to any economic activity.

The final Part, "People's behaviors to address climate change: proactive and conflicting actions", examines how our behaviors can influence environmental policies and how those policies can trigger positive or negative coping mechanisms in the administered community.

Chapter 17, on "Climate Change and consumer behavior", starts the analysis from a recent survey conducted across 28 countries worldwide, which reports that 70% of the total sample states to have made changes to their consumption behaviors out of concern about climate change. Of these, 17% refers "a lot of changes" and 52% refers "a few changes." Countries where consumers seem to be more prone to modify their behaviors to counteract climate change are India (88% of total interviewed), Mexico (86%), Chile (86%), China (85%), Malaysia (85%) and Peru (84%). Japan is the only surveyed country where almost 50% of people declared that they have not changed their behaviors in the attempt to impact less on the environment. The research work presented demonstrates that inducing behavior change is essential to the effectiveness of climate policies.

Chapter 18, "Climate change litigation: losing the political dimension of sustainable development" investigates the causes and consequences of conflicts brought before the courts to conquer the effective right to the environment. Precisely it investigates whether disputes brought by individuals and associations against States to challenge environmental mitigation's policies, the so-called climate litigation, are consistent with the fundamental structures of procedural rules and how they impact on the role of jurisdictions and on environmental protection policies. The analysis goes so far as to fear the risk that a purely subjective connotation of the environmentalist claims may diminish the awareness of the environment as a common good, which requires protection policies, and disarticulate the rules of actions in front of the courts of law. Finally, the reflection introduces a theory to reconnect the individual and collective dimensions of the environment, based on the overlapping of legal relations and rights and the complexity of contents and dimensions of sustainability, which requires an analysis of a proper political nature.

Chapter 19 explores "The judicial review of administrative decisions with environmental consequences." Starting from the analysis of the environmental protection as a juridical issue, the chapter explores the main characters of judicial review in this domain. Exploring the significance of the principle of sustainable development as a parameter of legality of administrative action, and analyzing selected case-law drawn from the US, the EU, Italy and the UK, the chapter conceptualizes judicial review as a fundamental instrument to guarantee a balanced application of a national model of growth.

Chapter 20, "Mediation in environmental disputes", describes the mediation in the subject at stake, as a problem-solving tool based on dialogue, facilitated by a third neutral, the mediator. The thesis is that encouraging public entities to settle environmental disputes through mediation will offer great opportunity to find solutions that allow communities to negotiate with public actors specific actions that successfully mitigate the impacts of invasive anthropogenic activities on the environment, thus obtaining a shared framework that results from a dialogue capable of including all the three dimensions of sustainability.

The four Parts of this volume present 20 chapters that marshal scholarly arguments and recent evidence on important aspects of climate change policy formation in the EU and US. Collectively, the sequence and accumulation of these disciplinary perspectives also demonstrates the complex interaction of policy subsystems and the importance of their interdependencies in the face of climate change mitigation and adaptation. These interdependencies are likely to increase in importance as climate change unfolds in uncertain ways that demand a robust capacity for adaptive policies. Given the critical interdependencies discussed throughout this volume, it is clear that further inter- and multi-disciplinary approaches are necessary to develop policies better prepared for the complexity and uncertainty of climate change in the coming decades.

Part I
Interpreting, Communicating and Composing a Complex Phenomenon

Chapter 2
The Ideological Trick of Climate Change and Sustainability

Gianluca Schinaia

Abstract This chapter analyzes how ideological bias plays a significant role in polarizing public debates on climate change policies. Urgent demands for global warming and sustainable development actions have become very popular among young people, international institutions, the scientific community, and public opinion. On the other hand, the right-wing ideology has endorsed the ideas of the climate denial countermovement worldwide, claiming to protect low and middle-class workers of Western countries, weakened by a protracted economic crisis. Ideological beliefs create disagreement on the perception of, and the solutions to tackle, the climate crisis. Disagreement causes mistrust of scientific data regarding climate change. So, the ideological clash between left and right-wing parties on the significance of global warming and its impacts and remedies reduces the solutions to tackle the climate crisis.

Specifically, the chapter presents how sustainable development issues have been pushed forward by entities traditionally far away from the conservative electorate. Furthermore, how the right-wing ideology has endorsed the ideas of the climate denial countermovement worldwide, focusing on Western countries and finally, to the communication problems and solutions to raise climate crisis awareness.

2.1 Introduction

Climate change is the hardest challenge humankind has ever faced. Solutions to mitigate the environmental, social and economic effects of global warming need to be shared. Nowadays, there are multiple answers based on interdisciplinary approaches combining sciences, economics, politics, technologies and social factors. However, ideological beliefs create disagreement on the perception of, and the

G. Schinaia (✉)
FPS Lab S.R.L.—University of Milan, Milan, Italy
e-mail: gianluca.schinaia@fpslab.it

© The Author(s), under exclusive license to Springer Nature Switzerland AG 2022
S. Valaguzza, M. A. Hughes (eds.), *Interdisciplinary Approaches to Climate Change for Sustainable Growth*, Natural Resource Management and Policy 47,
https://doi.org/10.1007/978-3-030-87564-0_2

solutions to tackle, the climate crisis. Disagreement causes mistrust of scientific data regarding climate change. Mistrust implies skepticism about most climate policies. In the end, ideological beliefs trump the value of scientific data. This condition is one of the greatest impasses to overcome for sustainable development, as described in this chapter.

Even though the climate change threat has enhanced international cooperation to slowly reach common solutions, the consequent decision-making process has been deeply divisive. Solutions adopted to face global warming imply massive changes, from corporate business models to human and social priorities. These remedies are modifying human perspectives on development and the means to achieve it. The characteristics of the climate system and the nature of the human influence on climate have profoundly challenged the governance, presenting a "truly complex and diabolical policy problem" (Steffen, 2011). Most of the efforts in terms of climate policies have proven to be disruptive, able to open several ideological disputes even on the very existence of a climate change issue. A poll presented by Yale University in 2019 showed that climate change is a more divisive issue for American electors than abortion or gun control (Milman, 2019).

Rarely scientific data on climate change are enough to overcome the social or individual belief in public debate. Even if the scientific consensus on human-caused contemporary climate change now exceeds 99% in the peer reviewed scientific literature (Lynas et al., 2021). It happens primarily because of the consequences of climate action on individuals and society, often stressed by organized corporate campaigns promoted to preserve the status quo. According to a 2016 research, more than 90% of dubious papers on climate change come from think tanks close to conservative ideologies. Relevant lobbies such as the Heartland Institute, the Cooler Heads Coalition, the Cato Institute or the Heritage Foundation. According to an analysis by sociologist Robert Brulle, 140 foundations have moved 558 million dollars to fund more than 100 denial organizations between 2003 and 2010. Where do these funds come from? Primarily from fossil fuel-related companies such as the Koch Family Foundation, British Petroleum, Shell or ExxonMobil (Levantesi, 2021). In many studies, scholars described how ideological filters strongly color public perception of climate change: facts are evaluated based on their fit to previously held beliefs (Denning, 2013). This mismatch between opinion and data has relevant implications regarding ideology in the public debate on climate change and sustainability issues. It involves media discussion and public perception of the issues investigated. In order to review the ways how ideology plays a role in the production, representation, and reception of climate change in media, five filters have been taken into account: economic factors, journalistic norms, political context, ideological cultures and citizen decoding (Maeseele & Pepermans, 2017). Alternatively, these filters can be classified into two views about climate change: is it seen as a post-ideological issue or not? If it is so considered, most people problematize the actual politicization of climate change, calling for its de-politicization. If it is not, the climate change topic is criticized for suppressing the role of democratic debate about its relevance and implications. Again, both views about climate change configure a truly complex and diabolical policy problem. The ideological

affects all public debate on the environmental policies for sustainable
lopment.

ually, science comprehension is linked to a higher level of education. When
nsider the climate change issue in developed countries, education is still
l to more pro-climate change beliefs among the political left and right, but its
are weaker among the political right than the political left (Czarnek, 2020).
nate skepticism is mainly linked to a right-wing ideology for many reasons.
others, climate change actions mostly require coordinated collective action,
the shadows of socialistic threats. Furthermore, actual climate movements
p roots with social ones protesting against neo-liberal principles and insti-
ack in the 1990s. In addition, climate skepticism—strengthened by the
lustrial Revolution tools—has opened to anyone the chance to question
assertions. These assertions remain for the few due to the technical nature
fic explanations. Finally, climate skepticism is empowered by actual and
reats to the present way of life for most people living in the (so-called)
countries. Skepticism is one of the main cornerstones supporting the rise
ive denial movement opposing deep transformative climate policies
n & Cook, 2011).

pter will analyze how sustainable development issues have been pushed
entities traditionally far away from the conservative electorate, such as
ments. Furthermore, how the right-wing ideology has endorsed the
climate denial countermovement worldwide, focusing on Western coun-
lly, to the communication problems and solutions to raise climate crisis

Is the New Black

gentle suggestion sounding all over the developed world. It feels like
d in many areas of daily life. A call is coming from public entities,
ies, media, arts, and entertainment. Just like the trees whispering to
pic nature novel, The Overstory by Richard Powers, awarded by the
or Fiction in 2019.

new black. In the fashion industry, the color black is regarded as a
it goes with everything. Nowadays, everything that sounds environ-
able—identified as "green"—is popular. At the moment, green can
f human activity. It is the message spread through marketing cam-
e specific service or product promoted. Attention is justified by the
ket share of green products and services that seem to be relentless.
ld of business opportunities in the promised lands, namely, circu-
onomy. In 2019, NYU Stern's Centre for Sustainable Business
f U.S. packaged-goods growth from 2013 to 2018 came from
ket products (Kronthal-Sacco & Whelan, 2019). A report
l (International Trade Centre, 2019) argued that: "Sustainable

product sourcing has become a top priority for retailers in key European Union markets as France, Germany, Italy, the Netherlands and Spain"—pointing out how 85% of European retailers reported increased sales of sustainable products over the past 5 years and 92% of them expect sales in sustainable products to increase in the next 5 years.

Shifting from products to services, the rising green trend seems specular. The sharing economy is changing the use of things, from ownership to rental. The energy sector is moving away from dependence on fossil fuels. In Europe, the share of renewable energy more than doubled between 2004 and 2019, representing 19.7% of the energy consumed in the EU-27 (Eurostat, 2020). In the United States, the growth of renewable energy remained significant under Trump's Administration. During the last decade, wind power tripled, and solar power increased fivefold in the United States: renewables now represent around 11% of all energy consumed. Moreover, US domestic renewable energy consumption will continue to increase until 2050 (U.S. Energy Information Administration, 2020).

Green is the new black even in the financial sector. In 2020, the world's biggest asset manager BlackRock declared it would make sustainability its "new standard" for investing.

The European Investment Bank recently took a similar decision. Environmental social, and governance (ESG) criteria are now a popular set of standards to evaluat companies and countries on how far they have progressed on these topics. Companie are increasingly adopting sustainability reports to show their corporate soci responsibility.

Despite the companies' move toward sustainability, the enthusiasm of mark players has been doubted too, criticized as "green capitalism": corporate gree washing of products, reinforcement of class oppression, and inefficient regulati of the industry to deceive consumers into being content with an illusion of progre (Engel, 2019). Still, this new green revolution is stepping into the light of the pub opinion, contaminating politics. The United States and European Union have set master plans on their respective sustainable development regional strategies: Green New Deal and the European Green Deal. The EU aims to be climate-neu by 2050, becoming the first continent to compensate for all its greenhouse gas en sions. The European Green Deal has also been built and approved because Europ green parties are becoming more popular, particularly among young electors. Yo politicians are very close to sustainability topics, too. Being sustainable is a pop new vision endorsed by Millennials and very young people belonging to the called Z and Alpha Generations. Nowadays, it is also embraced by youth m ments such as Extinction Rebellion or Fridays for Future. This new social acti has to do with the deep polarization of politics and ideology mentioned in the i duction. As authoritarianism, xenophobia, alt-right, and anti-immigrant forces risen all over the world, at the same time, environmental action and awareness a the impacts of climate change have grown significantly and mixed inside new y movements. It is not a brand new political phenomenon but a more comprehe evolution of the social protest movements of the last 30 years.

2.2.1 Brief Evolution of Main Protest Movement's Ideology

To some extent, youth eco-movements represent an answer to the criticism that sustainability is a "green myth" built up for market interests. These groups are targeting corporate and institutional behavior, instead of focusing on individual responsibilities. The neo-liberal market system relies on the power of individuals. Instead, climate action needs collectively shared plans. Nowadays, only mass movements have the power to alter the trajectory of the climate crisis, and it is time to stop obsessing with how personally green we individually live (Lukacs, 2017). Actual new transnational eco-movements have grown in public consideration during the last decades. These have both shared ideas and relevant differences from their ancestors. One of the most known is Fridays for Future, lead by Greta Thunberg. In the light of the evolving ideological tradition, let us consider another of the most emblematic eco-movement active today: Extinction Rebellion (XR).

XR was founded in the United Kingdom in May 2018, with around a hundred academics signing a call to action the following October. It draws inspiration from movements such as Occupy, Gandhi's independence movement and Martin Luther King, and the civil rights movement. Extinction Rebellion aims to support "a common sense of urgency to tackle climate collapse" worldwide. One of the inspirational movements for XR was Occupy Wall Street. Occupy and all its extensions worldwide, born from the occupation of Zuccotti Park in New York (September 17, 2011), intended to attack the financial system, protesting against economic inequality and the influence of international financial bodies, such as the European Central Bank or the International Monetary Fund. Occupy questioned the policies these institutions were taking to reduce wealth inequalities, as XR today also does, and the anti-globalization movement—namely "new global" movement—did before. The latter was a wider network of social organizations that defined the first international protest against international institutions in the late 1990s. In Seattle, November 30, 1999 registered one of the first most significant demonstrations of the new global movement against a World Trade Organization meeting. The WTO, World Bank, and the International Monetary Fund have always been the first targets of new global movements, getting millions of people out in the street to protest against the war in Iraq in 2003. Occupy has an ideological link to the new global movement critics. XR has a very close position to Occupy about the international financial system. XR, however, has a new focus on sustainability issues, sharing relevant points of the old new global movement agenda. Activists affiliated to the new global, Occupy, XR were primarily young people, politically oriented as left-wingers, anarchists, or independents.

Climate action claims presented by XR movement fall into disrepute for right-wing or conservative electors that historically do not support this range of social activism, which reflects a deep polarization of politics and ideology about sustainable development issues.

2.2.2 Students, the New Protest Vanguard

Something new was born after the subprime mortgage crisis. "A whole generation in 2010 were very motivated by slogans of 'no future'. They felt the future that they'd been promised—if they worked hard, got a good job, then they'd be able to achieve the level of economic stability and success which their parents had been promised—was no longer possible. Students were simply the first in a long line of groups who were going to have to pay for the crisis, and pay for austerity," says Matt Myers, author of *Student Revolt: Voices of the Austerity Generation* (Myers, 2017) Even though the anti-globalization movement faded out after 2002, the subprime crisis has reignited youth movements again. First with Occupy and after that through the Arab Spring revolts, students have become, step-by-step, the brand new movement vanguard. Undergraduate students have supported or even led many of the protests that have occurred in the last 10 years for democratic issues: Algeria, Bolivia, Britain, Catalonia, Chile, Ecuador, France, Guinea, Haiti, Honduras, Hong Kong, Iraq, Kazakhstan, Lebanon, Bangladesh, Germany, India, Italy, Malaysia, Quebec, South Africa, South Korea, Uganda, and many other countries. Usually, these protests start with a specific incident but afterward rise far beyond it: education fees (as in Chile), violation of democratic principles (as in Hong Kong), anti-racism protests (as by Black Lives Matter in the United States) or austerity measures (as in many European countries). Recently, economic inequality and global warming issues are gradually characterizing many kinds of youth protests all over the world. These tend to be leaderless, peaceful initially and occasionally deteriorating into street battles against the police. These protests are enhanced and sustained by social media, embracing undergraduate and high-school students (Larsson, 2020); these mainly focus on climate change but also on LGBT, feminist and anti-racist initiatives. "What we are witnessing in 2019 may not quite be a student revolution as it was in 1968; it may better be coined a youth (r)evolution" (Altbach & Luescher, 2019).

Young activists have gained two prominent supporters in recent years. First, international institutions. If WEF meetings were in the past targeted by "new global" protesters, in 2020, Greta Thunberg (an icon of the climate change youth movement), as a WEF guest in Davos, put the blame squarely on international leaders and tycoons. Now, European and U.S. plans on the environment share many initiatives proposed by the youth climate movements. This new shared vision by international institutions and young activists is viewed with suspicion by the conservative electorate. Furthermore, climate change movements have gained another key supporter: the scientific community.

2.2.3 The Scientific Community Approval

In 2019, a letter entitled "Concerns of young protesters are justified" was published in *Science* magazine (Hagedorn et al., 2019). Signed by over 20 scholars coming from many renowned universities in Europe and the United States, the letter clearly stated: "The enormous grassroots mobilization of the youth climate movement (…) shows that young people understand the situation (…) they deserve our respect and full support." In the same year, *Nature* published the following article: "Scientists worldwide join strikes for climate change" (Schiermeier et al., 2019) describing how researchers were among those protesting to urge action on global warming in North and Central America, Europe, Asia and Australia. These are just some examples of how generally the scientific community approves the ideas pursued by the climate movements. Approximately 97% of active publishing climate scientists believe in anthropogenic climate change, and it goes further to involve non-climate scientists in this position. On the other hand, only approximately half of the American public believes in anthropogenic climate change (Funk et al., 2016). The mismatch between data and public awareness has pushed the scientific community to endorse youth activist positions on climate requests. Thus, climate action movements now protest with the support of the scientific community.

Conservatives generally have respected scientists. However, the scientific community's endorsement of these new climate actions, led by groups traditionally far away from conservative ideology, has distanced right-wing supporters from scientific evidence on global warming. Cultural values are also significantly associated with certainty that climate change is occurring (Dunlap & Brulle, 2015). "According to the cultural cognition hypothesis, hierarchical individualists are more skeptical of environmental risks, including climate change, because accepting these risks would undermine hierarchical individualists' belief in commerce and industry. Those who hold more egalitarian and communitarian values tend to perceive environmental risks more acutely because they feel that commerce and industry (the drivers of many environmental risks) promote individuals over the community" (Carlton et al., 2015). Finally, the ideological bias on climate change beliefs sees, on the one side, movements led or formed by young students generally supported by the scientific community, with the international community appeasement; and, on the other, skeptics no longer relying on all scientific data released by their counterparts about climate change topics.

2.3 Climate Skepticism and Right-Wing Populism

In 2007, U.S. Republican representatives sounded quite convinced about the relevant role of global warming and its links with human actions. Several public declarations released that year by the party's major delegates were confirming it. "Technology will help us to confront the serious challenge of climate change"

(Former U.S. President George W. Bush). "I think that we have to accept the views that scientists have: there is a global warming and human condition contributed to that" (Former NY Mayor Rudolph Giuliani). Well-known Republican representatives such as Mitt Romney or John McCain endorsed the views expressed, too. Barack Obama's election in 2008 changed everything. Climate change actions and sustainable development plans played a pivotal role in Obama's program and agenda. Seven years after his election, former U.S. presidential candidate John McCain was ironically criticizing Obama's administration's shyness for taking part in the Syria conflict: "President of USA is saying the biggest enemy we have is climate change…". In 2015, during his electoral campaign talking about global warming, former U.S. President Donald Trump declared: "I think there'll be a little change here. It'll go up, it'll get a little cooler, it'll get a little warmer, like it always has for millions of years." During Trump's presidency, the climate change issue suffered from the ideological clash between left and right-wing political forces. Since 2019, President Trump singled out Greta Thunberg and indirectly climate activism (Noor, 2019). In November of the same year, the United States started the process of withdrawing from the Paris Agreement.

Nowadays, numerous studies have linked climate change skepticism to right-wing populisms. Considering conservative and alt-right perception of climate change, green is not the new black. According to their overall consideration of sustainability issues, it can be argued that green is the new red. As Simon Montlake wrote in an article on The Christian Science Monitor: "Climate denial in the U.S. is deeply rooted in an anti-government ideology that sees virtually all regulations, including curbs on carbon emissions, as leftist attacks on free enterprise" (Montlake, 2019). So the reactions to new climate change policies "appear to be rooted in fears of socialism." It's no coincidence that "Green is the new Red — Stop Enviro-Communism!" was the title of the manifesto of Norwegian Andres Breivik, the right-wing extremist who killed 69 kids at a summer camp in 2011. Naomi Klein also reported an ideological bias over climate change by right-wing. Author of *No Logo*, Klein was one of the inspirational leaders of the new global movement back in the 1990s. She published the book *"This Changes Everything"* (Klein, 2015), which investigates neo-liberal policies as a genetic factor of the climate crisis. The first chapter ("The Right is right: the revolutionary power of climate change") deals with the distance between conservative values and bodies with the needed transformative climate policies. As she wrote, "It is not opposition to the scientific facts of climate change that drives the deniers, so much as opposition to the concrete implications of these facts. Green economy, renewable energy, sharing mobility: saving humanity on the planet implies overcoming the extractive economy and the neo-liberal model."

Youth movements specifically (historically linked to anarchist, independent, and left-wing electorate) support all sorts of climate change actions. These actions are mostly targeting corporate activities and policies that prioritize individual needs (linked to neo-liberal ideals) over social demands. Furthermore, these movements are supported by the scientific community, academia, and some international institutions. The new alt-right groups consider these as intellectual elites, so weak, compromised, corrupted, and guilty of socialist aspirations.

Conservative politicians take into account new right-wing positions on global warming during their public speeches. For example, in 2020, Australian Deputy Prime Minister Michael McCormack shot down the prospect of a goal on net emissions in 2050: "IPCC is not governing Australia (…) Of course you've got to listen to the scientists but what you've also got to do is listen to the workers. Listen to those workers who (…) work hard in a coal-mine every day, those people in central and northern Queensland who rely on the resources sector for a job, for a future" (Martin, 2021). Thus, climate policies in political debates started to clash with job resilience for a vast majority of the population in Western countries. The subprime mortgage crisis, job automation, Covid-19: the last two decades have dramatically decreased the number of job opportunities available for lower and middle classes. Considering this scenario, climate policies are perceived as a radical-chic diversion, dangerous for "ordinary people life" (when it comes true). One of this bias's main consequences is the rejection of scientific evidence on climate change by many conservative groups living in developed countries. This rejection is driven by motivated cognition because people typically refuse evidence that can refute their fundamental beliefs. Studies show that rejection of scientific findings by the U.S. public is more prevalent on the political right than the left (Lewandowsky & Oberauer, 2016).

2.3.1 *Right-Wing Ideology Reduces the Effect of Education on Climate Change*

As many studies have shown, the questioning attitude regarding climate change is higher in the United States than in other countries. In less than 5 years, the United States passed from Obama's presidency, who considered global warming as the "greatest threat," to Trump's administration, who declared climate change a "hoax." A comparison across the last 30 years has revealed that in the past climate change concern was grounded by sociodemographic factors such as education, income, sex, age, or race: however, nowadays, this concern is better explained by political orientation (Driscoll, 2019).

This kind of evolution is also the result of a historical role played by corporations and politicians in deceiving the public about the risks of climate change, as scholars and investigative journalists demonstrated in several works (Dryzek et al., 2011). Sociologists link the rise of climate change denial to corporate and right-wing agents in a "top-down" social strategy (Dunlap & McCright, 2011) to influence public opinion on the potential dangers of climate policies that can disrupt its way of life. Researchers underline that consumption of conservative media can undermine the certainty that climate change is real. However, "bottom-up" factors push the denial of global warming's effects within right-wing voters, namely, ego or self-justification, group justification, and finally, system justification. The latter refers to the bias produced by the need to defend elements of the social, economic, and political systems on which one depends (Jacquet et al., 2014).

A study on web browsing histories from over 9000 participants in the United States, the United Kingdom and other four countries (Oxford Internet Institute, 2021) revealed that populist parties' supporters distrust the scientific consensus on climate change. The analysis highlights how supporters are mainly focused on promoting "the people needs" before climate change, choosing alternative explanations instead of scientific concerns on global warming. The same study shows that right-wing populist supporters are twice as likely to consume hyper-partisan media than left-wing supporters.

Another study demonstrates how general education and scientific literacy increase ideological bias's polarization along partisan lines and do not mitigate rejection of scientific findings (Lewandowsky & Oberauer, 2016). So, working to defeat these biases is firstly a challenge for media and communication rather than education. As seen in the introduction, education has positive effects on pro-climate change beliefs in countries that are considered to have low or mid-levels of development; in developed countries, right-wing ideology reduces the positive effects of education on these issues.

2.3.2 Out of the Palace: The Denial Countermovement

On January 6, 2021, thousands of people stormed the U.S. Congress in Washington, D.C. Most of them were Republican electors, alt-right militants, white supremacists, conservative evangelical Christians, and Qanon supporters. Nearly all of them were very politically active on social media, and a great majority supported a massive unofficial movement denying climate change. This topic is very popular in the United States, where the climate change denial industry is the widest of all developed countries. As already seen, the global warming denial position is strongly embraced by conservative voters and representatives: for instance, in the 2016 United States election cycle, every Republican presidential candidate questioned or denied climate change (Kaplan, 2015). The people that stormed the Capitol building were protesting against the last presidential election results and supporting Donald Trump, one of the most well-known politicians representing climate denial positions. Few of the rioters managed to get inside the Capitol building. Even though they used violence to enter, they appeared not to have further planned goals. The Capitol building intruders seemed lost and confused. Probably, they could have expected to feel a sense of belonging: being part of something broader, feeling part of a community of defenders and brave citizens; something similar to what animated people participating in climate movements demonstrations. Most of the Capitol building rioters of January 6 could be considered part of the climate change denial countermovement. Discrediting the scientific consensus over human-made global warming seems to have a philosophical inner link with the American conservative culture (Kane, 2007). First, climate change denial is based on an ideological commitment of conservatives and libertarians to neo-liberal policies, avoiding federal regulations. Second, climate change deniers present the need to defend the American way of life based on ever-expanding material growth (Collomb, 2014). Washington rioters could represent the desire to keep a position inside the palace, to preserve the status quo, and save the so-called American way of life.

However, the denial countermovement is spreading globally. It is unofficially represented by climate denial supporters on social media and officially endorsed by many political parties besides the United States and Australian right. Far-right European parties, such as the Finns Party in Finland, Vox in Spain, Freedom Party in Austria, or Alternative for Deutschland in Germany, share and promote ideas of climate change denial supporters. In Brazil, Ernesto Araujo, the former minister of foreign affairs in the Jair Bolsonaro government, called global warming a plot by "cultural Marxists," (Cockburn, 2018) and he dismissed the climate change division of the ministry. Under the right-wing Bolsonaro government, Amazon deforestation has dramatically increased.

In addition to politicians, people sharing denial countermovement ideas often refer to a generic global warming conspiracy theory. The claims that science behind climate change have been invented or distorted for ideological or financial reasons. Skeptical reports on global warming often support this theory: apparently, more than 90% of questioning papers on climate change originate from right-wing think-tanks (Xifra, 2016). So, climate change has gradually become a political opinion regardless of scientific findings. It has turned into an ideological battle for the new right-wing parties. These mainly target white people, worried about migrants, homosexuals, concerned about the effects of globalization on local communities, job markets, economic and social conditions. In the new right-wing storytelling, these factors do not affect international institutions, the scientific community, and the climate movements: these intellectual elites show a lack of empathy for the "ordinary people's needs." This right-wing political narrative has pushed climate change issues to the left-wing field. Sustainability and environmental topics are matters for "the people in the palace." The denial countermovement represents lower and middle-class people who felt abandoned by the society they knew in previous decades. These people still believe in the American Dream and the idea of an ever-expanding economy. The same people who now feel left out of the palace of power and are searching for a common battleground to fight for or against it. So, climate change has become a valuable field for a political battle.

2.4 Conclusions

Ideological bias plays a significant role in polarizing public debates on climate change policies. Urgent demands for global warming and sustainable development actions have become very popular among young people, international institutions, the scientific community, and public opinion. A new generation of international activists on climate issues have risen, pushing political demands to reform the global economy to make it fairer and more sustainable, just as independent and left-wing oriented groups (such as Occupy and new global movements) did in the past. On the other hand, a climate change denial countermovement was born to preserve the status quo, claiming to protect low and middle-class workers of Western countries, weakened by a protracted economic crisis. In addition, conservative parties around the globe are often representing the denial ideas of this countermovement. The ideological clash between left and right-wing parties on the significance of global warming and its impacts and remedies reduces the solutions to tackle the climate crisis.

A study on climate crisis denial (Häkkinen & Akrami, 2014) has identified three kinds of ideological variables: social-dominance orientation, right-wing authoritarianism, and left–right political orientation. The first variable considers the social group most motivated to reject evidence on climate change, but it is still very receptive to communication. This is a pivotal point because, as seen before, general education and scientific literacy were shown to increase the polarization of ideological bias along partisan lines and do not mitigate the rejection of scientific findings. For this reason, communication is a unique tool to contrast climate change denial. Among others, the Royal Swedish Academy of Science recently reported that the climate fight is being undermined by fake news and misinformation on climate issues (McKie, 2021). The University of Queensland in Australia issued guidelines on the five features of science denial arguments: fake experts, logical fallacies, impossible expectations, cherry-picking, and conspiracy theories. In order to overcome these arguments, scholars suggest using social media for climate news, even with a debunking role. Good use of these digital tools has led to a decrease in climate change skepticism, even if social context can moderate its effectiveness (Diehl et al., 2019). These elements require significant cooperation between professional communicators, journalists, social media managers, and scientists to improve communication on climate change and sustainability topics. Quoting Klaus Töpfer, the former executive director of the United Nations Environmental Program: "Communicating effectively about sustainable lifestyles is a challenge. One needs to consider not only what to communicate, but how to communicate it." (United Nations Environment Programme, 2005). Shifting from the means of communication to communication topics, studies suggest that specific knowledge of the mechanisms behind scientific findings can increase the acceptance of the data. It is crucial to easily deploy and present findings, describing these on a scientific basis.

Finally, the Italian environmentalist Alex Langer has drawn attention to a relevant point: "We will only attain ecological conversion when it becomes socially desirable," meaning that it is a priority to tear down ideological walls (Langer, 1994). There are many ways to pursue this, including both active policies and, again, communication strategies. For instance, it is working on one of the main pillars contributing to the growth of climate change denial: fear of change. This fear has been greatly exasperated by an extended economic crisis and the prospect of cross-sectoral job automation (National Research Council, 2008), and then further compounded by economic turmoil caused by Covid-19. Including new employment policies, reskilling and up-skilling plans, and social welfare for the weaker parts of society within climate policies would answer to the holistic revolution of sustainable development and most of the climate negationists' concerns. After these first measures, a green economy can boost new job opportunities and develop trust about the general idea of a sustainable future. The circular and sharing economy is already making room in relevant businesses, daily human life, and large cities: it is time to plan their applications in small towns and for little companies. Considering a communication perspective, it is also vital to open green economy innovations to ordinary people to make it free from an "elite stereotype." Here is necessary to expand the ways of disseminating climate change and sustainability issues to focus on new targets. For this very reason, a significant point concerns communication strategy about the climate crisis to engage "hostile audiences" (Denning, 2012).

Scholars show that ideological division on climate change is less pronounced among people whose close friends and family care about the issue. This research also underlines that conservatives are willing to shape global warming beliefs on the perceived social consensus (Goldberg et al., 2019), which means that it is essential to share and influence people to build a broader consensus on the climate crisis. In addition, researchers (Hornsey and Fielding, 2019) highlight strategies for promoting pro-environmental behaviors: favoring optimistic messages over pessimistic ones in describing the outlook for sustainability and evaluating in-group versus out-group messenger effects to deflate filter bubbles on social media (Chandler & Munday, 2016). To sum up, strategic communication can help to overcome ideological bias on climate change and sustainability.

References

Altbach, P. G., & Luescher, T. M. (2019). Students are the vanguard in the youth revolution of 2019. *University World News*, 7 December. Retrieved 6 April, 2021, from https://www.universityworldnews.com/post.php?story=20191204072235611.

Carlton, J., S., et al. (2015). The climate change consensus extends beyond climate scientists. *Environmental Research Letters*. https://doi.org/10.1088/1748-9326/10/9/094025

Chandler, D., & Munday, R. (2016). *A dictionary of social media*. Oxford University Press.

Cockburn, H. (2018). Brazil's foreign minister says climate change is a 'Marxist plot'. *The Independent*, 16 November. Retrieved from 8 April 2021, https://www.independent.co.uk/climate-change/news/brazil-climate-change-foreign-minister-ernesto-araujo-marxist-plot-global-warming-a8637281.html.

Collomb, J. (2014). The ideology of climate change denial in the United States. *European Journal of American Studies*. https://doi.org/10.4000/ejas.10305

Czarnek, G. (2020). Education, political ideology, and climate change beliefs around the world. In *Behavioral and social sciences*. Retrieved from 8 April, 2021, https://socialsciences.nature.com/posts/how-are-education-and-political-ideology-related-to-climate-change-beliefs-around-the-world

Denning S (2012) Effective engagement of hostile audiences on climate change. In *American Geophysical Union*. Retrieved 8 April, 2021, from https://ui.adsabs.harvard.edu/abs/2012AGUFMPA21A1957D/abstract.

Denning S (2013) I'm not a Warmist! Transcending ideological barriers in climate communication. In *American Geophysical Union*. Retrieved 2 April, 2021, from: https://ui.adsabs.harvard.edu/abs/2013AGUFMED31D..02D/abstract.

Diehl, T., Huber, B., Gil de Zúñiga, H., & Liu, J. (2019). Social media and beliefs about climate change: A cross-national analysis of news use, political ideology, and trust in science. *International Journal of Public Opinion Research*. https://doi.org/10.1093/ijpor/edz040

Driscoll, D. (2019). Assessing sociodemographic predictors of climate change concern, 1994–2016. *Social Science Quarterly, 100*(5), 1699–1708.

Dryzek, J. S., Norgard, R. B., & Schlosberg, D. (2011). *The Oxford handbook of climate change and society*. Oxford University Press.

Dunlap, R. E., & Brulle, R. J. (2015). *Climate change and society: Sociological perspectives*. Oxford University Press.

Dunlap, R. E., & McCright, A. M. (2011). Organized climate-change denial. In J. S. Dryzek, R. B. Norgard, & D. Schlosberg (Eds.), *The Oxford handbook of climate change and society* (pp. 144–160). Oxford University Press.

Engel, J. (2019). *An addiction to capitalism: A rhetorical criticism of mainstream environmentalism*. Retrieved 4 February, 2021, from https://digitalcommons.humboldt.edu.

Eurostat. (2020) *Renewable energy statistics*. Retrieved 2 April, 2021, from https://ec.europa.eu/eurostat/.

Funk, C. et al. (2016) The politics of climate. In *Pew Research Center*. Retrieved 7 April, 2021, from https://www.pewresearch.org/internet/wp-content/uploads/sites/9/2016/10/PS_2016.10.04_Politics-of-Climate_FINAL.pdf.

Goldberg, M., Van der Linden, S., Leiserowitz, A., & Maibach, E. (2019). *Reducing ideological bias on climate change*. Retrieved 8 April, 2021, from https://climatecommunication.yale.edu/publications/reducing-ideological-bias-on-climate-change.

Hagedorn, G., Kalmus, P., Mann, M., Vicca, S., Van Den Berge, J., et al. (2019). Concerns of young protesters are justified. *Science*. https://doi.org/10.1126/science.aax3807

Häkkinen, K., & Akrami, N. (2014). Personality and individual differences. *Elsevier, 70*, 62–65. https://doi.org/10.1016/j.paid.2014.06.030

Hornsey, M. J., & Fielding, K. S. (2019). Understanding (and reducing) inaction on climate change. *Social Issues and Policy Review, 14*(1), 3–35.

International Trade Centre. (2019). *The European union market for sustainable products*. Retrieved 2 February, 2021, from https://www.intracen.org.

Jacquet, J., Dietrich, M., & Jost, J. T. (2014). The ideological divide and climate change opinion: "Top-down" and "bottom-up" approaches. *Frontiers in Psychology, 5*, 1458. https://doi.org/10.3389/fpsyg.2014.01458

Kane, T. M. (2007). Hot planet, cold wars: Climate change and ideological conflict. *Energy and Environment, 18*(5), 533–547.

Kaplan, R. (2015). Where the republican candidates stand on climate change. *CBS News*, 1 September. Retrieved 3 April, 2021, from https://www.cbsnews.com/news/where-the-2016-republican-candidates-stand-on-climate-change/.

Klein, N. (2015). *This changes everything: Capitalism vs. the climate*. Penguin Books.

Kronthal-Sacco, R., & Whelan, T. (2019). *Sustainable share index: Research on IRI purchasing data (2013–2018)*. Retrieved 7 Apr 2021, from https://www.stern.nyu.edu.

Langer, A. (1994). *We will only attain ecological conversion when it becomes socially desirable*. Retrieved 5 April, 2021, from https://www.alexanderlanger.org/bs/279/1355.

Larsson, N. (2020). People thought I was too young to protest: The rise of student activism. *The Guardian*, 15 September. Retrieved 7 April, 2021, from https://www.theguardian.com/education/2020/sep/15/people-thought-i-was-too-young-to-protest-the-rise-of-student-activism .

Levantesi, S. (2021). *I bugiardi del clima*. Editori Laterza.

Lewandowsky, S., & Oberauer, K. (2016). Motivated rejection of science. *Current Directions Psychological Science, 25*(4), 217–222. https://doi.org/10.1177/0963721416654436

Lukacs, M. (2017). Neoliberalism has conned us into fighting climate change as individuals. *The Guardian*, 17 July. Retrieved 4 April 2021, from https://www.theguardian.com/

environment/true-north/2017/jul/17/neoliberalism-has-conned-us-into-fighting-climate-change-as-individuals.

Lynas, M. et al (2021). Environ. Res. Lett. 16 114005. https://iopscience.iop.org/article/10.1088/1748-9326/ac2966#references

Maeseele, P., & Pepermans, Y. (2017) *Ideology in climate change communication*. Retrieved 6 April, 2021, from https://oxfordre.com.

Martin, S. (2021). Michael McCormack dismisses claims he will stand down as National leaders. *The Guardian*, 16 February. Retrieved 7 April, 2021, from https://www.theguardian.com/australia-news/2020/feb/16/michael-mccormack-dismisses-claims-he-will-stand-down-as-nationals-leader.

McKie, R. (2021). Climate fight is undermined by media's toxic reports. *The Guardian*, 21 March. Retrieved 27 March, 2021, from https://www.theguardian.com/environment/2021/mar/21/climate-fight-is-undermined-by-social-medias-toxic-reports.

Milman, O. (2019). 'Climate crisis more politically polarizing than abortion for US voters, study finds. *The Guardian*, 22 May. Retrieved 2 April, 2021, from https://www.theguardian.com/us-news/2019/may/21/climate-crisis-more-politically-polarizing-than-abortion-for-us-voters-study-finds.

Myers, M. (2017). *Student revolt. Voices of the austerity generation*. Pluto Press.

Montlake, S. (2019). What does climate change have to do with socialism? *The Christian science monitor,* 5 August. Retrieved 6 April, 2021, from https://www.csmonitor.com/Environment/2019/0805/What-does-climate-change-have-to-do-with-socialism.

National Research Council. (2008). *Research on future skill demands: A workshop summary*. The National Academies Press.

Noor, P. (2019). Trump's latest attack on Greta Thunberg was sexist, ableist—and perhaps jealous. *The Guardian*, 12 December. Retrieved 8 April, 2021, from https://www.theguardian.com/environment/2019/dec/12/donald-trump-greta-thunberg-time-magazine.

Oxford Internet Institute. (2021). *Clear link between climate scepticism and support for right-wing populists, study find*. Retrieved 8 Apr 2021, from https://www.oii.ox.ac.uk.

Schiermeier, Q., Atkinson, K., Rodríguez Mega, E., Padma, T. V., Stoye, E., Tollefson, J., & Witze, A. (2019). Scientists worldwide join strikes for climate change. *Nature, 573*, 472–473. https://doi.org/10.1038/d41586-019-02791-2

Steffen, W. (2011). A truly complex and diabolical policy problem. In J. S. Dryzek, R. B. Norgard, & D. Schlosberg (Eds.), *The Oxford handbook of climate change and society*. https://doi.org/10.1093/oxfordhb/9780199566600.003.0002

United Nations Environment Programme. (2005). *Communicating sustainability, how to produce effective public campaigns*. Retrieved 2 April, 2021, from https://www.unep.org.

U.S. Energy Information Administration. (2020). *Renewable energy explained*. Retrieved 4 April, 2021, from https://www.eia.gov.

Washington, H., & Cook, J. (2011). *Climate change denials: Heads in the sand*. Earthscan.

Xifra, J. (2016). Climate change deniers and advocacy: A situational theory of publics approach. *American Behavioral Scientist, 60*(3), 276–287.

Gianluca Schinaia is head of sustainability and partner at the Italian media and digital agency FpS. As a professional journalist, he is a contributor for some Italian newspapers such as Wired, Avvenire, Il Fatto Quotidiano in the field of sustainability and he has won 4 journalistic prizes. As an author, he has contributed to an investigative book on environmental pollution in Italy. As a lecturer, he teaches climate change and sustainability communication at the Università degli Studi di Milano, principles of sustainability and brand journalism at the School of Journalism of the Università Cattolica, writing techniques and data visualization for sustainability at the European Institute of Design in Milan and he is a corporate trainer about SDGs topics. As a sustainability professional, he has a certificate on business sustainability management issued by Cambridge University and a Master's degree on interdisciplinary approaches to climate change issued by Università degli Studi di Milano.

Chapter 3
Words Count: The Role of Language in Overcoming Climate Inertia

Eleonora Ciscato and Marianna Usuelli

Abstract The meaning people attribute to climate change is closely related to the way it is formulated and presented by communicators. This, as research has widely proved, is highly influential on how citizens then take action to mitigate it. The chapter analyzes the climate change discourse in Italian newspapers. At first, it provides an overview on how it has been depicted in the last 30 years, essentially following the global discourse evolution and reflecting key international climate events. Then, the chapter introduces a more in-depth analysis of climate change metaphorical representation in one of the most popular national newspapers, *La Repubblica*. Through the critical analysis approach, we detected the illness, journey, and war metaphors in 1022 articles published between 2019 and 2020. A significant presence of war metaphors was observed, showing how the climate crisis—and the same happens with Covid-19 starting from 2020—tends to be described with a sense of emergency, fueling a divisive atmosphere.

Finally, our research shows how the Italian climate change narrative tends to follow global trends, adopting linguistic devices and linking the crisis mainly to global and far-away events, thus failing to intercept local experiences, possibly hindering citizens' and communities' proactive response.

E. Ciscato (✉)
University of Milan, Milan, Italy
e-mail: eleonora.ciscato@unimi.it

M. Usuelli
University of Milan, Milan, Italy
e-mail: marianna.usuelli@studenti.unimi.it; mariannausuelli@hotmail.it

3.1 Introduction

The number of people who experience the negative consequences of climate change is increasing, as it is the average global temperature. Despite the dire reality of the phenomenon, climate change cannot be seen, heard, or touched, thus the role of communication in spreading information and knowledge is essential in shaping public opinion and stimulating reactions from the public and politics.

Effectively, media and communication significantly contribute to building messages circulating in the public debate and shaping shared imaginaries (Appadurai, 2012). As mentioned in the previous chapter and shown by several researchers, the meaning people attribute to climate change is closely related to the way communicators formulate it: lexicon, discourse modalities, rhetorical figures, framing, storytelling are all filters influencing the interpretation and perception of climate change (Fløttum, 2017; Hulme, 2009; Lorenzoni & Pidgeon, 2006; Moser, 2016; Nerlich et al., 2010).

For example, James Painter holds that when journalists portray the consequences of climate change as "risks" rather than "uncertainties," a more robust response in terms of engagement and understanding can be reached (Painter, 2013).

Another example comes from the research at the *Yale Program on Climate Change Communication*, where it has recently been shown that the phrase "natural gas," even though referred to a fossil fuel composed almost entirely of methane (around 70–90%), evokes positive feelings in the U.S. population. The adjective "natural" attributed to "gas" distorts reality, and the source of energy is perceived as clean (Lacroix et al., 2020). This seems to be even more relevant in a polarized society such as the U.S. one, where the political discourse is imbued with denial instances, and climate change is a divisive issue.

One thing is what climate change is, another is how people interpret it. Climate change debate is thus far from being limited to science: as Hoffman writes (2012), "it's about values, culture and ideology." And, here, words matter.

That's why linguists, psychologists, anthropologists, historians, and sociologists have contributed to the prolific field of climate change communication, which saw a constant growth in the last 25 years (Moser, 2016).

Institutions, as well as news outlets, are adjusting to this, and the Intergovernmental Panel on Climate Change (IPCC) in 2016 reached the conclusion that "[IPCC] authors should be trained in writing and communicating, including the use of clear language" (Corner et al., 2018).

The British newspaper *The Guardian* then decided to adopt a firm editorial policy regarding articles on global warming. Graver and more marked expressions such as "climate emergency" or "climate crisis" should be used in place of "climate change" to increasingly convey the urgency and the devastating scope of the situation (Carrington, 2019).

However, finding the right balance in climate communication is no easy task: scholars widely agree that the pessimistic framing contributes to its failure. "The use of *alarmism* (...) has been much discussed in recent years, as research has shown that it might have the opposite effect to what was intended," write Nerlich et al. (2010). If faced just with irrevocable tipping points and extinction of animal species, people feel guilty, powerless, and less motivated to react collectively and individually. Apart from informing about the seriousness of the situation, it is essential then to present ways out: to stimulate a change, communicators should link climate change mitigation with positive desires and aspirations (Corner et al., 2018; Futerra, 2005; Nerlich et al., 2010; Stoknes, 2015).

Considering that the media are the main intermediators of scientific knowledge toward the public (Moirand, 2007), we decided to explore the Italian climate change debate in newspapers in the present chapter. How much do the media cover the issue? How has the press arena changed over the years? What are the words and the expressions mainly used to describe the phenomenon?

The first section of the present research aims at providing the reader with a general overview of the Italian media coverage on climate change from the beginning of 1990 until the end of 2020. For this reason, we selected a cluster of syntagms commonly used to discuss the topic, and we evaluated their presence and frequency in the titles of the most prominent Italian newspapers. This type of research provided us with the general trends of the Italian media coverage, which in the last 30 years has followed the global discourse evolution, being particularly connected to key international events.

The quantitative and diachronic section of the research also prompted us to frame and delimitate the second section, more qualitatively. Considering that one of the significant peaks in climate change media coverage concerns the time interval 2019–2020 (primarily related to the global climate movement), we decided to examine more in-depth climate change-related articles published over the 2 years and to analyze the use of metaphors in describing it.

As we will see, metaphors are powerful tools that convey knowledge and political messages. The analysis of metaphors used in the press is thus instrumental in exploring how climate change discussion is shaped, what political vision is embedded in the most common metaphors, and what types of responses rhetorical figures can stimulate. To carry out the research, we focused on one of the major Italian newspapers, *La Repubblica*, and adopted the critical analysis approach (Charteris-Black, 2004). Three metaphors have been searched for: the war, the illness, and the journey metaphors. International literature suggests that such images are often used in the political debate and have already been detected in different countries' press (Asplund, 2011; Atanasova & Koteyko, 2017; Cohen, 2011; Nerlich & Jaspal, 2013). To our knowledge, no specific research had been carried out in the Italian context, and in this chapter, we try to fill the gap. Our findings show that the war metaphor is predominant and that after the outbreak of the Covid-19 pandemic climate change coverage has decreased considerably.

As other scholars studied (Caimotto, 2021), this shift in the topic does not bring a decline in the use of war metaphors, which appear to be a solid and deeply rooted tool of Italian journalism. Finally, we considered the unclear and complex which in the last 10 years has been discussed not exclusively in scientific terms but also in economic, political, social terms.

Observing the diachronic trend of the graph, we note the presence of three main *discourse moment*s, which identify the emergence in the media of an intense discourse production on a specific issue in a defined temporal interval (Moirand, 2007). The first discourse moment can be seen in the period of the economic crisis (2007–2010). Some relevant climate-related events characterize these years: the publication of the Stern Review in 2006, the issue of IPCC's AR4, and Al Gore's "Unconventional Truth" in 2007, followed by the Nobel Prize recognition. The year 2009 was instead the year of the COP15 held in Copenhagen: the event received a great deal of media attention and was even amplified by the simultaneous Climategate controversial scandal.

It is also possible to notice on the graph the peak in newspapers' coverage occurred in 2015, concurrently with the negotiations of the Paris Agreement. Finally, the most recent discourse moment is trackable in 2019, when the global movement for climate inspired by the Swedish activist Greta Thunberg spread in all continents.

Far from being a merely Italian trend, our results find correspondence in the international literature (Boykoff et al., 2021a). Boykoff and colleagues have analyzed 120 newspapers from the five continents and found the same peaks in press production visible in the Italian scenario: 2007–2009, 2015, and 2019. Furthermore, similarly to what can be observed in our Italian graph, Boykoff and colleagues noticed that at a global level, media attention on climate change in 2020 dropped by 23% compared to 2019, still exceeding by 34% in 2018 of media coverage (Boykoff et al., 2021b). From a scientific perspective, this shows how the linguistic analysis of climate change communication cannot overlook the global dimension.

Indeed, our findings show that global events tend to overshadow local happenings and community experiences in the news coverage. While this naturally mirrors the global character of climate change, this adherence to international events influences readers' perception hindering the development of locally related solutions and proactive responses (Lorenzoni et al., 2007; Lorenzoni & Pidgeon, 2006).

Emotions and cultural beliefs count more than rational facts (Nerlich & Jaspal, 2012), and in several cases, inaction is not justified by a lack of knowledge and information, but rather by a communication failing to connect the fight against climate change to desires and shared values of communities (Giddens, 2009).

In this field, rhetorical figures, especially metaphors, are essential tools that can help, on one side, to transform scientific concept about climate change into simpler and tangible ones, and on the other, to connect global phenomena to local references and cultural values (Asplund, 2011; Corner et al., 2018; Fløttum, 2017).

3.3 Living in Metaphors

To analyze the power of communication devices, linguistics studies can be helpful. One of the major contributions came from the American scholar George Lakoff (2010), who explained how human beings think in terms of "frames," that is to say, people understand reality and its complexity by developing unconscious structures that create relations between elements, by assembling systems. While communicating, we activate neutrally such frames that trigger the emotional sphere of the brain, largely influencing action.

In this peculiar scope, studying linguistic devices in discourse such as metaphors can be insightful. According to Lakoff (2016), conceptual metaphors are "the main mechanism through which we comprehend abstract reasoning," and they work as cognitive devices that deconstruct complex problems, connecting a conceptual source domain that is known to be a conceptual target domain that is unknown. Through this "trick," one aspect of a concept can be understood in terms of another (Lakoff & Turner, 1980, 10), making it easier to grasp phenomena that are abstract or that have not been experienced directly by the subject in question.

Because they work as "messengers" of meanings between domains and, in this case, between scientific knowledge and the public (Maasen & Weingart, 2000), they play a decisive role: while simplifying complex issues, they naturally privilege some features over others. In doing so, they determine which side of the story to stress, leaving the rest in the shade or even assigning new nuances of meaning to the issue described (Schön, 1993, 41). This ambiguity assigns a potentially strong political power to metaphors that can be highly influential at the collective level (Lakoff & Turner, 1980, 57), and as underlined by Shaw and Nerlich, they "enable as well as constrain the ways we think about policy issues, especially with regard to largely abstract, complex and seemingly intractable problems like climate change" (Shaw & Nerlich, 2015, 35).

Especially when repeated over time and reinforced by different actors and voices, the stories created by metaphors make what has been defined as '"normative leap" from data to recommendations, from fact to values, from "is" to "ought" (Schön & Rein, 1994, 26).

The role of metaphors in describing environmental issues has been studied at large. As highlighted by Atanasova and Koteyko (2017, 72), when in the 1980s much of the attention was devoted to GHG emissions, the "greenhouse effect" metaphor had its momentum (Romaine, 1997), being then progressively abandoned after the release of the first IPCC report when media's attention diverted toward mitigation and adaptation (Nerlich & Jaspal, 2013). This change brought an increase in war metaphors that identified the global challenge as a "battle" and the rise in temperature as the "enemy"—indeed communicating urgency while also being divisive (Romaine, 1997).

Identifying a pattern in metaphor use over time and in different texts is no easy task because of the contingency of instances that can appear. The critical analysis approach (Charteris-Black, 2004) can be instrumental in guiding the research. It consists of three phases: the identification, the interpretation, and the explanation phases. The first phase requires distinguishing the corpus of texts to be studied: delimiting the period to cover, the types of texts to analyze, and the precise categorization of the keywords/expressions searched for. Once the corpus is delimited, the interpretation phase starts. It mainly consists of a case-by-case assessment of the word usage, intending to distinguish the instances where words are used metaphorically or literally, as the boundary can often be quite blurred. Finally, the explanation phase requires elaborating a convincing interpretation of the results, shading light over the political implications of a given metaphor, the message conveyed and, when apparent, the actors that more openly appropriate them.

3.4 Discussing the Italian Case: The Use of Metaphors in the Newspaper *La Repubblica*

Our research is grounded on existing literature covering metaphor use in the climate discourse (Asplund, 2011; Atanasova & Koteyko, 2017; Cohen, 2011; Nerlich & Jaspal, 2013). Specifically, we intended to integrate Atanasova and Koteyko's research on the presence and impact of three different metaphors: the war, the journey, and the illness metaphor. The work we referred to investigated *theguardian. com*, *Sueddeutsche.de*, and *NYTimes.com* that can be classified as widely read, left-center liberal newspapers. To complement it, we analyzed how *La Repubblica*, an Italian newspaper that takes a comparable political stand, portrayed climate change through metaphors.

In the first phase of the critical analysis approach, the identification phase, we proceeded by selecting the corpus of articles from *La Repubblica*, one of the major newspapers in Italy (ADS, 2021). Specifically, we limited the research to those published between January 1, 2019 and December 31, 2020. The time interval allowed us to cover the discourse moment of 2019 (see paragraph 2.2) and, at the same time, to examine the change due to the pandemic outbreak in 2020. Indeed, the literature shows the Covid-19 pandemic has stimulated an intense use of war metaphors in Italy and abroad (Caimotto, 2021), and we hypothesized the development of interconnections and overlappings with the climate change narration. In this phase, we also circumscribed the analysis by only selecting the articles containing the syntagm "climate change" (both in the singular and plural version, *cambiamento climatico* and *cambiamenti climatici*) together with one of the keywords covering the semantic area of the three metaphors (see Table 3.1).

In the interpretation phase, articles that only indirectly mentioned climate change were discarded, and a total of 1022 articles focusing on climate change were instead considered, as broader use of metaphors could be expected (Wallis & Nerlich, 2005). Among them, 430 (42%) made use of clear climate-change-related metaphors.

Table 3.1 Keywords for war, journey, and illness metaphors

War	War, danger, threat, battle, fight, enemy, defeat, struggle, win, defend, ally, lose, weapon, on the front line, victory, hit
Journey	Journey, road, evolution, passage, direction, route, transition
Illness	Disease, fever, patient, recover, symptom, drug, cure, sick, ill, take care of

Note: although the semantic field of illness/journey is wider than the one presented, the selected words are the ones we found more frequently as related to climate change metaphors

3.4.1 War Metaphor

The war metaphor is the most frequently used metaphor, present in 60.9% (see Table 3.2) of the articles analyzed. The political communication literature has widely studied it, and it is also frequent in the climate change debate of many countries (Asplund, 2011; Atanasova & Koteyko, 2017; Cohen, 2011; Nerlich & Jaspal, 2013). It conveys the notion of emergency, triggering and catalyzing national efforts, as well as stimulating unifying sentiment. Consequently, it is often used in the political discourse to legitimize the imposition of extraordinary or ambitious measures: it helped make greenhouse gas reduction proposals a reality (Cohen, 2011) and was also largely used to justify pandemic restrictions (Caimotto, 2021).

For a notion to be conceptualized in war terms, however, an enemy to fight is essential. The divisive climate that emerges from war metaphor use can, therefore, with repetition, lead to "dulling the critical faculties rather than awakening them" (Mio, 1997, 119) and contribute to climate fatigue.

Most of the war metaphors found in our corpus regarded articles with the exclusive presence of expressions like "fight to climate change," "combat climate change," "war to climate change" (*lotta al cambiamento climatico, combattere il cambiamento climatico, battaglia del clima, in prima linea*). Other common terms used were "threat," "existential threat," and "danger," reformulating climate change.

In articles where the war metaphor was more systematically and widely present, it was possible to proceed with the explanation phase. For instance, we found parallelisms between climate change and war. An example is the sentence

1. *Il cambiamento climatico è come la **terza Guerra mondiale**.*
   Climate change is like a **Third World War.**
   ("Se immagino l'Apocalisse mi viene in mente una data. Il 19 settembre 1991", 23/11/2019)
   Alternatively, again, we read:

2. *L'unica differenza tra l'immaginario presentato dal film e quello attuale è che allora la **minaccia** con la M maiuscola si chiamava **bomba H**, adesso cambiamento climatico.*
   The only difference between the imagery presented in the movie and the current one is that back then, the main **menace** was called **H bomb**, now it is climate change.
   (*"Il dottor Stranamore ed io"*, 2/3/2019).

However, the war metaphor can take different shapes and express opposite messages. Who is the enemy when we talk about climate change? In our corpus, the enemy is always different. As in this case, it can be the phenomenon itself:

Table 3.2 Frequency of metaphor use

War	Journey	Illness
60.9%	33.9%	5.1%

3. *La coalizione World War Zero è una **chiamata alle armi** non solo per **combattere una guerra contro il nemico comune** del cambiamento climatico.*
   The World War Zero coalition is a **call to arms** not only to **fight a war** against the common **enemy** of climate change.
   (*"Clima: Kerry guida il nuovo esercito della salvezza"*, 12/12/2019)
   The enemy can also be the CO2 or other physical consequences of climate change:

4. *Le foreste assorbono il 40 per cento di quell'anidride carbonica e bisogna **combattere il nemico nel suo campo di battaglia**.*
   Forests absorb 40 per cent of [that] carbon dioxide, you need to **fight the enemy on his battlefield**.
   (*"Stefano Boeri: Comincia in città il rinascimento dei nostri boschi"*, 13/9/2019)

5. *Ondate di caldo e microplastiche mettono a rischio la vita degli oceani. I nostri mari sono sotto scacco a causa degli effetti del cambiamento climatico, dell'inquinamento ambientale e dell'uso improprio delle sue risorse. Le notizie che arrivano dagli scienziati suonano come **bollettini di guerra** nei quali i numeri quantificano i danni: la frequenza delle ondate di calore marino.*
   Heatwaves and microplastics endanger the life of the oceans. Our oceans are in check due to the effects of climate change, environmental pollution, and the misuse of its resources. The news shared by scientists sounds like **dispatches** where numbers quantify the damage: the frequency of marine heatwaves.
   (*"I nostri mari sono in pericolo: salviamoli"*, 13/2/2020)

Very often, the war is against the fossil fuel companies and the negationists, sometimes personified by the former U.S. President Donald Trump, as in this case:

6. *Il **guanto di sfida** della Santa Sede alle «lobby del petrolio» che fanno profitto inquinando il globo è lanciato dall'inizio di questo pontificato. **Donald Trump è un nemico anche pubblicamente dichiarato**. Non a caso molti degli interventi in Italia di Steve Bannon sono diretti proprio contro Francesco e la sua **difesa del clima**.*
   Since the beginning of this papacy, the Holy See has **thrown down the gauntlet** to the "oil lobbies" that profit by polluting the globe. Donald Trump is a **publicly declared enemy**. It is no coincidence that many of Steve Bannon's speeches in Italy are directed against Francesco and his action in **defence** of the climate.
   (*"La benedizione del Papa a Greta 'Vai avanti nella tua battaglia'"*, 18/4/2019)

The metaphor can also take the form of a conflict between rich and poor, those who emit and those who suffer the consequences of the emissions:

7. *John Sauven di Greenpeace UK rincara: «Il mercato in crescita dei jet privati mette il pianeta **in pericolo**. Per essere chiari, ora la questione è: **o noi, o loro**».*
   John Sauven from Greenpeace UK adds: "The growing market for private jets puts the planet in **danger**. To be clear, the question is: **either us or them**."
   (*"Bill, Oprah e Meghan, quanto inquinano i vip sempre in volo"*, 18/11/2019)

Finally, the war of climate change can also be formulated as a battle against ourselves, our habits:

8. *E poi siamo in **guerra anche interiore**: causiamo questi fenomeni anche perché abbiamo stili di vita, a volte millenari, che si basano per esempio sulla carne.*

    And we also have an **internal war**: we cause these phenomena also because we have lifestyles, sometimes very ancient, based, for example, on meat.

    ("*È come una guerra: L'esempio di Greta ci può aiutare*", 25/8/2019)

What is worth highlighting here is that choosing who the "enemy" is is necessary to frame the phenomenon precisely, thus fostering a given vision and interpretation of reality.

In some cases, the war metaphor maintained a positive meaning: the verbs "to win" and "to lose" or the terms "defeat" and "victory" were used not to frame the climate change in a military context, but rather to formulate it as a challenge, a game, a race. Such shift in perspective is worth noticing, because it brings some analogies with the "journey metaphor," and it puts the phenomenon into a context of positive competition, not abusive struggle (Asplund, 2011).

9. *Quella contro i cambiamenti climatici è la sola **sfida da non perdere**.*
    Climate change is the only **challenge not to be missed**.
    ("*Ti importa del pianeta o fai finta?*" 4/9/2019)

10. «*La **corsa** contro il riscaldamento globale è una **corsa** che possiamo e dobbiamo **vincere***».
    «*The **race** against global warming is a **race** that we can and must win*».
    ("*Seguendo Leonardo DiCaprio su Instagram, il social network che vive della...*" 28/09/2019)

The outbreak of Covid-19 has led to a rapid decrease in the number of articles about climate change and war: if in 2019 we found 170 articles with war metaphors, in 2020, we detected just 92. The decline is probably due to increasing discussion on the pandemic which, as shown by Caimotto (2021), has also been dominated by war metaphors. As found in the corpus, the two battles (against climate change and Covid-19) have often been discussed together and put in relation: this has occurred mainly during the U.S. election campaign, where the two issues dominated the public debate and the expectations on Joe Biden's mandate on the adhesion to the Paris Agreement were high.

3.4.2 Illness Metaphor

The illness metaphor was the rarest in the corpus. We found it just in 5.1% of the articles. The most common image used is that of the "planet with the fever," or the "sick planet," to evoke global warming:

11. *Domani i giovani di 1.325 città in 98 Paesi diserteranno le aule per dire agli adulti: fate qualcosa per fermare la **febbre della Terra**. La più grande mani-*

festazione studentesca che si ricordi. Lei, Greta Thunberg, è soddisfatta ma non si accontenta.

Tomorrow, young people from 1325 cities in 98 countries will skip school to tell adults: do something to stop the **Earth's fever**. The largest student demonstration ever. Greta Thunberg is satisfied but not ultimately settled for [what has been done].

("Greta Thunberg '*Ragazzi per il clima non c'è più tempo anche gli adulti devono agire*'", 14/3/2019)

In the explanation phase, we noticed that environmental activists and Pope Francesco mainly use this metaphor. Some scholars (Arrese & Vara-Miguel, 2016; Sontag, 1989) highlighted that this metaphor is commonly associated with a Manichaeistic vision of the world, which opposes good and bad. This would be in line with the moralistic tone of the articles we found, underlining the "responsibility for the future" and the necessity to "take care" of the planet.

Concerning this topic, environmental communication research suggests that anchoring climate change in fear of disease, metaphors of illness might demotivate action (Höijer, 2010). The media is said to contribute to a culture of fear with its daily reports on scare stories, and, therefore, frequent exposure to messages conceptualizing climate change as an illness may lead to it becoming part of the expanding sphere of fear appeals in the media, to which the public gradually becomes indifferent (Höijer, 2010).

3.4.3 Journey Metaphor

The journey metaphor frames climate change in a context of a path to be travelled:

12. *È indubbio che il mondo si **muova verso** la **transizione** energetica. Forse se ne parla più di quanto non ci si stia lavorando, ma è solo una questione di tempo: l'**evoluzione** delle tecnologie permetterà di abbassare i prezzi dell'energia. Quindi: qual è la logica che sta dietro alle valutazioni di società come Saipem? Partiamo col dire che la **transizione** andrà a **velocità** diverse a seconda delle aree geografiche. Non è pensabile che Cina, India, ma anche l'Africa si possano permettere un **passaggio** in tempi brevi alla green economy. [...] Come management non possiamo che continuare a **spingere in questa direzione**. Il mondo dell'energia ha individuato il gas come la fonte energetica che accompagnerà la **transizione** energetica verso le rinnovabili e le tecnologie a emissione zero.*

    There is no doubt that the world is **moving toward** the **energy transition**. Perhaps we talk about it more than how much we are working on, but it is only a matter of time: the **evolution** of technologies will allow us to lower energy prices. So, what is the reason to go public for companies like *Saipem*? Let's start by saying that the **transition** will move at different **speeds** depending on the geographical areas. It is unthinkable that China, India, but also Africa can afford a **transition** to the green economy in a **short time**. [...] As part of the management, we can only continue to **push in this direction**. The energy sector has identified gas as the energy source that will accompany the energy **transition** toward renewables and zero-emission technologies.

(*"Saipem non è una società petrolifera i nostri progetti guardano al futuro"*, 28/9/2020)

We found this metaphor in 33.9% of the articles containing the journey keywords we selected (see Table 3.2). The term "transition" is the most common one, often formulated in different ways: *transizione energetica, transizione equa, transizione alle rinnovabili.*

Proceeding with the explanation phase, we find that this metaphor is often used to describe the decarbonization process, especially in articles on firms' pathway to becoming more sustainable and discussing the EU Green Deal targets. Many examples then show the use of normative adjectives associated with the pathway: *giusta direzione, strada sensata, strada della speranza*. This metaphor highlights the direction to take in a linearly evolutive scheme and often frames climate change as an "opportunity" to start a "green economy" and new "sustainable business." At the same time, as also shown by Atanasova and Koteyko (2017), it avoids the discussion of which destination to reach: "Journey metaphors seem to close down criticism (...), only appearing to engage with the problem at hand and embrace change while reinforcing business-as-usual."

This metaphor is usually accompanied by the use of consensual and conflictless expressions, such as "green economy," "sustainable path," "need for change," also typical of what has been called by some French linguists the *langue de coton* (Steiner, 2002). This type of language, spread from the beginning of the 2000s in international organizations and politics, is characterized by conflictless terms and rhetorics that tend to eliminate the political connotations in speech and legitimate wished practices. The *langue de coton,* drawing from the modernization ideology in which change is a synonym of positive evolution, helps us interpret the massive use of the journey metaphor in articles about companies and European politics.

3.5 Conclusions

This research allowed us to identify the main peculiarities and trends of the Italian climate change debate since its birth in the early 1990s.

First, our research showed how climate change media coverage evolved in the last 30 years, following global trends and shedding light on international events such as IPCC reports or the Conference of the Parties. Moreover, we detected an increasing trend in the diversification of syntagms used to discuss the topic, with an increasing role in the term "sustainability" in the last years.

We also found evidence of the global nature of the climate change debate in the second section of the research, where we qualitatively analyzed the types of metaphors used in articles. Considering the war, the illness, and the journey metaphors, already studied in international literature, we observed the clear prevalence of the war metaphor in the Italian articles by *La Repubblica*, conveying a sense of urgency and seriousness needed to face the issue. We also remarked a rapid decline in the number of articles about climate change in 2020, compared to 2019, following the Covid-19 outbreak. Concerning that, we underlined that the war metaphor has been widely used worldwide to describe the pandemic situation (Caimotto, 2021), revealing, as well as for climate change debate, the global nature of this communication.

Interestingly, some researchers have advanced the idea of collecting alternative metaphors to describe the pandemic, launching the initiative #ReframeCovid (Semino, 2020). They suggest that the war metaphor has been nourishing a divisive climate, constantly searching for an enemy to blame and unable to spread solidarity. Such need for different metaphors is not new to the climate change debate: several scholars (Russill, 2015) suggest that new environmental metaphors—maybe even less anthropocentric ones—could stimulate a powerful and more diversified reaction on readers.

As shown in our research, metaphors often end up being appropriated by specific interest groups. For example, even if quite rare in our corpus, the illness metaphor carries a moralistic and Manichaeistic understanding of climate change. The values of care and safety enshrined in this metaphor are consistent with the vision of activist groups and the Pope, who mostly utilize it. The exact appropriation mechanism occurs if we analyze the journey metaphor: the idea of a transformation and pathway toward a brighter future seems to be fitting the business and international institutions narrative.

The emergence of new metaphors could therefore stimulate different interpretations of the phenomenon, not divisive as the war metaphor suggests and, at the same time, not so deeply related to a specific group's values. New metaphors could express more diversified cultural references and values, helping to engage a wider part of the population to take an active response to climate change.

Intimately linked to this, both the terms "sustainability" and "climate change" tend to be conceptualized globally, leaving little space for local experiences and communities. However, as discussed in the article, communication strategies are only effective when adapting to local needs and contexts.

On the contrary, the concept of "sustainability" was conceived and developed in an international environment and scarcely adheres to local experience because of its ambition for a universal understanding. For this reason, some scholars discussed the "emptiness of sustainability" underlying how such a manipulable concept risks being co-opted by different powerful interests.

Therefore, the global nature of the climate change debate could constitute a limit for creating stronger responses by the public and a call for radical change. Creating a climate change narrative connected with local communities and concrete experiences and using, at the same time, diversified metaphors spreading different points of view could contribute to building a fertile ground for stimulating innovative and creative collective responses. Effective communication that considers all this is crucial in developing and implementing truly and desired sustainable policies.

References

Accertamenti Diffusione Stampa (2021). Retrieved 14 April, 2021, from http://www.adsnotizie.it/index.asp.
Appadurai, A. (2012). *Modernità in polvere*. Raffaello Cortina.
Arrese, Á., & Vara-Miguel, A. (2016). A comparative study of metaphors in press reporting of the euro crisis. *Discourse & Society, 27*(2), 133–155. https://doi.org/10.1177/0957926515611552
Asplund, T. (2011). Metaphors in climate discourse: An analysis of Swedish farm magazines. *Journal of Science Communication, 10*(4), 1–8. https://doi.org/10.22323/2.10040201
Atanasova, D., & Koteyko, N. (2017). Metaphors in online editorials and Op-Eds about climate change, 2006-2013: A study of Germany, the United Kingdom, and the United States. In K. Fløttum (Ed.), *The role of language in the climate change debate*. Routledge.
Boykoff, M., Aoyagi, M., Ballantyne, A. G., Benham, A., Chandler, P., Daly, M., Doi, K., Fernández-Reyes, R., Hawley, E., McAllister, L., McNatt, M., Mocatta, G., Nacu-Schmidt, A., Oonk, D., Osborne-Gowey, J., Pearman, O., Petersen, L. K., Simonsen, A. H., & Ytterstad, A. (2021a). World newspaper coverage of climate change or global warming, 2004–2021. In *Media and climate change observatory data sets*. Cooperative Institute for Research in Environmental Sciences, University of Colorado. Retrieved 15 April, 2021, from https://sciencepolicy.colorado.edu/icecaps/research/media_coverage/world/index.html
Boykoff, M., Church, P., Katzung, J., Nacu-Schmidt, A., & Pearman, O. (2021b). A review of media coverage of climate change and global warming in 2020. In *Media and climate change observatory*. Cooperative Institute for Research in Environmental Sciences. University of Colorado. Retrieved 15 April, 2021, from https://sciencepolicy.colorado.edu/icecaps/research/media_coverage/summaries/special_issue_2020.html
Caimotto, M. C. (2021). Siamo in guerra o sulla stessa barca? Le metafore della pandemia. In M. Cuono, F. Barbera, & M. Ceretta (Eds.), *L'emergenza Covid-19. Un laboratorio per le scienze sociali*. Carocci Editore.
Carrington, D. (2019). Why the Guardian is changing the language it uses about the environment. In *The Guardian, UK.*. Retrieved 14 April, 2021, from, https://www.theguardian.com/environment/2019/may/17/why-the-guardian-is-changing-the-language-it-uses-about-the-environment
Charteris-Black, J. (2004). *Corpus approaches to critical metaphor analysis*. Palgrave Macmillan.
Cohen, M. J. (2011). Is the UK preparing for "war"? Military metaphors, personal carbon allowances, and consumption rationing in historical perspective. *Climatic Change, 104*, 199–222. https://doi.org/10.1007/s10584-009-9785-x
Corner, A., Shaw, C., & Clarke, J. (2018). rinciples for effective communication and public engagement on climate change: a handbook for IPCC authors. In *Climate outreach. Oxford*. Retrieved 28 May, 2021,from https://www.ipcc.ch/site/assets/uploads/2017/08/Climate-Outreach-IPCC-communications-handbook.pdf
Fløttum, K. (2017). *The role of language in the climate change debate*. Routledge.
Futerra Sustainability Communications Ltd. (2005). *The rules of the game: Principles of climate change communications*. London. Retrieved 15 April, 2021, from https://www.stuffit.org/carbon/pdf-research/behaviourchange/ccc-rulesofthegame.pdf
Giddens, A. (2009). *The politics of climate change*. Polity.
Hoffman, A. (2012). Climate science as culture war. *Stanford Social Innovation Review, 10*(4), 30–37. https://doi.org/10.2139/ssrn.2944200
Höijer, B. (2010). Emotional anchoring and objectification in the media reporting on climate change. *Public Understanding of Science, 19*(6), 717–731. https://doi.org/10.1177/0963662509348863
Hulme, M. (2009). *Why we disagree about climate change*. Cambridge University Press.
Lacroix, K., Goldberg, M., Gustafson, A., Rosenthal, S., & Leiserowitz, A. (2020). Should it be called "natural gas" or "methane"? In *Yale program on climate change communication*. Retrieved 15 April, 2021, from https://climatecommunication.yale.edu/publications/should-it-be-called-natural-gas-or-methane/.

Lakoff, G., & Turner, M. (1980). *Metaphors we live by University of Chicago Press.* University of Chicago Press.

Lakoff, G. (2016). Language and emotion. *Emotion Review, 8*(3), 269–273. https://doi.org/10.1177/1754073915595097

Lakoff, G. (2010). Why it matters how we frame the environment. *Environmental Communication, 4*, 70–81. https://doi.org/10.1080/17524030903529749

Lorenzoni, I., & Pidgeon, N. (2006). Public views on climate change: European and USA perspectives. *Climatic Change, 77*, 73–95. https://doi.org/10.1007/s10584-006-9072-z

Lorenzoni, I., Leiserowitz, A., De Franca, D. M., Poortinga, W., & Pidgeon, N. (2007). Cross-national comparisons of image associations with "global warming" and "climate change" among laypeople in the United States of America and Great Britain. *Journal of Risk Research, 9*(39), 265–281. https://doi.org/10.1080/13669870600613658

Maasen, S., & Weingart, P. (2000). *Metaphors and the dynamics of knowledge.* Routledge.

Mio, J. S. (1997). Metaphors and politics. *Metaphor & Symbol, 12*(2), 113–133. https://doi.org/10.1207/s15327868ms1202_2

Moirand, S. (2007). *Les discours de la presse quotidienne. Observer, analyser, comprendre.* Presses Universitaires de France.

Moser, C. S. (2016). Reflections on climate change communication research and practice in the second decade of the 21st century: What more is there to say? *WIREs Climate Change, 7*(3). https://doi.org/10.1002/wcc.403

Nerlich, B., Koteyko, N., & Brown, B. (2010). Theory and language of climate change communication. *WIREs Climate Change, 1*, 97–110. https://doi.org/10.1002/wcc.2

Nerlich, B., & Jaspal, R. (2012). Metaphors we die by? Geoengineering, metaphors and the argument from catastrophe. *Metaphor and Symbol, 27*(2), 131–147. https://doi.org/10.1080/10926488.2012.665795

Nerlich, B., & Jaspal, R. (2013). UK media representations of carbon capture and storage: Actors, frames and metaphors. *Metaphor and the Social World, 3*(1), 35–53. https://doi.org/10.1075/msw.3.1.02ner

Painter, J. (2013). *Climate change in the media: Reporting risk and uncertainty.* I. B. Tauris.

Romaine, S. (1997). War and peace in the global greenhouse: Metaphors we die by. *Metaphor & Symbolic Activity, 11*, 175–194. https://doi.org/10.1207/s15327868ms1103_1

Russill, C. (2015). Climate change tipping points: Origins, precursors, and debates. *Climate Change, 6*, 427–434. https://doi.org/10.1002/wcc.344

Schön, D. (1993). Generative metaphor: A perspective on problem-setting in social policy. In A. Ortony (Ed.), *Metaphors and thought* (pp. 137–163). Cambridge University Press.

Semino E (2020) 'A fire raging': Why fire metaphors work well for Covid-19. In: Lancaster.ac.uk. Retrieved 15 April, 2021, from https://www.lancaster.ac.uk/linguistics/news/a-fire-raging-why-fire-metaphors-truly-fan-the-flames-of-covid-19.

Shaw, C., & Nerlich, B. (2015). Metaphor as a mechanism of global climate change governance: A study of international policies, 1992–2012. *Ecological Economics, 109*, 34–40. https://doi.org/10.1016/j.ecolecon.2014.11.001

Sontag, S. (1989). *AIDS and its metaphors.* Farrar.

Schön, D., & Rein, M. (1994). *Frame reflection: Toward the resolution of intractable policy controversies.* Basic Books.

Steiner, B. (2002). De la langue de bois à la langue de coton: les mots du pouvoir. In G. Rist (Ed.), *Les mots du pouvoir. Sens et non-sens de la rhétorique internationale. Cahiers de l'IUED* (pp. 193–208). Graduate Institute Publications. https://doi.org/10.4000/books.iheid.2470

Stoknes, P. E. (2015). *What we think about when we try not to think about global warming.* Chelsea Green Publishing.

Wallis, P., & Nerlich, B. (2005). Disease metaphors in new epidemics: The UK media framing of the 2003 SARS epidemic. *Social Science & Medicine, 60*, 2629–2639. https://doi.org/10.1016/j.socscimed.2004.11.031

Eleonora Ciscato is a PhD candidate in International and Public Law, Ethics and Economics for Sustainable Development at the University of Milan, Department of Italian and Supranational Public Law. Her research is mainly focused on environmental common goods, climate change governance, restoration practices and sustainable development policies. She is the co-founder of Diciassette, an association working on environmental justice and climate crisis awareness raising, both with civil society and Italian local institutions.

Marianna Usuelli holds a Master's degree in Anthropology obtained at the University of Turin and a Master's degree in Interdisciplinary Approaches to Climate Change at the University of Milan. She is a contributor to the Italian monthly magazine Altreconomia and to the newspaper Milano Finanza, covering environmental, climate change, and social themes. Currently, she covers the position of communication expert at Comune di Milano, holding workshops in highschools about climate change and sustainability awareness.

Chapter 4
Measuring Complex Socio-economic Phenomena. Conceptual and Methodological Issues

Filomena Maggino and Leonardo Salvatore Alaimo

Abstract The object of this book is to analyze how climate change can be a driver for sustainable growth by considering different disciplinary approaches. The starting point for such a reflection is to determine what is climate change and what sustainable growth means. Only then will we be able to investigate whether and how climate change may affect sustainable growth. The answer to these questions, which will be addressed in the following chapters, is not an easy task. The difficulty in understanding socioeconomic phenomena such as those examined in this volume is linked to their *complexity*.

4.1 Introduction

This book aims to analyze how climate change can be a driver for sustainable growth by considering different disciplinary approaches. The starting point for such a reflection is to determine what is climate change? And what is meant by sustainable growth? Only then will we be able to investigate whether climate change may affect sustainable growth and how. The answer to these questions, which will be addressed in the book, is not easy. The difficulty in understanding socio-economic phenomena such as those examined in this volume is linked to their complexity.

People have always felt an innate drive to know the reality to understand it and obtain information on the phenomena; this exercise helps achieve goals, satisfying needs and aspirations. In simple terms, we could say that knowledge is one of the

---

F. Maggino
Sapienza University of Rome, Pontificia Academia Mariana Internationalis, (Formerly) Control Room Benessere Italia at the Italian Prime Minister's Office, Rome, Italy
e-mail: filomena.maggino@uniroma1.it

L. S. Alaimo (✉)
Sapienza University of Rome, Italian National Institute of Statistics (Istat), Rome, Italy
e-mail: leonardo.alaimo@uniroma1.it

© The Author(s), under exclusive license to Springer Nature Switzerland AG 2022
S. Valaguzza, M. A. Hughes (eds.), *Interdisciplinary Approaches to Climate Change for Sustainable Growth*, Natural Resource Management and Policy 47, https://doi.org/10.1007/978-3-030-87564-0_4

components of our lives. Thus, the relationship between people and knowledge has always been a crucial topic in the reflections of scholars of every scientific discipline. Knowing reality means measuring it; measuring reality means dealing with its complexity.

In this chapter, we will address the topics of complexity and measurement of socio-economic phenomena.

4.2 Complexity and Knowledge

In recent years, complexity has become a mainstream topic in both the natural and social sciences. This term is used in different contexts and disciplines (e.g., physics, chemistry, biology, sociology, psychology). This often makes its definition problematic. What is complexity? What does complex mean?

Complexity in science has no precise meaning and no unique definition (Érdi, 2008). "Complex" is, sometimes, an abused term, used instead of other more appropriate ones (Maggino & Alaimo, 2021; Alaimo, 2021a, b). This term is often taken as a synonym for complicated, to refer to the difficulty in handling a situation or understanding a concept. Often, when we encounter challenging situations or we deal with phenomena whose meaning escapes us or for which we cannot find an immediate explanation, we tend to classify them generically as "complex" or "complicated." In this way, we give these two concepts the same meaning. But is that true? If we consider the etymology of the two terms, we can understand the difference in their meanings (De Toni & Comello, 2007; Letiche et al., 2012). Complicated comes from the Latin cum *plicum*, in which the term *plicum* indicates the fold of a sheet. Complex comes from the Latin cum *plexum*, where *plexum* means knot, weave. Consequently, complicated refers to something folded that can be explained and understood by unfolding its folds. Complex indicates something woven, knotted, interweaving, composed of many interconnected parts; compound; composite. An intricate association or assemblage of related things, parts, or units: an interrelated system. Dealing with a complicated problem requires adopting an analytic approach: the solution can be found by unfolding the problem in its creases, by identifying its basic components. In other words, the solution to the entire problem derives from the solution of its individual parts. As difficult as the problem may seem, it is always possible to find a solution. For instance, think of an embroidered tablecloth on a laid table and napkins with the same embroidery, but it is not visible, because they are folded. The embroidery on the latter will be immediately evident when we open them up by undoing the folds. The same thing happens when we try to solve a complicated problem: in order to understand it in its entirety (the embroidery hidden between the folds of the napkin), we have to identify its components (the folds of the napkin) and understand them (unfold them). Dealing with complexity requires a synthetic or systemic approach. It is not possible to understand the *plexum* by analyzing the individual components, because one would lose the whole. Think of a nice jumper, with an intricate weave and many different colors. If we split up the jumper weave in its basic threads, we obtain a set of threads whose analysis does not help recreate the original jumper. In other words, if we consider the single threads taken individually (adopting an analytic approach), we do not have a vision of the jumper, which comes from their interweaving. The solution to the complex problem must be found by trying to understand it as a whole. In brief, if we understand the single elements of a complicated problem or phenomenon, nothing prevents us from fully understanding it. It may take more or less time, more or less effort, but it will always be understood based only on the analysis of its individual components. On the contrary, focusing only on the analysis of its individual components does not provide a full understanding of a complex phenomenon; we must have a global perception of it. We need different approaches to the complicated and the complex. "The properties of the parts can be understood only within the context of the larger whole (...). Systems thinking is contextual, which is the opposite of analytical thinking. Analysis means taking something apart in order to understand it; systems thinking means putting it into the context of a larger whole" (Capra, 1996, 29–30). Contextuality is one of the main characteristics of complex systemic thinking: we must search for the sense of things, their meaning, within the context in which they are observed, in relation to the reality that surrounds them. The transition from analysis to synthesis, closely linked to the irruption of complexity in sciences, represents one of the most critical advances in twentieth-century science. This transition coincides with the awareness of understanding complexity employing analysis. "In the shift from mechanistic thinking to systems thinking, the relationship between the parts and the whole has been reversed. Cartesian science believed that in any complex system the behavior of the whole could be analyzed in terms of the properties of its parts. Systems science shows that living systems cannot be understood by analysis. The properties of the parts are not intrinsic properties but can be understood only within the context of the larger whole" (Capra, 1996, 37). This statement by Capra underlines that the synthetic approach does not aim at reducing complexity. A meaningful synthesis must be a *stylization of reality* (Maggino, 2017), presenting those characteristics that arise from the particular and often unique interconnections among its elements.

Complex is frequently used as a synonym for *difficult*. A complex problem is sometimes considered difficult because we are unable to understand or explain it. However, this difficulty is not inherent in the complex nature of the problem, but in trying to study it with an analytical approach, breaking it down into its essential components, rather than understanding it as a whole. It should also be highlighted that complexity is different from *completion*. Having a complex view of reality does not mean having a complete view of it. The latter indicates that all components of a phenomenon are included, with nothing missing. However, as clearly analyzed in the previous pages, having all the elements available and even analyzing them is not sufficient to understand a complex phenomenon. In understanding complexity, everything is interdependent; we cannot isolate the elements from one another. Having a sense of complexity means having the sense of solidarity—the sense of the multidimensional nature of reality (Morin, 2008).

We cannot approach the study of complexity through a preliminary definition: there is no such thing as one complexity but different complexities (Morin, 1985). This term can assume profoundly different meanings because it has been influenced by the contribution of many disciplines. Complexity does not belong to a particular theory or discipline but rather to a *discourse about science* (Stengers, 1985). Its importance coincides with a transformation in the relationship with knowledge. These two concepts are closely linked. Humanity has always aimed to reflect on its existence and investigate the possibilities and limits of knowledge. The more prominent the limits of human understanding become, the more critical complexity becomes in sciences: each increase in knowledge corresponds to an increase in ignorance and inability to know. Thus, the growing attention to complexity coincides with natural evolution in science corresponding to the transition from *classical* to *modern science*.

With the term "classical science," we indicate the scientific revolution of the seventeenth century. The idea behind this approach is that it is possible to know any object or phenomenon of reality by breaking it down into its elementary components. The objective is to simplify things and make the phenomena widely predictable, reducing them to their simple elements. In this way, it is believed that it is possible to achieve objective knowledge, idest, achieved by means of separation between subject and object, between beings and nature. The aim is to search for a model, an ideal representation of the phenomena that encloses all their characteristics. It is necessary to achieve the Platonic *hyperuranium*, explain reality in terms of generalization and immutability. The true and correct understanding of phenomena should seek their stability and unchangeability, considered essential characteristics of their objective nature. This approach to knowledge is based on two main concepts (for a complete analysis, please see Alaimo, 2020):

- **Causal explanation**: it is based on the assumption that by finding the cause of a phenomenon, we explain its behavior. The cause is the factor of an object. If the cause occurs, it inevitably determines the occurrence of the object. Consequently, the object is perfectly predictable. In this perspective, uncertainty is not admitted. Complex objects, being unpredictable, are considered *non-scientific*.
- **Experimental method**: this method, introduced by Galileo Galilei, is based only on what is expressly observable and measurable. It focuses on the observation of phenomena, on the use of mathematics, and on the reproducible experiment. Through observation and repeated experimentation, we can interpret the mathematical relationships that underlie and determine natural phenomena. Scientific hypotheses are then formulated and subjected to the control of the experimental method. The confirmed hypotheses become scientific laws.

From these two principles derives the *deterministic* view typical of classical science, id est, the idea that it is possible to predict the future (the effect) from the present (the cause). *Linearity* is the characteristic of the laws that describes reality: to specific causes correspond certain effects, which vary following linear laws. Time loses its meaning in classical science: it is an ideally reversible series of homogeneous events referable to quantitative laws. Everything is in equilibrium; if something does not seem to be in equilibrium, it is because of human limits. Objects are closed systems isolated from the environment. A deterministic, in equilibrium, linear reality is *ordered*, governed by precise rules and laws, in which there is no uncertainty. Reductionism is the concept that sums up this approach to knowledge. It entered scientific thinking between the seventeenth and eighteenth centuries, linked to the spread of the Newtonian mechanistic model, according to which reality can be *reduced* in terms of elementary particles and their movements. The importance of this principle for classical science lies in the idea that all phenomena can be explained rationally through mathematical models and laws. Knowledge is achieved by searching for a two-way correspondence between reality and a mathematical model capable of grasping its order that can be obtained only by reducing the heterogeneous to the homogeneous. All phenomena must be reduced to a purely quantitative and measurable level. This approach inevitably leads to the mechanistic construction of reality. Based on these assumptions, complexity is absolutely rejected. Conceived as a synonym for uncertainty, it concerns only superficial or illusory appearances, since the criterion of the truth of classical science is expressed by simple laws and concepts.

The principles of classical science are absolutely valid, and their importance is also acknowledged by scholars of complexity. The question is that those principles and criteria are insufficient to explain phenomena. Complexity does not contradict classical science but can be considered a complement to it, adding principles and concepts to those already present (Fig. 4.1). Starting with the birth of thermodynamics in the nineteenth century, several scientific contributions in different fields (e.g.,

	Classical Science	Complexity
Equilibrium	X	X
Non-equilibrium		X
Closed systems	X	
Open systems		X
Determinism	X	X
Fate		X
Linearity	X	X
Non-linearity		X
Reveribility	X	X
Irreversibility		X
Order	X	X
Disorder		X

Fig. 4.1 Main principles of classical science and complexity. *Source*: Alaimo (2020)

the Gestalt psychology, the theory of relativity, the uncertainty principle) formed the basis for the advent of complexity.

The main concepts of complexity derive from the research work of Ilya Prigogine, the best-known scholar in that field. He examines in depth the principles elaborated in other sciences and systematizes them, creating the complexity science. The starting points of his thought are the concepts of open *system* and *entropy*. The latter, introduced by Rudolf Clausius in 1865, states that in every mechanical process, part of (or all) the energy is dissipated in the form of heat. In simple words, entropy can be considered the impediment to transforming all the energy contained in a system. If a system has a limited amount of energy and is isolated (*closed system*), it is destined to exhaust the amount of transformable energy. All phenomena have a specific trend (*principle of irreversibility*), which is the one that tends to increase entropy. The latter grows until it reaches the state of *thermal equilibrium*, where changes in the system are no longer possible. Entropy is also interpreted as the amount of disorder in a system, because heat is the random movement of system's elements. Therefore, the second law of thermodynamics describes the irreversible movement of closed systems toward a state of disorder. In every closed system, energy is constant, while entropy tends to a maximum. According to Prigogine, systems cannot be conceived as closed, because they are contained within other systems and can exchange energy and information with them. For such open systems, the principle of entropy is not always valid without limitations because systems can exchange energy with the environment and other systems, varying their entropy. The variation of entropy ΔS in a system is given by:

$$\Delta S = \Delta_i S + \Delta_e S$$

where $\Delta_i S$ is the entropy produced within the system and $\Delta_e S$ is the one that the system receives from the environment. The latter can be null (in a closed system), positive, or even negative. The last situation is defined as *negentropy* (Schrödinger, 1944), the reverse concept of entropy, which describes the order that can emerge from chaos. This does not conflict with the second law of thermodynamics. In practice, systems tend to evolve between two opposing tendencies: entropy (disorder) and negentropy (order). They tend toward entropy, but they can also tend to a state of minimal entropy by importing energy from the outside. Thus, order and disorder, structure and change are linked together. Prigogine's work shifts the attention from stability to instability, *from being to becoming*, like the title of one of his famous books (Prigogine, 1982).

Phenomena are systems that can be in equilibrium but also *in non-equilibrium*. When equilibrium prevails, there is determinism, while when non-equilibrium prevails, *Fate* has an essential role. Uncertainty enters science, not in opposition to determinism: some phenomena can be predicted (determinism) and others cannot. Periods of linearity are followed by periods of *non-linearity*, where small changes can generate significant effects (the so-called *butterfly effect*). Reality is characterized by the presence of concepts, which only apparently seem to exclude each other. Contradiction is a purely complex concept. "In the classical view, when a contradiction appears in reasoning it is a sign of error. You have to back up and take a different line of reasoning. However, in a complex view, when one arrives via empirical rational means at contradictions, this points not to an error but rather the fact that we have reached a deep layer of reality that, precisely because of its depth, cannot be translated into our logic" (Morin, 2008, 45).

Complexity gives a new interpretation of the concept of *time*. It is conceived as a theoretically reversible set of homogeneous states in classical science, explainable by mathematical laws, and connected by causal links. Irreversibility becomes a key element of complexity: the past does not imply a certain future; the latter cannot be known from analyzing a series of conditions. In short, the future is open.

Reductionism is profoundly reshaped, because complexity requires a new way of looking at the world, a complex, non-intuitive, and non-linear causality, alongside the simple one derived from classical science. In order to understand complex phenomena, it is necessary to accept that there is a *circular relationship* between different and interconnected aspects of reality.

The idea of objective knowledge, obtained through detachment from observed reality, disappears. The static and immutable image of reality is lost, and it becomes clear that it is dynamic, temporal, in perpetual becoming. It is impossible to study phenomena isolated from their context, conceptualizing them as ideal entities. Reality is an entity that grows and develops over time, not a static object regulated by immutable laws. Each phenomenon manifests itself in an articulated way and presents the fundamental characteristic identified by Aristotle: from the interaction of the parts emerge new properties not present in the single parts. Morin (1977) defines them *emergencies*, new properties of a system with respect to those of the individual elements taken alone or linked by different interactions in another system. We cannot photograph reality as it is: the researcher builds a series of levels of reality, the result of his cultural preferences and cognitive abilities (Maturana & Varela, 1980). The idea of objective knowledge and the researcher distinct from the object of his investigation fails. The only way to create knowledge is by means of a dialogue between the researcher and reality, a dialogue that necessarily presupposes the subjective component. In this sense, we can affirm that complexity is *subjective*: the observer, based on his knowledge and experience of phenomena, establishes whether reality is more or less complex. In this vision, our knowledge is always relative to, and is conditioned by, a point of view. It is a product of our mind.

The concept of complexity has led to a number of important innovations in the relationship with knowledge. In particular, the need for a new way of looking at reality emerges: the importance of going beyond empirical evidence, trying to grasp at the same time the whole and the individual components that make it up.

4.3 Complexity and Systems

Complex is often associated with the term system. Before analyzing the main characteristics of complex systems, we must define a system. This term is used both in the common language and in many scientific disciplines and has several meanings and definitions. For instance, von Bertalanffy (1968), considered the father of systemic theory, defines it as a set of elements standing in interaction. This definition does not formally clarify which are the elements themselves. Furthermore, there is no reference to the criterion for choosing either objects or relations that are given a systemic character. The criterion of choice, specific to the observer, appears in the definition of Miller (1995): a system is a region delimited in space-time. The term *delimited* refers to an observer who chooses. A more precise definition of a system is that of Morin (1977): an organized global unit of interrelationships between elements, actions, or individuals. Thus, a set of elements, to be a system, must be governed by an organizational principle that establishes the rules of interaction between the elements.

An essential contribution to the theory of systems and the development of systemic thinking has been made by Donella H. Meadows. A system is "an interconnected set of elements that is coherently organized in a way that achieves something" (Meadows, 2009, 11). Meadows identifies the main components of a system: elements, interconnections, and functions. A system is not just a set/collection of things; they must be interconnected and have a purpose, an objective. The purpose of a system is often difficult to identify. "The best way to deduce the system's purpose is to watch for a while to see how the system behaves" (Meadows, 2009, 14). From this statement of Meadows, it can be deduced that a system has its behavior, different from its parts and that, like any behavior, it can change over time. Meadows highlights the dynamism of the systems, their adaptation over time. Obviously, the change can concern both the system as such and one or even all of its essential components. Change can also be traumatic and unexpected. Most systems can withstand the impact of drastic changes thanks to one of their fundamental characteristics, resilience, *id est*, "system's ability to survive and persist within a variable environment" (Meadows, 2009, 76).

Therefore, a system can be defined as an organic, global, and organized entity, made up of many different parts, aimed at performing a certain function. If one removes a part of it, its nature and function are modified; the parts must have a specific architecture, and their interaction makes the system behave differently from its parts. Systems evolve, and most of them are resilient to change.

What are the characteristics that make a system *complex*? Complex systems are made up of a great variety of elements, which have specialized functions. Therefore, elements are different from one another, and it is precisely this diversity that makes it difficult for their understanding. Elements often are other systems, which are in turn formed by systems and so on. This progressive encapsulation forms a *systemic hierarchy*, an essential characteristic of complex systems. Such a hierarchical structure allows the control of the elements, ensuring that they act in a coordinated and harmonious way. This type of structure is governed by the *slaving principle* (Haken, 1983): the elements at a lower hierarchical level are slaved to the upper level and to the overall (holistic) behavior. In order to understand a complex system, it is not necessary to analyze all its hierarchical levels, all the subsets that constitute it. In a complex system, the *interconnections* among the elements are more important than the elements themselves. The high density of interconnections is typical: the various elements are connected by a great variety of non-linear links. This is a fundamental characteristic. In simple systems, the whole is strictly equal to the sum of its parts; the connections do not bring any added value. Non-linear connections are essential in the definition of the structure and the organization of the system. Simple systems are characterized by few elements and few linear relationships between them; they can be analyzed analytically. Complex systems, on the contrary, are made up of many elements and many relations, linear and non-linear; they can be analyzed only in a synthetic way. In a complex system, elements and connections, besides being numerous, are various and different.

A particular type of complex system is *Complex Adaptive System* (CAS). It can be defined as an open system made up of numerous elements interacting with each other, in a linear and non-linear way, that constitute a unique and organic entity *capable of evolving and adapting to the environment*. Thus, a CAS adds to the other characteristics typical of complex systems, the ability to adapt and *learn*. CASs can adapt to the environment by processing information and building models capable of assessing whether adaptation is useful. The elements of such a system have the primary purpose of adapting and, consequently, they constantly look for new ways of doing things and learning.

The main characteristics of complex adaptive systems are typical of social organizations and phenomena. Each of them is made up of a network of elements, which interact both with one another and with the environment. They are multidimensional, and their different elements or dimensions are linked together in a non-linear way. They evolve over time, modifying both their dimensions and the links between them. Understanding complex socio-economic phenomena involves the ability to measure them using a synthetic approach. Their measurement requires the definition of systems of indicators capable of capturing their different aspects. As can be easily understood, these systems are dynamic, since they have to adapt to the changes in the measured phenomena. In simple terms, they are CASs and can be monitored and measured employing systems of indicators that are CASs themselves.

4.4 Measurement in Social Sciences

4.4.1 General Aspects of Measurement

The topic of measurement is often ignored by researchers and considered a niche field in the academic debate. However, scientific knowledge develops as *a dialogue between logic and evidence*, through two steps of analysis, linked together even if analytically distinct (Maggino, 2004):

- a theoretical-formal level, in which theories and hypotheses are developed and abstract concepts with their mutual relations are specified,
- an empirical level, in which hypotheses are verified through empirical data.

Each observation evaluated within a theoretical framework represents a datum. Consequently, we can obtain different types of data from the same empirical observation based on different theoretical frameworks. The framework allows comparing observation with one or more models, identified by a dimensional system based on an unambiguously defined unit. The relationship between the model and the observation is the product of the measurement. Therefore, knowledge is the result of an interaction between theory and observations represented and realized by measurement. This interaction is necessary and unavoidable (Alaimo, 2020).

Thus, we can consider measurement as the application of a formal model to a property of a series of empirical objects. That model can represent reality at different levels of accuracy. If it provides a faithful image of an empirical system, then the logical implications must be comparable with the observable behavior of the objects. If empirical observations are consistent with model-based predictions, then it can be concluded that the model provides an acceptable description of that segment of reality.

Generally speaking, measurement can be defined as the evaluation of the extension of something (an object, a property, etc.) in relation to a unit of measurement. According to Blalock (1982), measurement is a process by which numbers are assigned to objects so that it is also understood which types of mathematical operations can be legitimately used. Measuring involves a sort of *translation* (Alaimo, 2020), a shift from the plane of reality in which we observe phenomena to the plane of numbers in which we try to encode them. This translation must be:

- *meaningful*, it must reproduce as faithfully as possible the phenomenon in terms of numbers;
- *necessary*, it is the only way to know reality, which speaks to us with the *language of numbers*.

Of course, it is also possible to make errors, differences between the *true* value and the *measured* value; in particular, we can distinguish two types of error. The *random error* refers to all those factors that confuse and disturb the measurement of any phenomenon. The higher this error, the lower the level of reliability of the measuring instrument, *id est*, the overall consistency of a measure, its capacity to produce similar results under consistent conditions. Variables always contain a random

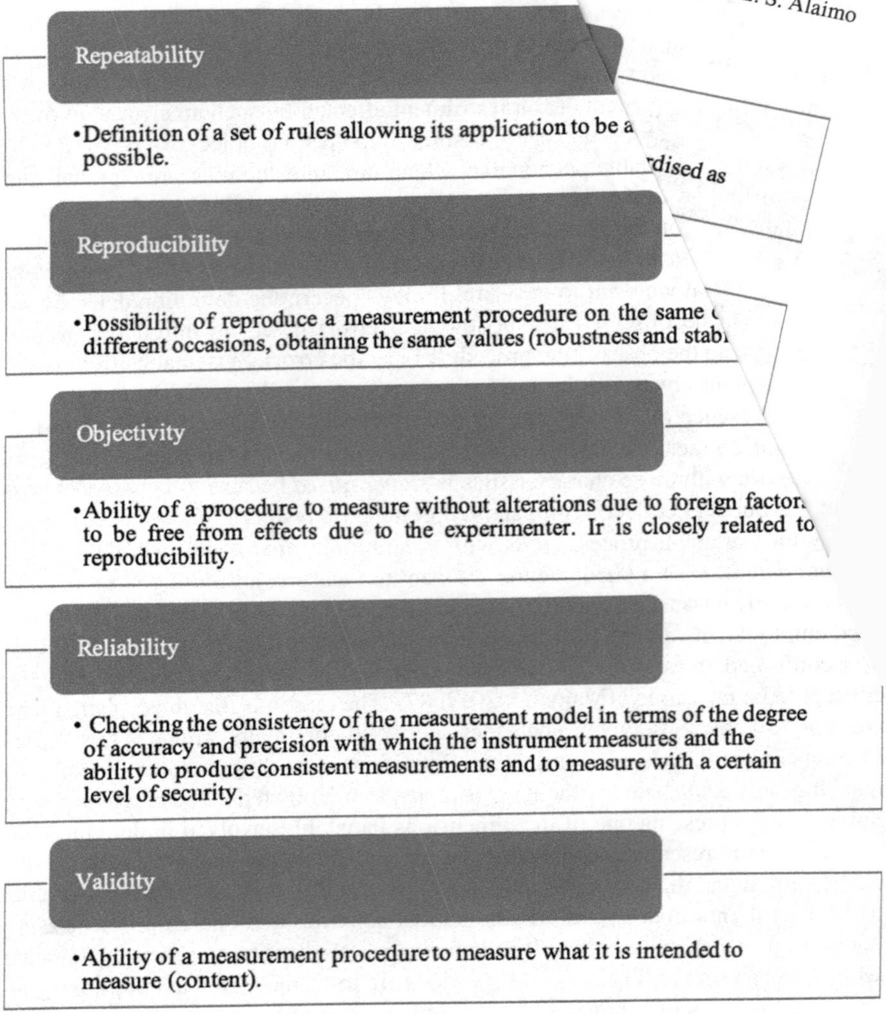

```
┌─ Repeatability
│  • Definition of a set of rules allowing its application to be a
│    possible.
│
┌─ Reproducibility
│  • Possibility of reproduce a measurement procedure on the same
│    different occasions, obtaining the same values (robustness and stab.
│
┌─ Objectivity
│  • Ability of a procedure to measure without alterations due to foreign factor.
│    to be free from effects due to the experimenter. Ir is closely related to
│    reproducibility.
│
┌─ Reliability
│  • Checking the consistency of the measurement model in terms of the degree
│    of accuracy and precision with which the instrument measures and the
│    ability to produce consistent measurements and to measure with a certain
│    level of security.
│
┌─ Validity
│  • Ability of a measurement procedure to measure what it is intended to
│    measure (content).
```

Fig. 4.2 Main characteristics of the measurement process. (Reproduced from Alaimo, 2020)

4.4.2 Measuring Socio-economic Phenomena: Conceptual and Methodological Aspects

Measuring socio-economic phenomena, such as wellbeing, sustainable development, climate change, and so on, requires acknowledging their specificity. Almost all social phenomena are complex; their measurement will also have to consider this complexity. This is the reason why we started this chapter with the concept of complexity. To correctly understand those phenomena, we must take into account their nature. They are something different from the simple sum of their parts. Knowing these phenomena means measuring them. Knowledge is the result of a complex interaction between theory and observations realized by means of measurement. This interaction is the basis of *scientific research*, which we can define as a creative process of discovery that develops according to an established itinerary and procedures that are consolidated within the scientific community. There is no contradiction between creativity and the presence of established procedures. As Reichenbach (1938) states, these are different steps in the process of knowledge. The first stage, called context of discovery, is not subject to rules and procedures: it is not possible to define logical and invariable rules that allow the creative function. Scientific work is not just about producing new theoretical hypotheses. The scientist must also test them. The *context of justification* consists precisely in the empirical verification of theories, which must be done following specific rules.

Sociologist Paul Felix Lazarsfeld made an essential contribution to the study of measurement in social sciences. He starts from the consideration of the specificity of measuring social phenomena. "When social scientists use the term measurement it is in a much broader sense than the natural scientists do" (Lazarsfeld, 1958, 100). Measurements in the social field have a typical character, requiring a specific procedure. Lazarsfeld defines this procedure as allowing the empirical translation of the theory, the so-called *operationalization*. The first step is the so-called *imagery of the concept*: the researcher must create a quite generic image of phenomenon (construct). The construct may often result from the perception of many heterogeneous phenomena having some underlying characteristic in common; the researcher tries to account for them. "In any case, the concept, when first created, is some vaguely conceived entity that makes the observed relations meaningful" (Lazarsfeld, 1958, 101). The second step is the concept specification, in which the construct is breaking down into components, called *dimensions*. They can be derived logically from the overall concept or one aspect can be deduced from another. "Every concept we use in the social sciences is so complex that breaking it down into dimensions is absolutely essential in order to translate it into any kind of operation or measurement" (Lazarsfeld, 1958, 102). The third step is the *selection of indicators* for each dimension identified. At this stage, the researcher has to address some problems. The first problem is understanding what an indicator is. Lazarsfeld affirms that indicators are directly suggested to researchers by common experience, and "each indicator has not an absolute but only a probability relation to our underlying concepts" (Lazarsfeld, 1958, 103). This means that the relationship depends on the definition of the concept. In this sense, an indicator is *purposeful statistics* (Horn, 1993): it is not simple crude statistical information but represents a measure organically connected to a conceptual model. Consequently, a wide variety of possible indicators can be identified to measure a specific dimension of a concept. This raises another question: how many indicators should we consider? There is no *correct* answer to that question. Generally speaking, we need to choose a number of indicators that allow us to adequately represent the desired conceptual dimension, avoiding redundancy and ensuring the reduction of error. The last step is the *combination of indicators* into indices. The researcher must "put Humpty Dumpty together again" (Lazarsfeld, 1958, 104). The concept needs to be reconstituted. All the indicators that we have collected and used have produced data; at this point, a synthesis of the indicators must be made. Synthesis is the only way that allows us to have a meaningful view of complex phenomena.

This *operationalization* process translates (abstract) concepts into (measurable) variables. Therefore, the variable is an operationalized concept; more precisely, it consists of the operationalized property of an object. It is important to underline the highly subjective nature of this process: the way in which the researcher decides to operationalize a concept is questionable. There is no absolutely correct definition: the decision on how to operationalize depends only on the researchers' choices. Finally, it should be noted that the process described by Lazarsfeld is typical of quantitative sociological research. In fact, in qualitative research, there is no equivalent to the operationalization of concepts. Qualitative research moves differently. The concept is used as a sensitizing concept (Blumer, 1954), it is an orientation toward research.

In social sciences, dealing with measurement means dealing with indicators. They represent the result of the translation of reality to the plane of numbers. The term is different from index. As stated in Horn's definition, an indicator is a purposeful statistics; thus, a statistical index becomes an indicator only when its definition and measurement occur in the ambit of a conceptual model. "Indicator is what relates concepts to reality" (Maggino, 2017, 92). It represents what is actually measured with reference to a specific dimension or aspect of a phenomenon (Maggino, 2015). Several *subjectivities* are involved in this process. The definition of phenomena is subjective. Describing reality always depends on the researchers' point of view, on the *small windows* through which he observes reality. The definition of the hypotheses on reality is pervaded by subjectivity: researchers can change perspective in a path of knowledge in continuous evolution. Subjectivity also refers to the kind of information defined in the ambit of a conceptual framework and subsequently observed. In any case, this process cannot be considered arbitrary since it always involves a relationship with reality. Given the complexity of such a reality, we can consider data as a *fragmented text*; the researcher must read this text looking for a sense. This *sense structuring* process is not an arbitrary one but necessarily involves some subjectivity (Maggino, 2017).

Indicators should be developed employing a step-by-step, *hierarchical design*, a specification of the Lazarsfeld's model described in the previous pages. The first step is *defining the concept*, id est, the question *what is the phenomenon to be studied*? It is not a simple question. A good starting point is not to rely on common sense but to seek out what other researchers have done. However, evaluating the objectivity and quality in the definition of the phenomenon only by considering the reference to the literature selected by the researcher is not entirely correct. Also, literature selection is a subjective activity. The second step is the *identification of latent variables*. Each of them represents an aspect to be observed and confers an explanatory relevance onto the corresponding defined concept. They reflect the nature of the phenomenon consistently with the conceptual model. Based on its level of complexity, the variable can be described by one or more factors. The different factors of each variable are referred to as *dimensions*. The last step is the selection of basic indicators to measure the defined variables. Each latent variable could be defined and measured by a single indicator. This *single indicator approach* is weak and assumes the existence of a direct correspondence between one latent variable and one indicator. It is preferable to adopt a *multi-indicator approach*, consisting of using several indicators for each conceptual dimension to cover the conceptual dimension's variability. This approach increases measurement accuracy and precision, allowing to compensate for the random error.

The correct application of the hierarchical design produces a *system of indicators*. This is not a simple collection of measures. It is a complex system, and as such, it has all its fundamental characteristics. Indicators within a system are interconnected, and new properties typical of the system emerge from these interconnections. Using such a system, we can measure (and understand) a complex phenomenon that would not otherwise be measurable.

4.5 Final Remarks

Measuring in social sciences requires a robust conceptual definition, a consistent collection of observations, and a consequent analysis of the relationship between observations and defined concepts. Managing indicators introduces at the same time (Maggino, 2017, 111):

- a challenge, represented by the need of dealing with complexity;
- a need, given by the need of making indicators relative; and,
- a risk, given by the reductionism.

Indicators are the tools to understand complexity. They play a crucial role in describing, understanding, and controlling complex systems. An indicator is, therefore, a tool for understanding reality. It is not necessarily a number. It can be an object, a map, an image. It allows us to grasp the complexity and guide us in understanding it (Maggino et al., 2021).

The *soft power* of numbers and indicators is characteristic of our time (Porter, 1996). If we want to use indicators and other measures to make the world navigable in simpler terms, let us be careful what we wish for. We must be sure to give as meaningful as possible a representation of reality, preserving the systemic nature of phenomena.

References

Alaimo LS (2020) *Complexity of social phenomena: Measurements, analysis, representations and synthesis*. Dissertation, University of Rome, La Sapienza.
Alaimo L.S. (2021a). Complexity and knowledge, in: F. Maggino (Ed.), Encyclopedia of Quality of Life and Well-being Research, Cham: Springer- pp. 1–2. https://doi.org/10.1007/978-3-319-69909-7 104658-1. 744.

Alaimo, L.S. (2021b). Complex systems and complex adaptive systems, in: F. Maggino (Ed.), Encyclopedia of Quality of Life and Well-being Research, Cham: Springer, 2021, pp. 1–3. https://doi.org/10.1007/978-3-319-69909-7_104659-1

von Bertalanffy, L. (1968). *General system theory: Foundations, development, applications.* George Braziller.

Blalock, H. M., Jr. (1982). *Conceptualization and measurement in the social sciences.* SAGE Publications.

Blumer, H. (1954). What is wrong with social theory? *American Sociological Review, 18*, 3–10. https://doi.org/10.2307/2088165

Capra, F. (1996). *The web of life: A new scientific understanding of living things.* Anchor Books.

De Toni, A. F., & Comello, L. (2007). *Viaggio nella complessità.* Marsilio Editori.

Érdi, P. (2008). *Complexity explained.* Springer.

Haken, H. (1983). *Synergetics: An introduction.* Springer Verlag.

Horn, R. V. (1993). *Statistical indicators for the economic and social sciences.* Cambridge University Press.

Lazarsfeld, P. F. (1958). Evidence and inference in social research. *Daedalus, 87*(4), 99–130.

Letiche, H., Lissack, M., & Schultz, R. (2012). *Coherence in the midst of complexity: Advances in social complexity theory.* Palgrave Macmillan.

Maggino, F. (2004). *La misurazione nella ricerca sociale: teorie, strategie, modelli.* Firenze University Press.

Maggino, F. (2015). Assessing the subjective wellbeing of nations. In W. Glatzer, L. Camfield, V. Møller, & M. Rojas (Eds.), *Global handbook of quality of life* (pp. 803–822). Springer.

Maggino, F. (2017). Developing indicators and managing the complexity. In F. Maggino (Ed.), *Complexity in society: From indicators construction to their synthesis* (pp. 87–114). Springer.

Maggino, F., & Alaimo, L. S. (2021). Complexity and wellbeing: Measurement and analysis. In L. Bruni, A. Smerilli, & D. De Rosa (Eds.), *A modern guide to the economics of happiness* (pp. 113–128). Edward Elgar Publishing.

Maggino, F., Bruggemann, R., & Alaimo, L. S. (2021). Indicators in the framework of partial order. In R. Bruggemann, L. Carlsen, T. Beycan, C. Suter, & F. Maggino (Eds.), *Measuring and understanding complex phenomena. Indicators and their analysis in different scientific fields* (pp. 17–29). Springer.

Maturana, H. R., & Varela, F. J. (1980). *Autopoiesis and cognition: The realization of the living.* D. Reidel Pub. Co.

Meadows, D. H. (2009). *Thinking in systems: A primer.* Earthscan.

Miller, J. G. (1995). *Living systems.* University of Colorado Press.

Morin, E. (1977). *La méthode.* Tome I. La nature de la nature.

Morin, E. (1985). Le vie della complessità. In G. Bocchi & M. Ceruti (Eds.), *La sfida della complessità* (pp. 49–60). Feltrinelli.

Morin, E. (2008). *On complexity. Advances in systems theory, complexity, and the human sciences.* Hampton Press.

Porter, T. M. (1996). *Trust in Numbers: The pursuit of objectivity in science and public life.* Princeton University Press.

Prigogine, I. (1982). *From being to becoming: Time and complexity in the physical sciences.* W.H. Freeman.

Reichenbach, H. (1938). Experience and prediction: An analysis of the foundations and the structure of knowledge. Chicago: The University of Chicago Press.

Schrödinger, E. (1944). *What is life? The physical aspect of the living cell and mind.* Cambridge University Press.

Stengers, I. (1985). Perché non puó esserci un paradigma della complessità. In G. Bocchi & M. Ceruti (Eds.), *La sfida della complessità* (pp. 61–83). Milano.

Filomena Maggino is a professor of social statistics at the Sapienza University of Rome and coordinator of the Department of Integral Wellbeing, Pontifical Academy of Mary (Pontificia Academia Mariana Internationalis), Holy See. She is past president of the "Control Room Benessere Italia," Italian Presidency of the Council of Ministers (Conte's Cabinets). She serves as editor in chief of Social Indicators Research journal (Springer), editor in chief of Encyclopedia of Quality-of-Life and Well-Being Research (Springer), and president and co-founder of the Italian Association for Quality-of-Life Studies (AIQUAV). She was president of the International Society for Quality-of-Life Studies (ISQOLS) and chair and organizer of two ISQOLS conferences (Florence 2009 and 2012). She is advisor of several governmental organizations and member of ad-hoc committees about issues related to indicators, well-being, quality of life, and sustainable development.

Leonardo Salvatore Alaimo is aggregate professor of social statistics at the Department of Social Sciences and Economics, Sapienza University of Rome. He holds a Ph.D. in social statistics. He was researcher at the Italian National Institute of Statistics—Istat. He was expert on statistics at the Prime Minister Office—Italian Government (Conte's cabinet). He took part in several national and international conferences and he is author of several papers published in a wide range of international peer-reviewed journals. His research interests are mainly the measurement of complex phenomena, the synthesis of systems of statistical indicators and the multivariate statistics.

Part II
The Scientific Debate

Chapter 5
Glaciers: Vanishing Elements of Our Mountains and Precious Witnesses of Climate Change

Guglielmina Diolaiuti, Maurizio Maugeri, Antonella Senese, Veronica Manara, Giacomo Traversa, and Davide Fugazza

Abstract This chapter focuses on the clear reduction in glacier extension that has been observed worldwide in the global warming period: this reduction has strongly affected the Alpine region that has been subjected to a much higher temperature increase than the average Earth's one.

Specifically, the chapter presents different techniques of remote-sensing analysis of glaciers applied by the University of Milan (Italy) which in the recent past have led to the compilation of regional, national, and international glacier inventories, to the description of phenomena of great interest such as glacier darkening, and to the study of dynamic evolution of glaciers with impacts on environmental risk and danger conditions and therefore on tourist presence.

5.1 Introduction

Climate change is mainly represented by an increase in air temperature that has significant effects on many terrestrial environmental systems. Among them, the elements of the cryosphere are probably the most fragile and sensitive (IPCC, 2013). Glaciers, in particular, are among the best witnesses of climate change. They are not the unique elements of the natural landscape able to answer to climate and its variations but they are the ones that respond most clearly and unambiguously.

The average Earth's surface air temperature has increased globally by about 1 °C in the last 150 years (IPCC, 2013). However, some areas have experienced a more

---

G. Diolaiuti (✉) · M. Maugeri · A. Senese · V. Manara · D. Fugazza
University of Milan, Milan, Italy
e-mail: guglielmina.diolaiuti@unimi.it; maurizio.maugeri@unimi.it;
antonella.senese@unimi.it; veronica.manara@unimi.it; davide.fugazza@unimi.it

G. Traversa
University of Siena, Siena, Italy
e-mail: giacomo.traversa@student.unisi.it

substantial temperature increase: Italy has experienced a temperature increase of about 3 °C from the beginning of the nineteenth century (Brunetti et al. (2006)—updated record available at https://www.isac.cnr.it/climstor). In the same interval, glaciers worldwide have significantly reduced their length and extension (Frezzotti & Orombelli, 2014). In Alaska, some glaciers have diminished their length up to 50 km, while in Italy, the Forni Glacier, in the Stelvio National Park, has retreated its tongue by more than 2 km (Diolaiuti & Smiraglia, 2010). A length reduction of this magnitude within such a short period is undeniable, and it represents a clear signal that the climate equilibrium has been broken.

Several valleys, which in the recent past were nesting glaciers, are now abandoned by ice, and here forest and vegetation are rapidly rising, regaining the territories left free by glaciers (D'Agata et al., 2019). Therefore, the wood and the forest have risen in the last decades tens of meters of altitude, even though only a careful and trained eye is able to detect such changes. Moreover, animal and plant species are in crisis and at risk of extinction, but few people can see and understand the evidence of such phenomena. Instead, a glacier tongue that in the past occupied a whole valley and that today has retreated by 2 or more kilometers is a witness of the climate change impact, detectable and recognizable by everyone (Figs. 5.1–5.4).

Glaciers are not only witnesses of climatic variations but also of air and water quality (Miner et al., 2017). To understand why, we must trace back glaciers' origin. Specifically, glacier ice is different from sea ice that forms over the sea of extreme latitudes. Sea ice is simply frozen seawater. Differently, the ice of glaciers derives from snow transformation, also known as snow metamorphism. Layer by layer, the snow that falls in winter accumulates on the surface of the glacier, it compacts, melts, and freezes, and finally, it turns into ice. As soon as it falls, the "fresh" snow is light, featuring a density of about 100 kg/m^3. It is, in fact, very porous and rich in air. With the passing of weeks and months, the compaction and transformation of the snow crystals cause it to become denser and heavier, and at the end of the season, the "old" snow (named "firn") has a density of about 500–600 kg/m^3. Glacier ice is denser than firn, and it reaches the density value of 917 kg/m^3. The transformation from snow to ice is very slow: it takes about 10 years in the Alps and even 100 years in Antarctica, and it requires not only an abundant winter snow accumulation but also cool summers and adequate altitudes in order not to completely melt the snow coverage. This slow process causes many substances that travel into the atmosphere to remain trapped in the glacier ice. There are many pollutants, volatile substances that travel for thousands of km and then fall with the snow and remain in the ice until it melts (Fig. 5.5). This is the case of dichloro-diphenyl-trichloroethane (commonly known as DDT) used in the past, pesticides, flame retardants, flavorings. Everything remains trapped in the glacier ice, which also represents an archive of our chemical impact on the environment and a document to read, to understand, and hopefully to change!

Glacier ice cores can therefore also be used to document the time evolution of air pollution levels. This topic is particularly interesting for the Southern Alps residents as they are close to the Po Plain, which has exhibited big changes in air pollution loads over the past decades (e.g., Manara et al., 2019).

5 Glaciers: Vanishing Elements of Our Mountains and Precious Witnesses of Climate...

Fig. 5.1 The Forni Glacier (Italy). Photo by V. Sella 1890

Substances trapped in snow and ice are released when they melt, and are therefore present in the freshwaters of streams and rivers of glacial origin (Miner et al., 2017). The most abundant melting of glacier ice in recent years (Cannone et al., 2008) also impacts the amount of pollutants released and the release rate and, therefore, climate change impacts not only on the extent of glaciers but also on the quality of the freshwater they release.

Fig. 5.2 The Forni Glaciers (Italy). Photo by Casati Archive, 1929

Once released by glacier melt, pollutants enter rivers, lakes, and seas and reach the food web as well. Among the emerging pollutants on glaciers also micro- and macro-plastics are found (Ambrosini et al., 2019). By studying glacier meltwater and cataloging all the substances it contains, we can describe the history of the circulation of pollutants in the atmosphere, and therefore of the human impact on air and water quality.

Fig. 5.3 The Forni Glaciers (Italy). Photo by A. Desio, 1947

Different techniques are available to describe in detail and continuously glacier changes, ranging from field surveys to remote investigations with drones and satellites. This latest technique is called Remote-Sensing, and it is the science that allows to studying an object from afar, acquiring information and images repeatedly, over large surfaces, and safely. Remote-sensing is developing thanks to the strong development of satellite technologies quickly.

For glaciers, Remote-Sensing is particularly suitable because it allows acquiring information on the extent and characteristics of the glacier surface (e.g., presence of crevasses, surface debris distribution, and features, the occurrence of the supraglacial lake and water ponds, collapsed areas, etc.) without the need to reach the glacier or to approach the most fragile and hazardous areas. The comparison of

Fig. 5.4 The Forni Glacier (Italy). Photo by C. Smiraglia, 2019

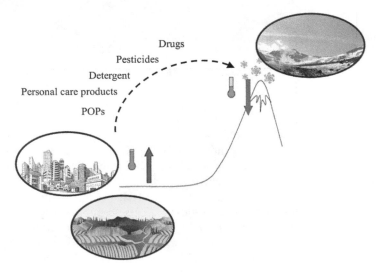

Fig. 5.5 Sketch showing the origin of the pollutants present in snow and ice and consequently in the resulting meltwaters

multitemporal remote sensed data permits quantifying changes on a weekly, monthly, or seasonal scale. With airplanes and satellites, glacier inventories can be developed, thus describing the glacier coverage of an entire region or an entire country. Remote-sensing not only allows observing glaciers directly, but it also permits to monitor meteorological variables (e.g., cloudiness and solar radiation), which have a substantial impact on the energy balance at the glacier surface and then on the snow and melt magnitude and rate (Manara et al., 2020; Senese et al., 2020).

In this chapter, we will summarize different methods of remote-sensing analysis of glaciers applied by the University of Milan (Italy), which in the recent past have led to the compilation of regional and national glacier inventories, to the description of phenomena of great interest such as glacier darkening and the study of the dynamic evolution of glaciers with impacts on environmental risk and dangerous conditions, and therefore on tourist presence.

5.2 Glacier Inventories: Useful Tools to Describe Glacier Evolution in a Changing Climate

Glacier inventories are suitable tools to investigate mountain glaciation in a changing climate (Paul et al., 2004). They should be carried out at intervals compatible with the characteristic response time of mountain glaciers (few decades or less in the case of small glaciers), even if the currently observed glacier downwasting calls for more frequent updates (Paul et al., 2007, 2011; Pfeffer et al., 2014).

For the last two decades, the University of Milan has been considered as a national reference research group to produce glacier inventories. The inventories have been developed by analyzing different types of remote-sensing data and referring to glacial areas of different extensions (from regional to national and extranational scale). In the inventories, glacier area data, the most used characteristic, are reported together with other geometry information (e.g., maximum length, minimum and maximum elevations, aspect, and slope) to evaluate glacier shrinkage and its magnitude.

5.2.1 The New Italian Glacier Inventory

To contribute to the knowledge of the rate and magnitude of Alpine glacier involution, researchers from the University of Milan published in 2015 the New Italian Glacier Inventory (Diolaiuti et al., 2019; Smiraglia et al., 2015). To detect Italian glaciers, hence mark their boundaries and calculate their area together with other crucial information (e.g., glacier name, id code, coordinates, slope, aspect, etc.), the fundamental data sources in the New Italian Glacier Inventory were color orthophotos used as base layer in a GIS (Geographic Information System) environment. The orthophotos were derived from high-resolution aerial photos featuring low or absent cloud coverage and acquired at the end of the summer when glaciers show the minimum snow mantle, and then their limits appear clearer and more detectable. They were made available kindly by regional and local administrations of Italy. Specifically, the analyzed orthophotos were surveyed in 2005 (Valle d'Aosta, RAVA Flight); 2007 (Lombardia, digital color orthophoto BLOM-CGR S.p.A.-IIT2000/VERS.2007); 2008 (Provincia Autonoma di Bolzano-Alto Adige, PAB Flight); 2009 (Veneto: LIDAR survey performed by Regione Veneto-ARPAV); 2009–2011 (Regione Piemonte, ICE Flight); 2011 (Trentino, PAT Flight); 2011 (UM flight Friuli-Venezia Giulia). The orthophotos are purchasable products featuring a planimetric resolution specified by 1 pixel (pixel size: 0.5 m × 0.5 m). The planimetric accuracy stated by the manufacturers is ±1 m. In some cases, to improve the glacier mapping, satellite images were also used (Valle d'Aosta, 2009 SPOT images featuring a resolution of 6 m) and field and literature data as well (Friuli-Venezia Giulia and Abruzzo). A cross-check of the glacier data thus obtained was performed considering already existing regional or local inventories, published maps and cartography, and dedicated field surveys. To assess the potential error affecting data entered in the glacier inventory, the approach introduced by Vögtle and Schilling (1999) was adopted. This method was largely applied to evaluate the area error of Lombardy glaciers (Citterio et al., 2007; Diolaiuti et al., 2012a) and Aosta Valley glaciers (Diolaiuti et al., 2012b), and it is based on the calculation of the area buffer for each mapped glacier. The buffer extent depends on the glacier boundary, the pixel size, and the uncertainty of the applied mapping method (this latter is due to the manual operator and evaluated for each glacier in relation to the experience of the operator and his/her knowledge of the surveyed glacier area). The final precision of the whole glacier coverage was determined by taking the root of the squared sum of all the buffer areas. Thanks to the high quality resolution of the analyzed orthophotos and the accurate manual mapping, the obtained glacier area data featured an error of less than ±2% of the actual value. Few exceptions occur in the case of supraglacial debris presence (i.e., debris-covered glaciers). These conditions are becoming even more frequent over recent years, and they are making it more difficult and uncertain to detect and map glacier outlines; particularly, the glacier snout mapping is complex since at lower elevations debris coverage can reach higher depth. In such conditions, the mapped glacier area can be underestimated by up to 10% of the actual value. From the orthophoto analysis also information on glacier aspect and type were derived (Paul et al., 2011; Pfeffer et al., 2014).

The New Italian Glacier Inventory thus obtained (Fig. 5.6, see also Diolaiuti et al., 2019) is an inventory that describes the 903 glaciers of Italy. This inventory has been compared with the previous one, which dated back to 1962. It was found that together the 903 Italian glaciers currently extend for about 369 km^2 of area (which is more or less the surface of Garda Lake), and they have lost over the last

Fig. 5.6 The New Italian Glacier Inventory, a bilingual open access work describing the 903 glaciers of Italy

50 years about 157 km² of the area, which means more or less an extension like that of the Como Lake.

The glacier regression is not equally intense on all 903 glaciers of Italy: some (the largest ones) have shrunk by about 25–30% in half a century and others (the smallest ones) have reduced even by 50% in the same time interval.

Against this intense area reduction, the number of Italian glaciers between 1962 and 2010 increased by 68 units. This increase is due to the fragmentation of the glaciers, and as a consequence, two or more smaller ice bodies can be derived from a single large glacier (Fig. 5.7).

It is, therefore, evident that the increase in the number of glaciers is not a positive signal but is an indicator of the great crisis of alpine and global glacialism.

Such an intense reduction is unfortunately generalizable to the whole of Europe since similar data were found by studying French, Austrian, and Swiss glaciers (Diolaiuti et al., 2019). Among the others, Maisch et al. (2000) analyzed area data of a comprehensive and representative sample of European glaciers and reported a general Alpine decrease of 27% from the mid-nineteenth century to the mid-1970s, and losses even more substantial in some subregions of the Alps.

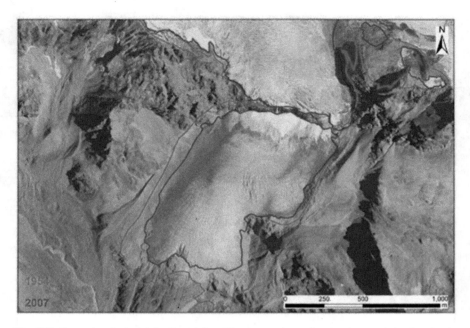

Fig. 5.7 Example of glacial fragmentation. The comparison of orthophotos of the Lombardy Region allowed identifying a case of fragmentation (red perimeters and blue perimeters) which occurred in this case in the time interval 1954–2007. The background images is the 2007 color orthophoto (courtesy Lombardy Region)

5.2.2 The European Glacier Inventory

The availability in the first decade of the twenty-first century of many national glacier inventories (e.g., Swiss since 2004, Austrian since 2005, French since 2014, Italian since 2015) has prompted European researchers to join forces to build a single Alpine glacier inventory based on homogeneous and coeval data to describe the state of health of glaciers throughout Europe. Researchers from the University of Milan participated in this international project. Specifically, the collaboration between the University of Milan, the University of Zurich, the University of Grenoble, and the Austrian company ENVEO IT Gmbh led to the publication of the entire glacier inventory open access mode (Paul et al., 2020). The study is based on data acquired by Sentinel-2 satellites in the period 2015–2017, made available free of charge by the European Space Agency (ESA). The researchers processed the data through an algorithm that allows the automatic recognition of ice and subsequently made corrections starting from the glaciological and geomorphological evidence to better delineate debris-covered glaciers. The application of an exclusively automatic technique is more problematic. The study results are also supported by a detailed analysis of the precision in the realization of the glacier outlines, which is around 5%.

Fig. 5.8 Area distribution (km²) of alpine glaciers from the European Glacier Inventory (Paul et al., 2020)

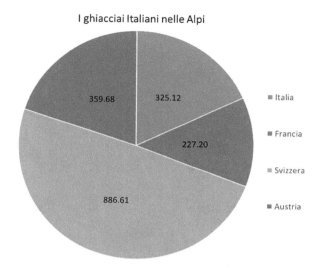

From the Alpine glacier inventory (Paul et al., 2020), it results that there are 4395 glaciers in the Alps, with a total area of 1.806 km², distributed for 49.4% in Switzerland, 20% in Austria, 12.6% in France, and 18% in Italy, with 325 km² of glacier coverage (Fig. 5.8). There are giant glaciers such as Aletsch, with its 77 km², and many glaciers are featuring a size minor than 0.1 km², which represent the majority of Alpine glaciers. Most of the Alpine glaciers are exposed to the North, where the lower solar radiation ensures a more prolonged survival, while the median altitude is around 3000 m a.s.l.

Comparing the European glacier inventory data published in 2020 with those of the previous Alpine inventory relating to 2003, for a selection of glaciers, the losses were approximately 13.2%. This corresponds to an annual reduction rate of about 1.1% and indicates how the retreat of glaciers has continued without pause from the 1980s to the present day.

If we focus on the Italian glaciers and compare the data obtained from the analysis of Sentinel Images (325 km²) with the surface of the Italian glaciers surveyed in the previous national inventory (the New Italian Glacier Inventory above described, see also Smiraglia et al. (2015)) which is based on data acquired in the period 2005–2011 (369 km²), we obtain a loss of the glacier surface of about 44 km². This occurred in less than a decade and an annual retreat rate that exceeds 1.6% for the Lombard glaciers.

The case of the Forni Glacier (Stelvio national park, Lombardy) is paradigmatic. Forni was the largest Italian valley glacier. It is now divided into three parts no longer communicating with each other (Fig. 5.9).

If we then compare these new data of Italian Glaciers with those of the previous century, published with the first Italian Glacier Inventory compiled in 1960 by the Italian Glaciological Committee (CGI 1959–1962), the reduction of Italian glaciers is equal to 200 km², a surface slightly minor than that one of the Maggiore Lake.

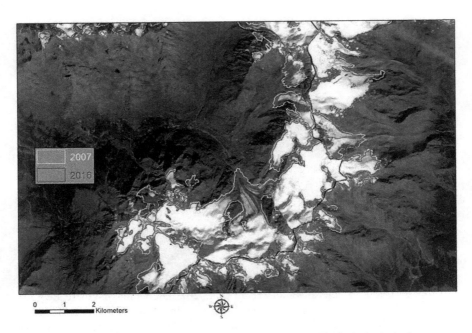

Fig. 5.9 The retreat of the Ortles-Cevedale glaciers, of which the Forni glacier is the largest one, seen from satellite (Sentinel images). All glaciers appear dramatically reduced in the 2007–2016 time frame, and glacier fragmentation is evident. From the comparison of the glacier boundaries, the expansion of outcropping rocks and nunataks is also evident

5.2.3 The Karakoram Anomaly Investigated by Satellite

The Karakoram Range is one of the most glacierized mountain regions globally, and glaciers there are an important water resource for Pakistan. The attention paid to this area is increasing, because its glaciers remained relatively stable in the early twenty-first century, in contrast to the general glacier retreat observed worldwide on average. This condition is also known as "Karakoram Anomaly." Within this context, the Central Karakoram National Park (CKNP) Glacier Inventory was developed by the University of Milan in cooperation with the Ev-K2-CNR Pakistan association and the Pakistan Meteorological Department. The inventory has been developed within the framework of the Project "Social Economic Environment Development (SEED) in the Central Karakoram National Park (CKNP) Gilgit Baltistan Region" Phase II, funded by the Government of Italy and the Government of Pakistan in the framework of the Pakistan-Italian Debt for development Swap Agreement (PIDSA). The inventory is available at the link: https://sites.unimi.it/glaciol/index.php/en/cknp-glacier-inventory/.

The CKNP Glacier Inventory is a fundamental achievement, providing an updated picture of the status of Pakistan–Karakoram glaciers based on a standardized analysis of recent satellite images. Considering that 70% of Pakistan's freshwater resources come from glacier melting, the CKNP Glacier Inventory represents the key baseline information for the scientific community and policy makers in climate change, water resources assessment, and sustainable management.

For the compilation of the CKNP glacier inventory, Landsat images were used despite the previously mentioned inventories (i.e., orthophoto and Sentinel-2) following Paul et al. (2009). More precisely, Level 1 T Landsat Thematic Mapper (TM) and Enhanced TM Plus (ETM+) scenes of 2001 and 2010 were processed and analyzed (Minora et al., 2016).

In the CKNP, there are 608 glaciers (among which are some of the largest Karakoram glaciers: Baltoro, Biafo, and Hispar). Glaciers span a broad range of sizes, types (i.e., mountain glaciers, glacierets, hanging glaciers, compound-basin valley glaciers), and surface conditions (i.e., debris-free and debris-covered ice). Their total area in 2001 was 3681.8 ± 27.7 km^2, ~35% of the CKNP area. This area represents ~24% of the Karakoram Range's glacier surface within Pakistan (Bajracharya & Shrestha, 2011). The most significant ice body is Baltoro Glacier, with an area of 604.2 km^2, while the mean glacier size results in 6.1 km^2. Glacier minimum elevation (i.e., ~glacier terminus elevation) ranges between 4000 and 5000 m a.s.l. on average, with few larger glaciers reaching farther down (between 3000 and 3500 m a.s.l.). Smaller glaciers (<1 km^2) show higher termini location, similarly to what is observed in other glaciated regions. Finally, more than 60% of glaciers feature a length of 1–5 km, and we observe that glaciers range in elevation from 2250 to 7900 m a.s.l.

Figure 5.10 shows the 2010 glacier distribution covering an area of 3682.1 ± 61.0 km^2, slightly more than in 2001. The analysis of the area changes during 2001–2010 reveals general stability, evidence of the peculiar behavior of glaciers in the Karakoram in contrast to a worldwide shrinkage of most mountain glaciers outside the Polar Regions. The total area change is $+0.3 \pm 67.0$ km^2, given by an area gain of $+7.7 \pm 40.1$ km^2 and a loss of -7.4 ± 53.0 km^2. The Baltoro Glacier is the glacier with the largest loss (-2.1 km^2: from 604.2 km^2 in 2001 to 602.1 km^2 in 2010). On the other hand, a relatively large debris-free glacier (i.e., Shingchukpi Glacier) experienced the maximum area gain (+1.7 km^2: from 11.8 km^2 in 2001 to 13.5 km^2 in 2010).

Despite the overall stable situation, some glaciers showed considerable changes. Some of these are surge-type glaciers. The Karakoram hosts several surge-type glaciers: this type of ice body displays cyclically short-term active phases involving rapid mass transfer from high to low elevations and long-term quiescent phases of low mass fluxes. The most prominent surge example is the Shingchukpi Glacier with the largest surge advance (ca. 2220 m). It is now in touch with the Panmah Glacier. The overall contribution of the advancing surge-type glaciers to the CKNP area gain is 2.6 km^2, about 33% of the total area gain in 2010 with respect to 2001. This net area gain was evaluated without considering glacier tributaries; for these latter, the area increase is already accounted for in the extent of the main glaciers. Neglecting the surge-type advances, the remaining glacier surface is still more or less stable, even if slightly negative. Despite the relatively large length and area changes, and the high flow velocities during the active phase of a surge (up to

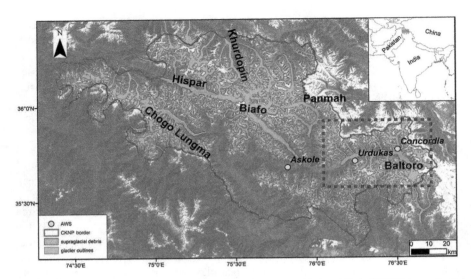

Fig. 5.10 The map shows the border of the Central Karakoram National Park (CKNP) in northern Pakistan (red line), the glacier boundaries (blue lines), and the glacier areas covered by supraglacial debris (orange). Names of the most expansive glaciers are shown, with Baltoro glacier highlighted by a box with a blue dashed line

5 km yr^{-1}) for the Khurdopin Glacier in the 1970s, according to Quincey and Luckman (2014), it is difficult to connect such advances to changes in mass balance. Previous works on surging glaciers in the Karakoram have suggested that climatically induced changes in glacier thermal conditions may be linked to observed exceptional surging (Hewitt, 2005). Others indicate that a change in subglacial drainage is the dominant control (Mayer et al., 2011).

A supervised classification applied to the Landsat images allowed the spatial analysis of the supraglacial debris, which can be brought by landslides from the steep rock-walls surrounding the glaciers, rock falls, and debris-laden snow avalanches. The supraglacial debris coverage was found equal to 765.5 ± 25.7 km^2 in 2001 and 919.1 ± 58.6 km^2 in 2010, i.e., about 21% of the total ice-covered area. According to our supraglacial debris occurrence calculation, the debris cover increased by 153.6 ± 64.0 km^2. In general, 27.3% of the CKNP glaciers were found to be debris-covered. Therefore, if CKNP glaciers are divided into debris-free and debris-covered types, we can immediately recognize two patterns. On the one hand, debris-covered glaciers are mostly larger (Baltoro and Hispar Glaciers belong to this group), and they reach the lowest elevations (even below 3000 m a.s.l.). Moreover, they are covered by debris almost entirely up to about 4000 m a.s.l.: the maximum supraglacial-debris cover is found at 4300 m a.s.l. On the other hand, debris-free glaciers are smaller, and their termini are found higher up on average (4500 m a.s.l., almost 700 m above the mean termini of debris-covered glaciers).

5.3 Remote-Sensing to Describe Glacier Darkening

Although we are used to thinking of glaciers as white, immaculate bodies, they are covered by light-absorbing particles, which make them appear darker. The effect is not simply aesthetic but also impacts on the glacier albedo, which is a measure of the amount of solar radiation that glacier can reflect. Indeed, it governs its surface energy balance and, therefore, the amount of meltwater produced.

Light-absorbing particles can be of autochthonous or allochthonous origin and include a mineral and organic component. The mineral component is made up of debris and dust and can be transported from the lateral moraines, from the rock walls, subjected to degradation through more frequent freeze-thaw cycles as temperatures increase, and from places far away such as deserts. The organic component includes the product of bacterial decomposition of organic material, black carbon (BC), emitted by forest fires or anthropogenic activities, such as combustion of diesel engines, residuals of pollen or vegetation, and living organisms such as algae (Azzoni et al., 2016). While a thick mantle of rock debris insulates the ice and protects it from melting, with a thin layer of dust or other light-absorbing material, the prevailing effect is that of decreasing the glacier albedo, and melt is enhanced. Furthermore, increasing summer temperatures lead to earlier snowmelt, the shrinking of the glacier accumulation areas, and uncovering of bare ice, which further promotes melt (Fugazza et al., 2019) in a feedback loop.

It is therefore not surprising that an increase in the amount of light-absorbing impurities and, in general, glacier darkening has been reported on glacier or ice sheets in different regions of the world. However, the extent of darkening and its effect on albedo is still largely unknown. To study this phenomenon, we adopted a methodology first described by Klok et al. (2003) to estimate glacier albedo from images of the Landsat family of satellites (TM, ETM+, and OLI). The approach was first tested on the Forni glacier, the largest in the Ortles-Cevedale group (11.34 km^2; Smiraglia et al., 2015) using images from 2011 to 2013 and then extended to its mountain group, the Ortles-Cevedale, using the entire record of Landsat images from 1984 to 2011 to assess trends in glacier albedo and therefore the extent of glacier darkening. The Ortles-Cevedale is part of Stelvio National Park, a protected area in the Central Italian Alps, where glaciers also play an important role in producing hydroelectric energy (D'Agata et al., 2018).

To estimate the albedo from satellite data, the process requires a series of steps, as satellite sensors measure radiance, a directional quantity observed by the field of view of the sensor in a discrete spectral band. In contrast, the albedo is defined over the entire hemisphere and includes radiation between 300 and 3000 μm. Starting from two spectral bands, the methodical steps include radiometric calibration, conversion to reflectance, atmospheric and topographic correction, anisotropic correction, and conversion from narrowband to broadband albedo. A number of additional inputs are required, including a digital elevation model for topographic correction and a vertical profile of temperature, precipitation and ozone, and aerosol optical thickness for atmospheric correction. The uncertainty of the albedo estimation was

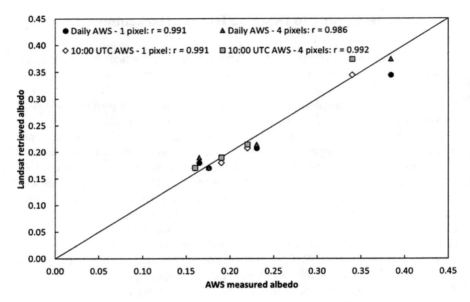

Fig. 5.11 Validation of satellite albedo against ground observations from the AWS (automatic weather station) located on Forni glacier (Fugazza et al., 2016)

quantified by Fugazza et al. (2016) as 0.07, which mainly stems from the anisotropic correction. To validate the method, we performed a comparison between satellite-albedo from four Landsat scenes and ground observations from the automatic weather station located on the tongue of the Forni glacier, by using data acquired in 2011–2013 (Fig. 5.11).

The validation showed a very good agreement in the comparison between hourly ground albedo and the satellite estimate (which is an instantaneous value from ca. 10:30 AM), as well as the average daily observations from the ground station.

By first testing the method on these four Landsat scenes, we were also able to identify the main drivers of temporal and spatial variability, i.e., summer air temperatures and snowfall: although the images were all acquired during the summer period, the mean albedo of the entire glacier largely depends on the presence of snow and the air temperatures since the last snowfall, which can melt snow and leave bare ice on the glacier tongue. For this reason, in our subsequent analysis of albedo trends, we selected the glacier ablation area, identified as the area which is not covered by snow in any of the summer images. An issue that needs to be properly solved is represented by cloud cover, which can be high even in summer and partly obscure the glaciers. While cloud cover itself can be recognized automatically using the Landsat near-infrared and thermal infrared bands, cloud cover shadows are more difficult to identify, and often the automatic procedure fails. Thus, we inspected all images manually to select glaciers that were cloud-free in all scenes. Our final database for assessing albedo trends included 15 glaciers larger than 0.2 km^2, corresponding to 86% of the glacier area in the Ortles-Cevedale group.

For these glaciers, we calculated the mean albedo of the ablation area and assessed its trends. Out of all 15 glaciers, 14 (13) had statistically significant negative albedo trends at the 95% (99%) confidence level throughout the investigation. The mean trend was -0.003 ± 0.001 y^{-1}, ranging between -0.001 and -0.006 y^{-1}.

Lastly, we performed a comparison of the negative albedo trends against the increase in debris cover ($+0.32$ km^2 y^{-1}) reported by Azzoni et al. (2018). The two variables show a significant correlation, although the magnitude of the correlation is weak ($r = -0.52$). This might be caused by the different years of analysis in the study of Azzoni et al. (2018), which only considered the years from 2003 to 2012. Furthermore, the latter study only considered thick, continuous debris cover, while by estimating glacier albedo, the effects of both thin/sparse and thick/continuous debris are included. It is likely a combination of both factors, as well as the increasing summer temperatures reported in the area ($+0.5$ °C between 1988 and 2006; Cannone et al., 2008), which play a role in determining the observed decrease in glacier albedo and which will likely lead to a further reduction in the coming years.

5.4 Unmanned Aerial Vehicles (UAVs) for High-Resolution Mapping of Glacier Hazards and Thickness Changes

Glaciers are very dynamic features, continuously flowing and sculpting the landscape through bedrock erosion and redepositing rocks and sediments. On glaciers themselves, the action of meltwater can rapidly change the surface by creating moulins, *bédières*, and cavities. Glacier changes are even more evident now that the increasing temperatures are causing their widespread thinning and retreat, calling for the frequent update of glacier inventories.

The constant monitoring of glaciers requires innovative techniques which do not rely on field sampling, as they are often in remote locations and hard to reach. Satellites already play a critical role by allowing the acquisition of several images with a daily (e.g., MODIS), weekly (Sentinel-2), or biweekly revisit time (Landsat). However, this theoretical temporal resolution can be much lower in reality, as optical satellite sensors cannot observe the surface beneath clouds, thus hampering the application of satellite remote-sensing and the calculations of variables, such as glacier albedo. For such reasons, the past decade has seen a large interest in the applications of unmanned aerial vehicles (UAVs) to glaciology. In this field, UAVs are particularly favorable because they allow repeated measurements of the surface with the desired time interval. By allowing to choose the acquisition day and time, they are less sensitive to clouds and permit capturing images with the most favorable illumination conditions.

The other great advantage of UAVs over satellite sensors is the spatial resolution, which can be as high as a few cm. No satellite can rival this, and those that come close are commercial satellites whose images are rather expensive. Another advantage yet is the customizable payload: although most frequently UAVs are equipped with simple RGB digital cameras, it is generally possible for them to carry all types of sensors ranging from thermal radiometers to LiDARs to hyperspectral sensors.

The product of UAV surveys are orthophotos and digital elevation models (DEMs). With the former, the measurement of glacier retreat over short periods as well as a high-resolution classification of glacier features becomes possible, while the latter allows estimating the glacier thickness changes by differencing the surface elevation imaged at different times. With this method, it is also possible to estimate the glacier mass balance, which is considered the most direct response of the glacier to climate forcing.

Since 2014, we have carried out several surveys on the Forni glacier using various UAV equipment. This glacier was chosen because it represents a tourist attraction, with several routes leading to the summit of Mount S. Matteo (3673 m a.s.l.) both in winter and summer. Owing to the number of people visiting the glacier, studying dynamic features, such as crevasses, which constitute a risk for tourists and hikers, has become a priority. Beside crevasses, a peculiar feature observed in recent years (Azzoni et al., 2017) is collapsing areas, named ring faults in structural terms. Ring faults are circular or semicircular fractures with stepwise subsidence caused by englacial or subglacial meltwater. The water flow causes voids at the interface between the ice and bedrock, eventually leading to the collapse of the cavity roofs (Azzoni et al., 2017; Fugazza et al., 2018). These structures are particularly dangerous for hikers, and tourists visiting glaciers, and therefore studying them at a safe distance is paramount.

In 2014, a first survey was carried out using a commercial fixed-wing SwingletCam drone equipped with a compact RGB digital camera. In subsequent surveys in 2016, 2017, and 2018, we employed a UAV prototype built with a Tarot frame, a GPS/GLONASS navigation system and a different RGB camera. In the last survey in 2020, we used instead a commercial DJI Phantom 4RTK, equipped with a 20 Mp digital camera, a multi-constellation navigation system, and RTK (real-time kinematics) system for sub-cm positioning accuracy.

The goal of the first survey was to test the possibility of using drones on the glacier while also developing an automatic approach to detect the glacier outlines based on segmentation and classify supraglacial features, including epiglacial lakes and melt ponds, crevasses (using Gabor filters), and supraglacial debris (Fugazza et al., 2015).

The goal of the 2016 survey was to compare the point clouds obtained using the UAV against those from terrestrial photogrammetry and laser scanners, to assess whether UAV products could be accurate enough to detect the glacier thickness changes over a time span of 2 years, as well as repeating the mapping of hazardous features such as crevasses and ring faults. Over 4 days, we simultaneously acquired data using the three different approaches; the comparison between the different point clouds was carried out in five separate areas on the glacier, including mostly vertical or sub-vertical features (Fig. 5.12).

Fig. 5.12 Areas identified for comparison of point clouds from UAV, terrestrial photogrammetry, and laser scanning acquired in 2016 (Fugazza et al., 2018)

The comparison demonstrated that although UAV products are slightly less accurate compared to those from the other two approaches, with a lower point density and differences up to 15 cm compared to the laser scanner point clouds, their uncertainty is still more than an order of magnitude smaller than the thickness changes occurring to the glacier, which were quantified in an average of −5 m per year over the ablation tongue, with a maximum difference of −38.71 m (Scaioni et al., 2017). The combination of terrestrial and UAV photogrammetry proved particularly useful to detect vertical and sub-vertical surfaces and identify potential glacier hazards, which were mapped manually.

Subsequent surveys carried out in 2017 and 2018 were also used to test different equipment such as multispectral cameras and to continue monitoring the changes in glacier thickness over the ablation tongue, which were as high as 30 m in 2016–2017 at collapsing ring faults (Yordanov et al., 2019), as well as comparing different approaches of point cloud coregistration for change detection (Di Rita et al., 2020).

Despite the technological advancements and the improved ability of glacier monitoring with UAVs, some issues remain in high-altitude environments and particularly glaciers, where lower air temperature reduces battery duration, and katabatic winds can compromise the aircraft stability. In contrast, at present the mapping of a large glacier such as the Forni in its entirety is still a difficult task, even this drawback will likely be compensated in the future, and new generation UAVs will become the tool of choice to monitor the glacier mass balance and glacier hazards.

5.5 Remote-Sensing as a Tool for Studying Antarctica, the Ice Reserve of our Planet

Antarctica is the fourth continent by extent and is almost entirely covered by ice and snow, having about 90% of Earth's ice. The inner continent is almost entirely interested by the presence of snow, which covers an ice layer of about 2 km thickness on average, with peaks of over 4 km.

Antarctica is one of the widest and most reflective surfaces of the planet. A potential variation of its surface albedo could deeply affect the Earth's energy balance by controlling the absorption of surface solar radiation, leading to relevant effects on sea level. In addition, the continent is a heat sink for the Southern Hemisphere, and thus has an important control over the circulation of the Earth's atmosphere (King & Turner, 2007).

Throughout history, the albedo, as many other climatic parameters of relevance, has been acquired on ice sheets and ice shelves by instruments mounted on automatic weather stations or in-situ stations (Driemel et al., 2018). However, the surface covered by these punctual records cannot be representative of a large area, that is, an entire glacier or ice sheet. Many of these instruments would be required to cover such a wide area, for example, an ice sheet. In this context, remote-sensing acquires importance. NASA provides a series of albedo products obtained by combining observations from the Moderate Resolution Imaging Spectroradiometer (MODIS) sensors onboard the Terra and Aqua satellites.

MODIS data were used to study the ice sheets in Antarctica. A preliminary study (Traversa et al., 2019) analyzed the surface albedo of Antarctica and investigated eventual signals of variations in space and time using MODIS GLASS product in more than a decade (2000–2012) at approximately 6 km spatial resolution. No significant changes of summer averages over time were detected. At this coarse scale, the only variations were found by dividing the continent by means of its altitude, where an increasing heterogeneity was discovered moving from inner Antarctica towards the coastal zones. Additionally, dividing the continent into three subregions in correspondence with ocean basins (Fig. 5.13), the Atlantic area, which includes the Antarctic peninsula, presented higher variability in respect to the other zones. Effectively, it reaches the lowest latitude of the continent and thus shows the highest temperatures, and the presence of cyclones and easterly winds produces frequent snowfall and almost continuous drifting snow, which supply the surface with small and highly reflective snow grains (Pirazzini, 2004).

On the other hand, considering Antarctica in its entirety, the detected trends of inter-annual and intra-annual analysis were mainly due to the sun elevation variations in time, as observed by Pirazzini (2004), that found a relationship between albedo and the sun elevation angle in Antarctica. Nevertheless, the so-obtained albedo results presented good correlations with other climatic features, for example, temperature and precipitations, especially in the central summer period. However, MODIS spatial resolution resulted in being too coarse (i.e., at most, 500 m) to satisfactorily represent albedo variation of local features of the ice sheet (e.g., blue ice,

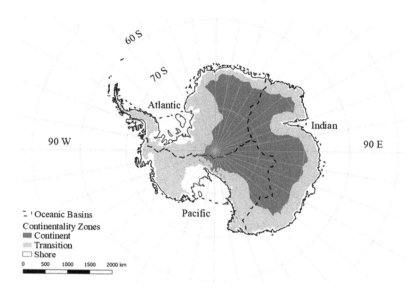

Fig. 5.13 Antarctica divided into the three oceanic basins by the dashed lines and into continentality zones by different colored areas (Traversa et al., 2019)

wind crust zones). For this reason, a methodology to retrieve reliable albedo measurements at higher spatial resolution was necessary, and it was possible starting from Landsat satellite imagery at 30 m resolution. Taking advantage of the existing model of albedo calculation validated by Fugazza et al. (2016) in the Alpine region, composed by different steps and corrections (e.g., atmospheric and topographic corrections), a new adjusted and updated model was applied on the ice sheets, including Antarctica (Traversa et al., 2021). Despite the remote location, a vast dataset of the Antarctic continent exists (Matsuoka et al., 2021), with data referring to a wide range of fields (geology, geography, biology, human sciences, etc.) that has been fundamental in these glaciological studies. Making a comparison of satellite data with field observations (7 automatic weather stations and 3 in-situ stations), excellent statistics were found.

On a 43-scene analysis in Antarctica, a 99.5% of correlation between field and remote data was calculated, and a final mean absolute error of 0.02 was provided. Thus, the resulted model could help construct a record to analyze albedo variations in Antarctica and, in general, in the Polar Regions.

5.6 Conclusions

In the case studies that we have reported and illustrated here, remote-sensing techniques have resulted in powerful, reliable, and indispensable tools for obtaining homogeneous and multitemporal data fields covering entire glacial regions. Such data fields are becoming more relevant every day because only with data extended over time and space, with high resolution and up-to-date continuously, is it possible to contribute to detailed knowledge and quantification of the impacts of climate change.

Besides high-resolution fields of glacier variables, remote-sensing techniques can also provide fields of meteorological variables. Such data fields are available from the EUMETSAT Satellite Application Facility on Climate Monitoring (CM SAF) that develops, generates, archives, and distributes high-quality satellite-derived products of the energy and water cycle in support of monitoring and understanding climate variability and climate change (https://www.cmsaf.eu). The goal of the last generation of satellite products is not only to capture the spatial distribution of meteorological variables but also to provide climate data records for essential climate variables suitable for climate monitoring. This is a very challenging goal, because the very rapid evolution of the satellite network and the high complexity of the procedure required to transform satellite observation into meteorological data can induce many inhomogeneities in climate data records derived from satellite observations. Another very important source providing fields of meteorological variables is reanalysis datasets. Also in this case, the quality and the accuracy of the data have improved quickly in the last decades, and the last generation products (see e.g., ERA5—Hersbach et al., 2020) have a much higher resolution than the ones which were available until a few years ago. For some regions, high-resolution reanalysis fields have also been produced, and these fields can have a resolution of a few km, even though the higher resolution does not always mean higher accuracy. The last source providing high-resolution fields of meteorological variables are models which combine high-resolution geographic information with lower-resolution networks of meteorological stations. Such models have recently allowed the reconstruction of high-resolution monthly temperature and precipitation fields for the Adda river catchment from the beginning of the nineteenth century (Crespi et al., 2021). Despite the greatest potential accuracy, the limit of such models is that they can produce high-resolution fields only when a high-density network of station records is available. Remote-sensing techniques, reanalysis data-sets, and a combination of high-resolution geographic information with lower-resolution networks of meteorological stations have allowed in the recent past a substantial increase in our ability of estimating meteorological variables over glacier areas.

The analyses carried out on the Italian and European Alps and in Antarctica highlight a clear involutionary phase of the cryosphere on a global scale that cannot be ignored and which is highly correlated with the global warming phase of the last century. The only exception is represented by the Karakoram, wherein the last decades there has been a phase of stasis. A slight glacier expansion known as "Karakoram Anomaly" seems to be linked to a peculiar climatic dynamics which took place here between the late 1990s of the twentieth century and the first decade of the twenty-first century, amplified by the very high altitudes at which the glaciers are located and favored by the consistent surface debris coverage that reduces the ice melting on many glacier tongues. Apart from this, exceptional situation that once again made it was possible to study in detail, thanks to remote-sensing techniques, the rest of the planetary cryosphere has been experiencing a phase of intense crisis and collapse.

The glacier-darkening phase, which is modifying the surface conditions of the planet's glaciers, reducing their reflectivity and increasing the solar energy they absorbed and then the intensity of the ice melting, has certainly modified magnitude and rates of glacier involution. Darkening is not directly caused by climate change, but it is in any case by human activity which, through pollution, black carbon emissions, and forest fires, modifies air quality and impacts on snow and ice albedo and solar absorption. Also this global phenomenon, which can accelerate the decline and extinction of many glaciers on the planet, has been observed and proved thanks to remote-sensing.

What future? What will be the aspect of our mountains in the next 50 or 100 years? Projections obtained applying physically based models running on average climate scenarios (Garavaglia et al., 2014) suggest that most glaciers in the Italian Alps could shrink to 20% of their existing surface by 2080. These values could become even more dramatic if the darkening intensifies by further increasing the ablative rates.

What impact will this glacial involution have on the Alpine landscape, on ecosystems, on the economy?

The ongoing glacier shrinkage is changing deeply the mountain landscape of the Alps (D'Agata et al., 2019), which are expected to show features and forms now visible within the Pyrenees (where the present glaciation is the relict of the previous one and is formed by small cirque glaciers). In a second phase, they are expected to resemble the Apennines (where only the Calderone Glacier can be found, actually classified as debris-covered glacier together with small snow fields).

Geodynamically, the transition from the glacial system to the paraglacial one is now occurring (Ballantyne & Benn, 1994, 1996). The areas wherein the recent past the main shaping and driving factors were glaciers are now subjected to the action of melting water, slope evolution, and dynamics and periglacial processes. Morphological changes develop at different rates in relation to shape and features of the newly exposed areas. Bare rock exposures (e.g., *roches moutonneé*, smoothed surfaces, etc.) accomplish meltwater runoff while unconsolidated till deposit are unstable and can be remobilized by running waters or by gravitative processes. Under such changing environmental features, the new territories are available for plant and tree colonization. The melting of glaciers not only has obvious impacts on the surrounding ecosystems, but it also has adverse consequences upon the value of the sites where they are located in the context of natural heritage. Heritage is an irreplaceable source of life and inspiration, it is humankind's legacy from the past, with which we live in the present and pass on to future generations (UNESCO, 2007).

Also, the GEO-Heritage (Bosson & Reynard, 2012) properties could be exposed to the unfavorable effects of changing climate, and this is particularly the case of mountain glaciers, among the most fascinating elements of the high elevation environment.

Last but not least, the contraction of the Alpine glaciers has an impact not only on the high mountain landscape (D'Agata et al., 2019) but will also have economic repercussions on tourism and on hydropower production. To evaluate these impacts, a study was carried out by the University of Milan, the Milan Polytechnic, and the Lombardy Region aimed at evaluating the volumetric contraction of all the glaciers in the Province of Sondrio between 1981 and 2007 (D'Agata et al., 2018). The volume variation on the glaciers in the province of Sondrio was -1.5 km^3 in 26 years. This value is equal to an average water release per year of 56 million m^3. It should be borne in mind that every year in Lombardy 27 billion m^3 of water fall as rain and snow. 56 million compared to 27 billion is a tear! However, we must also consider that this tear becomes available during the summer period, especially in July and August, to power the hydroelectric plants (and it is no coincidence that Lombardy accounts for 28% of the National Hydroelectric) and feeding rivers and streams during droughts. Finally, the role played on average in those 26 years by the glacial melt on the hydroelectric production of the Province of Sondrio, where numerous reservoirs are located, was also estimated. The Province of Sondrio in the study by D'Agata et al. (2018) has been divided into two sub-regions, one characterized by glaciers and artificial reservoirs for hydroelectric production and a second with the presence of reservoirs but not glaciers. It was evaluated that the water in the study period (1981–2007) fed the reservoirs, and it was found that in the subregion where the glaciers are present, their meltwater counted up to 15–20% of the total water available to produce power. This value is modest but not negligible, and we should take these values into account when the glaciers will no longer exist or when their extent will be considerably reduced.

References

Ambrosini, R., Azzoni, R. S., Pittino, F., Diolaiuti, G., Franzetti, A., & Parolini, M. (2019). First evidence of microplastic contamination in the supraglacial debris of an alpine glacier. *Environmental Pollution, 253*, 297–301. https://doi.org/10.1016/j.envpol.2019.07.005

Azzoni, R. S., Senese, A., Zerboni, A., Maugeri, M., Smiraglia, C., & Diolaiuti, G. A. (2016). Estimating ice albedo from fine debris cover quantified by a semi-automatic method: The case study of Forni glacier, Italian Alps. *The Cryosphere, 10*, 665–679. https://doi.org/10.5194/tc-10-665-2016

Azzoni, R. S., Fugazza, D., Zennaro, M., Zucali, M., D'Agata, C., Maragno, D., Cernuschi, M., Smiraglia, C., & Diolaiuti, G. A. (2017). Recent structural evolution of Forni glacier tongue (Ortles-Cevedale group, central Italian Alps). *Journal of Maps, 13*, 870–878. https://doi.org/10.1080/17445647.2017.1394227

Azzoni, R. S., Fugazza, D., Zerboni, A., Senese, A., D'Agata, C., Maragno, D., Carzaniga, A., Cernuschi, M., & Diolaiuti, G. A. (2018). Evaluating high-resolution remote sensing data for reconstructing the recent evolution of supra glacial debris: A study in the Central Alps (Stelvio Park, Italy). *Progress in Physical Geography, 42*(1), 3–23. https://doi.org/10.1177/0309133317749434

Bajracharya, S. R., & Shrestha, B. (Eds.). (2011). *The status of glaciers in the Hindu Kush-Himalayan region*. ICIMOD.

Ballantyne, C. K., & Benn, D. I. (1994). Paraglacial slope adjustment and resedimentation following glacier retreat, Fabergstolsdalen, Norway. *Arctic and Alpine Research, 25*, 255–269. https://doi.org/10.2307/1551938

Ballantyne, C. K., & Benn, D. I. (1996). Paraglacial slope adjustment during recent deglaciation and its implications for slope evolution in formerly glaciated environments. In M. G. Anderson & S. M. Brooks (Eds.), *Advances in hillslope processes* (Vol. 2, pp. 1173–1195). Wiley.

Bosson, J. B., & Reynard, E. (2012). Geomorphological heritage, conservation and promotion in high-alpine protected areas. *Eco Mont-Journal on Protected Mountain Areas Research, 4*(1), 13–22. https://doi.org/10.1553/eco.mont-4-1s13

Brunetti, M., Maugeri, M., Monti, F., & Nanni, T. (2006). Temperature and precipitation variability in Italy in the last two centuries from homogenised instrumental time series. *International Journal of Climatology, 26*, 345–381. https://doi.org/10.1002/joc.1251

Cannone, N., Diolaiuti, G., Guglielmin, M., & Smiraglia, C. (2008). Accelerating climate change impacts on alpine glacier forefield ecosystems in the European Alps. *Ecological Applications, 18*, 637–648. https://doi.org/10.1890/07-1188.1

Citterio, M., Diolaiuti, G., Smiraglia, C., D'Agata, C., Carnielli, T., Stella, G., & Siletto, G. B. (2007). The fluctuations of Italian glaciers during the last century: A contribution to knowledge about alpine glacier changes. *Geografiska Annaler: Series A Physical Geography, 89*, 164–182. https://doi.org/10.1111/j.1468-0459.2007.00316.x

Crespi, A., Brunetti, M., Ranzi, R., Tomirotti, M., & Maugeri, M. (2021). A multi-century meteo-hydrological analysis for the Adda river basin (Central Alps). Part I: Gridded monthly precipitation (1800–2016) records. *International Journal of Climatology, 41*, 162–180. https://doi.org/10.1002/joc.6614

D'Agata, C., Bocchiola, D., Soncini, A., Maragno, D., Smiraglia, C., & Diolaiuti, G. A. (2018). Recent area and volume loss of alpine glaciers in the Adda River of Italy and their contribution to hydropower production. *Cold Regions Science and Technology, 148*, 172–184. https://doi.org/10.1016/j.coldregions.2017.12.010

D'Agata, C., Diolaiuti, G., Maragno, D., Smiraglia, C., & Pelfini, M. (2019). Climate change effects on landscape and environment in glacierized Alpine areas: Retreating glaciers and enlarging forelands in the Bernina group (Italy) in the period 1954–2007. *Geology, Ecology, and Landscapes*. https://doi.org/10.1080/24749508.2019.1585658

Di Rita, M., Fugazza, D., Belloni, V., Diolaiuti, G., Scaioni, M., & Crespi, M. (2020). Glacier volume change monitoring from UAV observations: Issues and potentials of state-of-the-art techniques. In *ISPRS - international archives of the photogrammetry, remote sensing and spatial information sciences. XXIV ISPRS congress, commission II, Copernicus GmbH, XLIII-B2-2020* (pp. 1041–1048). https://doi.org/10.5194/isprs-archives-XLIII-B2-2020-1041-2020

Diolaiuti, G., & Smiraglia, C. (2010). Changing glaciers in a changing climate: How vanishing geomorphosites have been driving deep changes on mountain landscape and environment. *Géomorphologie: Relief, Processus, Environnement, 2*, 131–152. https://doi.org/10.4000/geomorphologie.7882

Diolaiuti, G., Bocchiola, D., Vagliasindi, M., D'agata, C., & Smiraglia, C. (2012a). The 1975–2005 glacier changes in Aosta Valley (Italy) and the relations with climate evolution. *Progress in Physical Geography, 36*(6), 764–785. https://doi.org/10.1177/0309133312456413

Diolaiuti, G., Bocchiola, D., D'agata, C., & Smiraglia, C. (2012b). Evidence of climate change impact upon glaciers' recession within the Italian alps: The case of Lombardy glaciers. *Theoretical and Applied Climatology, 109*(3–4), 429–445. https://doi.org/10.1007/s00704-012-0589-y

Diolaiuti, G. A., Azzoni, R. S., D'Agata, C., Maragno, D., Fugazza, D., Vagliasindi, M., Mortara, G., Perotti, L., Bondesan, A., Carton, A., Pecci, M., Dinale, R., Trenti, A., Casarotto, C., Colucci, R. R., Cagnati, A., Crepaz, A., & Smiraglia, C. (2019). Present extent, features and regional distribution of Italian glaciers. *La Houille Blanche, 105*(5–6), 159–175. https://doi.org/10.1051/lhb/201903

Driemel, A., Augustine, J., Behrens, K., Colle, S., Cox, C., Cuevas-Agulló, E., Denn, F. M., Duprat, T., Fukuda, M., Grobe, H., Haeffelin, M., Hodges, G., Hyett, N., Ijima, O., Kallis, A., Knap, W., Kustov, V., Long, C. N., Longenecker, D., ... Konig-Langlo, G. (2018). Baseline surface radiation network (BSRN): Structure and data description (1992–2017). *Earth System Science Data, 10*(3), 1491–1501. https://doi.org/10.5194/essd-10-1491-2018

Frezzotti, M., & Orombelli, G. (2014). Glaciers and ice sheets: Current status and trends. *Rendiconti Lincei—Scienze Fisiche e Naturali, 25*(1), 59–70. https://doi.org/10.1007/s12210-013-0255-z

Fugazza, D., Senese, A., Azzoni, R. S., Smiraglia, C., Cernuschi, M., Severi, D., & Diolaiuti, G. A. (2015). High-resolution mapping of glacier surface features. The UAV survey of the Forni glacier (Stelvio National Park, Italy). *Geografia Fisica e Dinamica Quaternaria, 38*, 25–33. https://doi.org/10.4461/GFDQ.2015.38.03

Fugazza, D., Senese, A., Azzoni, R. S., Maugeri, M., & Diolaiuti, G. A. (2016). Spatial distribution of surface albedo at the Forni glacier (Stelvio National Park, central Italian Alps). *Cold Regions Science and Technology, 125*, 128–137. https://doi.org/10.1016/j.coldregions.2016.02.006

Fugazza, D., Scaioni, M., Corti, M., D'Agata, C., Azzoni, R. S., Cernuschi, M., Smiraglia, C., & Diolaiuti, G. A. (2018). Combination of UAV and terrestrial photogrammetry to assess rapid glacier evolution and map glacier hazards. *Natural Hazards and Earth System Sciences, 18*, 1055–1071. https://doi.org/10.5194/nhess-18-1055-2018

Fugazza, D., Senese, A., Azzoni, R. S., Maugeri, M., Maragno, D., & Diolaiuti, G. A. (2019). New evidence of glacier darkening in the Ortles-Cevedale group from Landsat observations. *Global and Planetary Change, 178*, 35–45. https://doi.org/10.1016/j.gloplacha.2019.04.014

Garavaglia, R., Marzorati, A., Confortola, G., Bocchiola, D., Cola, G., Manzata, E., Senese, A., Smiraglia, C., & Diolaiuti, G. (2014). Evoluzione del Ghiacciaio dei Forni. *Neve & Valanghe, 81*, 60–67. https://issuu.com/aineva7/docs/nv81

Hersbach, H., Bell, B., Berrisford, P., et al. (2020). The ERA5 global reanalysis. *Quarterly Journal of the Royal Meteorological Society, 146*, 1999–2049. https://doi.org/10.1002/qj.3803

Hewitt, K. (2005). The Karakoram Anomaly? Glacier Expansion and the 'Elevation Effect,' Karakoram Himalaya. *Mountain Research and Development, 25*, 332–340. https://doi.org/10.1659/0276-4741(2005)025[0332:TKAGEA]2.0.CO;2

IPCC. (2013). Climate change 2013: The physical science basis. In T. K. Stocker, D. Qin, G. K. Plattner, M. Tignor, S. K. Allen, J. Boschung, A. Nauels, Y. Xia, V. Bex, & P. M. Midgley (Eds.), *Contribution of working group I to the fifth assessment report of the intergovernmental panel on climate change* (p. 1535). Cambridge University Press.

King, J. C., & Turner, J. (2007). *Antarctic meteorology and climatology*. Cambridge University Press. ISBN 0521039843, 9780521039840.

Klok, E. J., Greuell, W., & Oerlemans, J. (2003). Temporal and spatial variation of the surface albedo of Morteratschgletscher, Switzerland, as derived from 12 Landsat images. *Journal of Glaciology, 49*, 491–502. https://doi.org/10.3189/172756503781830395

Maisch, M., Wipf, A., Denneler, B., Battaglia, J., & Benz, C. (Eds.). (2000). *Die Gletscher der Schweizer Alpen. Gletscherhochstand 1850, Aktuelle Vergletscherung, Gletscherschwund Szenarien* (2nd ed., 373 pp.). VdF Hochschulverlag.

Manara, V., Brunetti, M., Gilardoni, S., Landi, T. C., & Maugeri, M. (2019). 1951–2017 changes in the frequency of days with visibility higher than 10 km and 20 km in Italy. *Atmospheric Environment, 214*, 116861. https://doi.org/10.1016/j.atmosenv.2019.116861

Manara, V., Stocco, E., Brunetti, M., Diolaiuti, G. A., Fugazza, D., Pfeifroth, U., Senese, A., Trentmann, J., & Maugeri, M. (2020). Comparison of surface solar irradiance from ground observations and satellite data (1990–2016) over a complex orography region (Piedmont-Northwest Italy). *Remote Sensing, 12*(23), 3882. https://doi.org/10.3390/rs12233882

Matsuoka, K., Skoglund, A., Roth, G., de Pomereu, J., Griffiths, H., Headland, R., Herried, B., Katsumata, K., Le Brocq, A., Licht, K., Morgan, F., Neff, P. D., Ritz, C., Scheinert, M., Tamura, T., Van de Putte, A., van den Broeke, M., von Deschwanden, A., Deschamps-Berger, C., ... Melvær, Y. (2021). Quantarctica, an integrated mapping environment for Antarctica, the Southern Ocean, and sub-Antarctic islands. *Environmental Modelling & Software, 140*, 105015. https://doi.org/10.1016/j.envsoft.2021.105015

Mayer, C., Fowler, A. C., Lambrecht, A., & Scharrer, K. (2011). Surge of north Gasherbrum glacier, Karakoram, China. *Journal of Glaciology, 57*(205), 904–916. https://doi.org/10.3189/002214311798043834

Miner, K. R., Blais, J., Bogdal, C., Villa, S., Schwikowski, M., Pavlova, P., Steinlin, C., Gerbi, C., & Kreutz, K. J. (2017). Legacy organochlorine pollutants in glacial watersheds: A review. *Environmental Science: Processes & Impacts, 19*, 1474–1483. https://doi.org/10.1039/C7EM00393E

Minora, U., Bocchiola, D., D'Agata, C., Maragno, D., Mayer, C., Lambrecht, A., Vuillermoz, E., Senese, A., Compostella, C., Smiraglia, C., & Diolaiuti, G. A. (2016). Glacier area stability in the Central Karakoram National Park (Pakistan) in 2001–2010: The "Karakoram anomaly" in the spotlight. *Progress in Physical Geography: Earth and Environment, 40*, 629–660. https://doi.org/10.1177/0309133316643926

Paul, F., Kääb, A., Maisch, M., Kellenberger, T., & Haeberli, W. (2004). Rapid disintegration of alpine glaciers observed with satellite data. *Geophysical Research Letters, 31*, L21402. https://doi.org/10.1029/2004GL020816

Paul, F., Kääb, A., & Haeberli, W. (2007). Recent glacier changes in the Alps observed from satellite: Consequences for future monitoring strategies. *Global and Planetary Change.* https://doi.org/10.1016/j.gloplacha.2006&07.007

Paul, F., Barry, R. G., Cogley, J. G., Frey, H., Haeberli, W., Ohmura, A., Ommanney, C. S. L., Raup, B., Rivera, A., & Zemp, M. (2009). Recommendations for the compilation of glacier inventory data from digital sources. *Annals of Glaciology, 50*, 119–126. https://doi.org/10.3189/172756410790595778

Paul, F., Frey, H., & Le Bris, R. (2011). A new glacier inventory for the European Alps from Landsat TM scenes of 2003: Challenges and results. *Annals of Glaciology, 52*(59), 144–152. https://doi.org/10.3189/172756411799096295

Paul, F., Rastner, P., Azzoni, R. S., Diolaiuti, G., Fugazza, D., Le Bris, R., Nemec, J., Rabatel, A., Ramusovic, M., Schwaizer, G., & Smiraglia, C. (2020). Glacier shrinkage in the Alps continues unabated as revealed by a new glacier inventory from Sentinel-2. *Earth System Science Data, 12*, 1805–1821. https://doi.org/10.5194/essd-12-1805-2020

Pfeffer, W. T., Arendt, A. A., Bliss, A., Bolch, T., Cogley, J. G., Gardner, A. S., Hagen, J. O., Hock, R., Kaser, G., Kienholz, C., Miles, E. S., Moholdt, G., Mölg, N., Paul, F., Radic, V., Rastner, P., Raup, B. H., Rich, J., Sharp, M. J., & The Randolph Consortium. (2014). The Randolph glacier inventory: A globally complete inventory of glaciers. *Journal of Glaciology, 60*(221), 537–552. https://doi.org/10.3189/2014JoG13J176

Pirazzini, R. (2004). Surface albedo measurements over Antarctic sites in summer. *Journal of Geophysical Research: Atmospheres, 109*, D20118. https://doi.org/10.1029/2004JD004617

Quincey, D. J., & Luckman, A. (2014). Brief communication: On the magnitude and frequency of Khurdopin glacier surge events. *The Cryosphere, 8*, 571–574. https://doi.org/10.5194/tc-8-571-2014

Scaioni, M., Corti, M., Diolaiuti, G., Fugazza, D., & Cernuschi, M. (2017). Local and general monitoring of Forni glacier (Italian Alps) using multi-platform Structure-From-Motion photogrammetry. In *The International Archives of the photogrammetry, remote sensing and spatial information sciences. ISPRS Geospatial Week 2017–18-22 September, Wuhan, China, Copernicus GmbH. XLII-2/W7* (pp. 1547–1554). https://doi.org/10.5194/isprs-archives-XLII-2-W7-1547-2017

Senese, A., Manara, V., Maugeri, M., & Diolaiuti, G. A. (2020). Comparing measured incoming shortwave and longwave radiation on a glacier surface with estimated records from satellite and off-glacier observations: A case study for the Forni Glacier, Italy. *Remote Sensing, 12*, 3719. https://doi.org/10.3390/rs12223719

Smiraglia, C., Azzoni, R. S., D'Agata, C., Maragno, D., Fugazza, D., & Diolaiuti, G. A. (2015). The evolution of the Italian glaciers from the previous data base to the new Italian inventory. Preliminary considerations and results. *Geografia Fisica e Dinamica Quaternaria, 38*, 79–87. https://doi.org/10.4461/GFDQ.2015.38.08

Traversa, G., Fugazza, D., Senese, A., & Diolaiuti, G. A. (2019). Preliminary results on antarctic albedo from remote sensing observations. *Geografia Fisica e Dinamica Quaternaria, 42*, 245–254. https://doi.org/10.4461/GFDQ.2019.42.14

Traversa, G., Fugazza, D., Senese, A., & Frezzotti, M. (2021). Landsat 8 OLI broadband albedo validation in Antarctica and Greenland. *Remote Sensing, 13*, 799. https://doi.org/10.3390/rs13040799

UNESCO (2007). *Case studies on climate change and world heritage* (82 pp). Retrieved from http://whc.unesco.org/en/activities/473

Vogtle, T., & Schilling, K. J. (1999). Digitizing maps. In H. P. Bahr & T. Vogtle (Eds.), *GIS for environmental monitoring* (pp. 201–216). Schweizerbart.

Yordanov, V., Fugazza, D., Azzoni, R. S., Cernuschi, M., Scaioni, M., & Diolaiuti, G. A. (2019). Monitoring Alpine glaciers from close-range to satellite sensors. In *The International Archives of the Photogrammetry, Remote Sensing and Spatial Information Sciences. ISPRS Geospatial Week 2019–10-14 June 2019, Enschede, The Netherlands, Copernicus GmbH, XLII-2/W13* (pp. 1803–1810). https://doi.org/10.5194/isprs-archives-XLII-2-W13-1803-2019

Guglielmina Diolaiuti professor (associate) of physical geography and geomorphology, is at Milan University since 2002. She has more than 20 years of experience in the field of cryosphere sciences and glaciology, area of expertise in which she has published more than 100 papers on international peer-reviewed journals. Her research activities are mainly focused on: (i) describing, evaluating, and quantifying climate change impacts on glaciers; (ii) surveying and describing glacier micro-meteorology; and (iii) projecting glacier evolution and meltwater discharge under different climate change scenarios.

Maurizio Maugeri full professor of atmospheric physics, is at Milan University, Department of Environmental Science and Policy, since autumn 1991. He has more than 30 years of experience in the field of atmospheric physics, area of expertise in which he has published more than 100 papers on international peer-reviewed journals. His research activities are mainly focused on the reconstruction of the evolution of the climate of the last two centuries for Italy, the Alpine Region and the Mediterranean area.

Antonella Senese is a researcher of glaciology and alpine climatology at University of Milan, Department of Environmental Science and Policy. Since 2010, she has published more than 50 papers in international peer-reviewed journals, divulgative works, books or individual chapters, conference proceedings and technical reports, as the result of studies carried out in collaboration with national and international research groups. Her research activities are mainly focused on monitoring the state of mountain glaciers (Alps, Himalaya, Karakoram, Patagonia) and their meteorological conditions.

Veronica Manara is a post-doc researcher at Milan University, Department of Environmental Science and Policy. Since 2014, her research activities are mainly focused on the reconstruction of the evolution of the climate of the last two centuries for Italy, the Alpine Region and the Mediterranean area. She published about twenty papers in international peer-reviewed journals, conference proceedings, and technical reports and she presented the results of her research at many international and national conferences.

Giacomo Traversa is a PhD student in environmental, geological and polar sciences and technologies at the University of Siena. After a master thesis on Antarctic glaciology, he specialized in polar glaciology and geomorphology using remote sensing applications on the Antarctic and Greenland ice sheets, focusing on ablation areas analysis. He has already published papers in international peer-reviewed journals and conference proceedings.

Davide Fugazza is a researcher in glaciology and remote sensing at University of Milan, Department of Environmental Science and Policy. Since 2015, he has researched innovative applications of remote sensing in the field of cryospheric sciences, by processing images taken from satellites, aircraft, and drones, focussing on the retreat and darkening of glaciers in the European Alps, Patagonia, and Himalayas. He has published more than twenty papers in international peer-reviewed journals, conference proceedings, and technical reports.

Chapter 6
Rural Revival and Coastal Areas: Risks and Opportunities

Felice D'Alessandro

Abstract The present chapter draws attention to the urgent need to support effective policy responses and identify appropriate nature-based adaptation measures that could be implemented, particularly in low-density coastal areas, for reducing the risk and enhancing the environmental and societal resilience of rural coastal communities to the adverse effects of climate change.

Special attention has been paid to rural revival, emphasizing the role of natural capital as a window of economic opportunity for vulnerable rural communities. Natural capital represents a pivotal contributor to encourage inclusive and sustainable growth and promote investments in protecting natural assets, improving the efficiency of natural resource use.

While there is considerable relevant research in this field, the debate is still open, and the problem remains challenging for coastal and marine sciences.

6.1 Introduction

Climate change is broadly recognized as a key environmental issue that involves compound interplays including hazards, exposure, and vulnerability of human and natural systems, resulting in growing risks (IPCC, 2014).

This topic is of major concern to the coastal and marine ecosystems, which, being highly exposed and vulnerable to extreme events, are experiencing the adverse consequences of hazards related to climate change.

Given the complexity of the coastal and marine ecosystems, comprehensive frameworks should require a multidimensional and integrated approach in which multiple hazards (e.g., storm surge, sea-level rise, sea temperature), triggering and

---

F. D'Alessandro (✉)
University of Milan, Milan, Italy
e-mail: felice.dalessandro@unimi.it

© The Author(s), under exclusive license to Springer Nature Switzerland AG 2022
S. Valaguzza, M. A. Hughes (eds.), *Interdisciplinary Approaches to Climate Change for Sustainable Growth*, Natural Resource Management and Policy 47,
https://doi.org/10.1007/978-3-030-87564-0_6

cascade-effect-threatening multiexposed elements changing over time (e.g., communities, agriculture, infrastructures) lead to a multivulnerability (Gallina et al., 2016).

Therefore, to reduce such vulnerability, it would be necessary to examine the exposure of the basins of interest, their sensitivity to the hazards-induced changes, and adaptive capacity to propose effective adaptation measures.

From an adaptation perspective, nature-based restoration of rural coastal systems is increasingly viewed as an effective strategy (e.g., Maza, 2020). Typical decision-making issues arising from climate change involve the management of impacts on environmental resources (e.g., biodiversity and habitats), encouraging natural, sustainable, and soft solutions (D'Alessandro & Tomasicchio, 2016; D'Alessandro et al., 2020).

In light of this context, there is an urgent need to support effective policy responses and identify appropriate nature-based adaptation measures that could be implemented particularly in low-density coastal areas, for reducing the risk and enhancing the environmental and societal resilience of rural coastal communities to the adverse effects of changing climate.

Rural communities are strongly dependent on nature for their subsistence. Most rural people rely directly on it for their daily energy, food, water, and income needs. For this reason, natural capital is a crucial component of rural livelihoods, representing a pivotal contributor to encouraging inclusive and sustainable growth and promoting investments in protecting and enhancing natural assets, improving the efficiency of natural resource use. Addressing rural revival can provide interesting opportunities also influencing sea-based economic activities, including tourism and recreation.

The present chapter examines the potential revival of rural coastal areas under changing climate identifying risks and opportunities. Specifically, the work is organized as follows. In Sect. 6.2, the requirements for multidimensional risk frameworks are discussed. In Sect. 6.3, bridges between risk and adaptation theory and practice are explored. Section 6.4 focuses on the role of natural capital as a window of economic opportunity for rural communities. The chapter concludes with some recommendations for future research.

Given the importance and complexity of the topic, this work cannot be fully encompassing, but it is limited to give a general overview outlining some key directions.

6.2 Assessing the Risks of Climate Change

According to the IPCC definition (IPCC, 2012), the risk is characterized as "the potential for consequences where something of human value (including humans themselves) is at stake and where the outcome is uncertain." At the same time, the risk is often represented as "the probability of occurrence of hazardous events or trends multiplied by the consequences if these events occur." Combining these two definitions, compound events can be embedded in the general risk framework linking hazards, exposure, and vulnerability. In particular, given that in the past most climate-related-risks studies focused on single drivers, and given the evidence that the particularly worrisome events are typically multivariate in nature, here it is specified that a better understanding of compound events may improve projections of potential high-impacts on specific targets and sectors (i.e., tourism, agriculture, health). Moreover, it can provide a bridge between climate scientists, engineers, social scientists, impact modelers, and decision-makers, who need to work closely together to lead to this complexity (Zscheischler et al., 2018).

6.2.1 Hazards

Coastal areas are especially demanding sites for climate change impacts, the most relevant being flooding and erosion because of exposure to multiple drivers, which can occur at different time scales (Ranasinghe et al., 2013). Other expected impacts include salt intrusion of surface and ground waters, loss of coastal natural protection due to coral bleaching, and the decline/loss of coastal wetlands (Nicholls et al., 2007; Wong et al., 2014). Furthermore, dependence among the climate drivers can lead to extreme compound events (e.g., Tavakol et al., 2020). There are different multivariate approaches to consider interdependencies among climate drivers in this field. Copulas are widely used since they represent an efficient tool to investigate the statistical behavior of dependent variables (Salvadori et al., 2014, 2015, 2020).

Recognizing sea level rise as the main climate driver in coastal areas has resulted in improved projections. Key aspects of making accurate projections include understanding long-term sea-level variability. In literature, extensive studies have been devoted to exploring sea level variability. IPCC warned that at current trends, the projected increments in mean sea level for the year 2100, relative to the 1986–2005 time series of tide gauges records, are 400, 470, 480, and 840 mm, for the *Representative Concentration Pathways* scenario indicated as *RCP2.6, RCP4.5, RCP6.0*, and *RCP8.5*, respectively. As a result, global mean sea levels have been rising through the past century and are projected to rise at an accelerated rate throughout the twenty-first century (Fig. 6.1).

Several authors have been motivated to search for already existing accelerations in observations, which would be, if present, vital for coastal protection planning purposes. No scientific consensus has been reached yet on how a possible acceleration could be separated from intrinsic climate variability in sea-level records. This has led to an intense debate on the existence of significant accelerations in regional and global sea levels, and, if absent, on the general validity of current future projections (Visser et al., 2015; Tomasicchio et al., 2018). Some researchers have reported accelerations (Douglas, 1991; Church & White, 2006; Jevrejeva et al., 2014), where others (Houston & Dean, 2011; Watson, 2011; Boretti, 2012) have found decelerations, sometimes based on the same data. Visser et al. (2015) tried to shed light on the controversial discussion of the quantification of the sea level rise by providing a

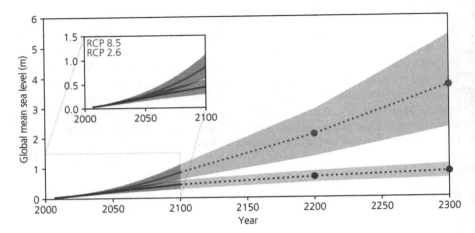

Fig. 6.1 Projected sea-level rise until 2300 (Oppenheimer et al., 2019)

comprehensive review of the trend methods used so far (for a total of 30 methods). The authors believe that much of the misunderstanding/controversies in the scientific community are due to the different mathematical or statistical characteristics of the considered models: a different approach may lead to contradictory acceleration/deceleration inferences. Similar conclusions have been reached in Thorarinsdottir et al. (2017).

6.2.2 Exposure

Coastal areas are complex environments where natural and socioeconomic systems are deeply intertwined (Fay et al., 2009). A first pass assessment should involve analyzing the exposure and sensitivity of such systems to climate change (e.g., sea level rises, the coast will be impacted, hence is vulnerable to erosion and/or flooding).

Exposure accounts for the whole inventory of elements, including the presence of people, livelihoods, species or ecosystems, environmental functions, services, and resources, infrastructure, or economic, social, or cultural assets in places and settings that could be adversely affected by an impact (IPCC, 2014). Although reducing exposure to physical assets, such as buildings and infrastructures, is common practice, information associated with indirect effects cannot be disregarded. For instance, when an industrial plant becomes flooded, consequences are not limited to the damages to structure and contents but can include loss of profits due to business interruption or delay (Toimil et al., 2020). However, socio-economic and natural indicators need to be allocated geographically and at the appropriate resolution. This is especially challenging when dealing with multiple sectors where information is heterogeneous in format, time, and space (Toimil et al., 2017).

Table 6.1 Coastal dimensions of exposure and sensitivity to climate hazards (Fay et al., 2009)

Climate hazards to which the natural system is exposed	Sensitivity determinants of the natural system (first level)	Outcome I (bio-geophysical impacts)	Sectors exposed to Outcome I	Sensitivity determinants of the natural system (second level)	Outcome II (socioeconomic impacts)
Extreme storms Sea level rise Sea temperature rise Sea water acidification	Geology Morphology Tidal range	Erosion Dune destabilization Flooding/inundation Saltwater intrusion	Ecological systems Biodiversity Economy Ag-forestry Fisheries Aquaculture Industry (e.g., tourism) Infrastructures Ports shipping Housing Roads Water Energy	Population density Number of marine/coastal protected areas Revenues from fishery and aquaculture Revenues from tourism Historic/cultural importance	Loss of lives Loss of property Damages to infrastructures Damages to agriculture, fisheries etc. Loss of cultural resources Forced migration Loss of ecosystems good and services Direct changes to sea: Impacts on biodiversity and fisheries

Given a stretch of coast at risk, a multidimensional and integrated approach requires that multiple hazards interact with multiple exposed elements. Table 6.1 focuses on different coastal dimensions of exposure and sensitivity. Proceeding from left to right, the climate hazards initially affect the natural system, and the magnitude of floods and/or erosion (outcome I) are mediated by the sensitivity of this system. For instance, the magnitude of erosion caused by sea-level rise depends on geological and morphological features of the coasts (e.g., sandy beaches versus rocky reefs, wave-dominated versus tide-modified beaches (Short, 2006)). The bio-geophysical impacts (Outcome I) triggered by climate hazards and mediated by the sensitivity of the natural system (first-level sensitivity) affect a range of natural and socioeconomic coastal sectors. The magnitude of the impacts of the socioeconomic system (Outcome II) depends both on the type and the magnitude of the bio-geophysical impacts (Outcome I) and on its second-level sensitivity. The latter is often calculated based on the social and economic importance of the coasts as measured by a range of indicators: population density, number of marine/coastal protected areas, revenues from fisheries, aquaculture, and tourism, and historic/cultural importance.

Outcomes II include, among the others, damages to housing, industrial, and transport infrastructures. Furthermore, ecosystems can be damaged: sea storms may impact wetlands as saltwater infiltration into aquifers has been proven to reduce the resilience of the coastal system; increases in sea temperature and acidification impact flora and fauna directly, causing consequences for biodiversity, fisheries, and aquaculture.

6.2.3 Vulnerability

While exposure represents the elements at risk that could be adversely affected, vulnerability characterizes the different elements at risk toward a given hazard intensity. In a broad sense, since the ability of the systems to cope with changes varies with time and across physical space and social frames, vulnerability should include economic, social, geographic, demographic, cultural, institutional, governance, and environmental factors (IPCC, 2012; Toimil et al., 2020).

A key international definition of vulnerability in climate change research is the one proposed by the IPCC (2014) for which the vulnerability of a system is a function of three main elements: (i) the magnitude and rate of climate variations to which a system is exposed (i.e., exposure); (ii) the degree to which a system could be affected by climate-related stimuli (i.e., sensitivity); (iii) the ability of a system to adjust or to cope with climate-change consequences (i.e., adaptive capacity). It is crucial to distinguish sensitivity from vulnerability. Sensitivity does not account for the moderating effect of adaptation measures; vulnerability can be viewed as the sum of the impacts remaining after adaptation has been considered (Torresan et al., 2008). Moreover, vulnerability is a multidimensional concept encompassing a wide range of bio-geophysical, socioeconomic, institutional, and cultural factors, and its integrated assessment requires combining quantitative and qualitative approaches to capture its multiple dimensions. The approaches used to assess the multivulnerability components depend on the geographic scale, data availability, and the models used, resulting in different levels of uncertainty.

As far as coastal systems are concerned, many important methods were proposed by different authors to assess vulnerability related to climate change, including participatory, model-based, agent-based, and index-based approaches (Hinkel, 2011; Pantusa et al., 2018).

6.3 Adaptation of Coastal Systems to Climate Change Impacts

Adaptation to climate impacts needs undertaking actions to minimize threats or to maximize opportunities resulting from changing climate and its effects.

As such, robust adaptation strategies specifically suited to the challenges of rural coastal areas need to be urgently considered, embedding interrelated aspects such as environmental change and associated risks to communities, coupled with the policy frameworks and financing. These kinds of actions are crucial to build the foundations for sustainable coastal development, mainly supporting socioeconomic services and biodiversity, increasing resilience.

Significant challenges for science and practice relate to the general requirements for developing and implementing adaptive solutions such as nature-based measures and their monitoring systems.

Restoring natural coastal environments is increasingly viewed as an effective option to adapt and enhance resilience to the adverse effects of climate change in rural coastal areas, especially if projected to experience severe impacts due to increased exposure to coastal flooding and erosion under rising sea levels.

Growing numbers of approaches have emerged as key nature-based solutions in recent years (e.g., Maza, 2020). For instance, dune recovery projects integrate vegetation efforts with natural, sustainable, and soft solutions. In many cases, these approaches have been implemented as demonstration projects in protected areas. As an example, Fig. 6.2 shows the results gained with an innovative and environmentally friendly technique for the consolidation of coastal sand dunes investigated within a recent field experiment conducted along the Adriatic coast in the Salento area (south of Italy) (D'Alessandro et al., 2020).

Although nature-based solutions can already be found along many coasts worldwide, in-depth investigations on efficiency, vulnerabilities, and natural dynamics are often lacking. To overcome these knowledge gaps, scientific insight from systematic research is urgently needed to identify options that combine spatial and statistical analysis, remote-sensing, field-monitoring techniques, and numerical modeling to generate a novel understanding of nature-based coastal adaptation to erosion and flood risks in low-density coastal areas.

Furthermore, to maximize the success and effectiveness of adaptation measures, there needs to be a joint understanding between people within communities, local authorities, and other key stakeholders. This will help all parties work together to develop adaptation pathways based on mutual knowledge of the issues relating to climate change to overcome risks and create opportunities.

Fig. 6.2 Visual observation of the restored dune 6 months after the intervention (D'Alessandro et al., 2020)

6.4 Coastal Natural Capital and Opportunities for Rural Communities

"Natural capital refers to the elements of nature that produce value or benefits to people (directly and indirectly), such as the stock of forest, rivers, land, minerals, and oceans, as well as the natural processes and functions that underpin their operation" (UK NCC, 2013).

Like the other forms of capital, natural capital provides today and in the future goods and services that are essential for individual and society wellbeing and development, and represent the foundation on which other types of assets are built. Specifically, marine and coastal ecosystems, such as watersheds, wetlands, mangroves, and coastal dunes, yield a wide range of "ecosystem services," from biodiversity and culture to carbon storage and flood protection.

Historically, rural communities have depended on nature for subsistence as part of strategies for coping with and recovering from natural and human-induced shocks. For instance, most rural people rely directly on nature for their daily energy, food, water, and income needs (TEEB, 2011). A significant decline in ecosystem services would directly affect these populations' energy, food, and water security. Land, water, and soil degradation, and the associated reduction in agricultural yield could drastically lower the earning capacity of vulnerable groups. For this reason, natural capital is a crucial component of rural livelihoods representing a key contributor to rapid economic growth.

Conventional approaches to measuring wealth and economic development, such as Gross Domestic Product (GDP), do not adequately take the values of these goods and services or the degradation of these ecosystems into account. The UN System of National Accounts, for example, does not fully capture the status and benefits of marine and coastal ecosystems (Essam & Milligan, 2016).

Until recently, for marine and coastal policymaking, the value of ecosystem services produced by natural assets to human wellbeing was largely invisible in economic and financial decisions, and nature's services were often viewed as "free" or "public goods." Conventional policy frameworks for marine and coastal development have tended to prioritize the conversion of marine and coastal ecosystems into other forms of wealth, such as hotels for tourists or aquaculture farms in place of mangrove forests. As a result, governments and other decision-makers have pursued economic development strategies that rely on the intensive use of natural capital. This approach has led to the over-exploitation of natural capital and the degradation and destruction of ecosystems.

In contrast, now there is growing recognition in both the public and private sectors that natural capital assets give rise to economically valuable goods and services. Natural capital has been equated with other forms of capital (built, social, and human). In this path, holistic frameworks can help policymakers to develop policies and to promote investments in protecting and enhancing natural assets, improving the efficiency of natural resource use, and mitigating the impact of economic activities on natural capital.

One of the most critical questions that policymakers raise is what kind of marine and coastal resource management maximizes economic, ecological, and societal net benefits.

In an era where carbon-trading and associated credit schemes are becoming increasingly important in combating climate change, growing attention needs to be given to marine ecosystems and their possibilities to store carbon. The latest scientific findings focus on the potential of natural coastal ecosystems, such as tidal marshlands, mangrove forests, or seagrass meadows, to act as carbon sinks. Ocean-based CO_2 sequestration technologies as potential climate change mitigation tools are also under investigation. Additionally, Posidonia meadows' stocks is a further clear example of natural capital that can provide many positive ecosystem services both to marine ecosystem functioning and to direct and indirect effect to human and economic activities, e.g., tackling seaboard erosion, maintaining nursery habitats of commercial fish species, climate regulation (INCC, 2018) (Fig. 6.3). As far as climate regulation, Posidonia meadows are one of the largest CO_2 sinks in the Mediterranean Area (IUCN, 2012).

Furthermore, mitigating the impact of land-based pressures on coastal water quality can encourage sustainable growth, thereby supporting the livelihood of the rural communities and increasing their access to opportunities when considering the economic viability of tourism and recreation activities. The viability of tourism and recreation activities in rural areas mainly depends on their adaptive capacity and less vulnerability to climate change allowing safe and resilient ecosystems. There is evidence that the development of natural assets, protection of biodiversity, and

Fig. 6.3 Posidonia meadows (concepts: F. Boero; art: A. Gennari – *source*: Danovaro & Boero, 2019)

managing hazards can improve the competitiveness of tourist destinations, which contributes to the positive relationship between tourism growth and economic expansion with additional cultural and social value.

6.5 Conclusions and Recommendations

The present work draws attention to the urgent need to support effective policy responses and identify appropriate nature-based adaptation measures that could be implemented, particularly in low-density coastal areas, for reducing the risk and enhancing the environmental and societal resilience of rural coastal communities to the adverse effects of climate change.

Special attention has been paid to rural revival, emphasizing the role of natural capital as a window of economic opportunity for vulnerable rural communities.

While there is considerable relevant research in this field, the debate is still open, and the problem remains challenging for coastal and marine sciences.

Below, an overall summary of recommendations for future research and practice, which have been organized in to four blocks following the paper structure, is presented:

In scientific practice, it should be recognized that the most severe climate-related impacts are caused by the compound effects of multiple hazards. A systematic research program is overdue and more comprehensive multidimensional risk frameworks are needed. Such approaches require considering the non-stationarity of the risk components and allowing to quantify uncertainty. Additionally, novel methodologies and tools are needed to assess the multivariate analysis of extremes.

Major challenges involve developing and implementing nature-based adaptation measures whose behavior requires better understanding. Improved understanding and application of such knowledge will form a critical part of coastal adaptation planning, likely reducing the need for expensive engineering options in some locations and providing a complementary tool in hybrid engineering design.

Furthermore, it seems necessary to (i) improve knowledge and enhance natural capital recovery measures, (ii) implement innovative financing tools, (iii) ensure natural ecosystems functionality and integrity, and (iv) establish synergies among green infrastructures and rural areas. Maintaining and enhancing coastal systems will also support the continued provision of other coastal services, including providing food and maintenance of coastal resource-dependent livelihoods.

Dealing with such complex issues, stronger collaboration and synchronization across multiple fields of research (e.g., coastal engineering, environmental sciences, economics, and policy) will be necessary to provide multidisciplinary and comprehensive approaches to climate change risk assessment and nature-based adaptation processes. At the same time, it is crucial to propose methodologies to assess the value of ecosystem services and promote investments in protecting and enhancing natural assets, improving natural resource use efficiency.

References

Boretti, A. (2012). Is there any support in the long term tide gauge data to the claims that parts of Sydney will be swamped by rising sea levels? *Coastal Engineering, 64*, 161–167.

Church, J. A., & White, N. J. (2006). A 20th century acceleration in global sea-level rise. *Geophysical Research Letters, 33*.

D'Alessandro, F., & Tomasicchio, G. R. (2016). Wave-dune interaction and beach resilience in large-scale physical model tests. *Coastal Engineering, 116*, 15–25.

D'Alessandro, F., Tomasicchio, G. R., Francone, A., Leone, E., Frega, F., Chiaia, G., Saponieri, A., & Damiani, L. (2020). Coastal sand dune restoration with an eco-friendly technique. *Aquatic Ecosystem Health & Management, 23*(4), 417–426.

Danovaro, R., & Boero, F. (2019). Italian seas. In C. Sheppard (Ed.), *World seas: An environmental evaluation* (2nd ed., pp. 283–306). Academic Press.

Douglas, B. C. (1991). Global sea level rise. *Journal of Geophysical Research, 96*, 6981–6992.

Essam, Y. M., & Milligan, B. (2016). *Using natural capital accounts to inform marine and coastal ecosystems policy*. Waves.

Fay, M., Block, R., Carrington, T., & Ebinger, J. (2009). *Adapting to climate change in ECA. Europe and Central Asia knowledge brief* (Vol. 5). World Bank Group.

Gallina, V., Torresan, S., Critto, A., Sperotto, A., Glade, T., & Marcomini, A. (2016). A review of multi-risk methodologies for natural hazards: Consequences and challenges for a climate change impact assessment. *Journal of Environmental Management, 168*, 123–132.

Hinkel, J. (2011). Indicators of vulnerability and adaptive capacity: Towards a need for sea level rise information: A decision analysis perspective. *Earth's Future, 7*, 320–337.

Houston, J. R., & Dean, R. G. (2011). Sea-level accelerations based on U.S. tide gauges and extensions of previous global-gauge analyses. *Journal of Coastal Research, 27*(3), 409–417.

INCC—Italian Natural Capital Committee. (2018). *Second report on the state of natural Capital in Italy*. Rome.

IPCC. (2012). In Field CB et al (ed) *Managing the risks of extreme events and disasters to advance climate change adaptation. A Special Report of Working Groups I and II of the Intergovernmental Panel on Climate Change. Cambridge Groups I and II of the Intergovernmental Panel on Climate Change*. Cambridge University Press

IPCC. (2014). Summary for policymakers. In C. B. Field et al. (Eds.), *Climate change 2014: Impacts, adaptation and vulnerability. Part A: Global and sectoral aspects. Contribution of Working Group II to the Fifth Assessment Report of the Intergovernmental Panel on Climate Change*. Cambridge University Press.

IUCN (2012). *Annual report. Nature + towards nature-based solutions*. Retrieved 7 April, 2021 from https://www.iucn.org/sites/dev/files/import/downloads/iucn_global_annual_report_2012.pdf.

Jevrejeva, S., Moore, J. C., Grinsted, A., Matthews, A. P., & Spada, G. (2014). Trends and acceleration in global and regional sea levels since 1907. *Global and Planetary Change, 113*, 11–22.

Maza M. (2020). Nature-based solutions for coastal defence: key aspects in the modelling of flow-ecosystem interactions. In *Proceedings of Virtual Conference on Costal Engineering, 2020*.

Nicholls, R. J., Wong, P. P., Burkett, V. R., Codignotto, J. O., Hay, J. E., McLean, R. F., Ragoonaden, S., & Woodroffe, C. D. (2007). In M. L. Parry et al. (Eds.), *Coastal systems and low-lying areas* (Climate Change 2007: Impacts, adaptation and vulnerability. Contribution of Working Group II to the Fourth Assessment Report of the Intergovernmental Panel on Climate Change) (pp. 315–356). Cambridge University Press.

Oppenheimer, M., Glavovic, B. C., Hinkel, J., van de Wal, R., Magnan, A. K., Abd-Elgawad, A., Cai, R., Cifuentes-Jara, M., DeConto, R. M., Ghosh, T., Hay, J., Isla, F., Marzeion, B., Meyssignac, B., & Sebesvari, Z. (2019). Sea level rise and implications for low-lying Islands, coasts and communities. In H.-O. Pörtner et al. (Eds.), *IPCC Special report on the ocean and cryosphere in a changing climate*.

Pantusa, D., D'Alessandro, F., Riefolo, L., Principato, F., & Tomasicchio, G. R. (2018). Application of a coastal vulnerability index. A case study along the Apulian coastline, Italy. *Water, 10*(9), 1218.

Ranasinghe, R., Duong, T., Uhlenbrook, S., Roelvink, D., & Stive, M. (2013). Climate-change impact assessment for inlet-interrupted coastlines. *Nature Climate Change, 3*, 83–87.

Salvadori, G., Tomasicchio, G. R., & D'Alessandro, F. (2014). Practical guidelines for multivariate analysis and design in coastal engineering. *Coastal Engineering, 88*, 1–14.

Salvadori, G., Durante, F., Tomasicchio, G. R., & D'Alessandro, F. (2015). Practical guidelines for the multivariate assessment of the structural risk in coastal and off-shore engineering. *Coastal Engineering, 95*, 77–83.

Salvadori, G., Tomasicchio, G. R., D'Alessandro, F., Lusito, L., & Francone, A. (2020). Multivariate Sea storm hindcasting and design: The isotropic buoy-ungauged generator procedure. *Scientific Reports, 10*(1), 20517.

Short, A. D. (2006). Australian Beach system: Nature and distribution. *Journal of Coastal Research, 22*, 11–27.

Tavakol, A., Rahmani, V., & Harrington, J. (2020). Probability of compound climate extremes in a changing climate: A copula-based study of hot, dry, and windy events in the Central United States. *Environmental Research Letters, 15*, 104058.

TEEB—The Economics of Ecosystems and Biodiversity. (2011). *The economics of ecosystems and biodiversity: Mainstreaming the economics of nature: A synthesis of the approach, conclusions and recommendations of TEEB*. http://www.teebweb.org/publication/mainstreaming-the-economics-of-nature-a-synthesis-of-the-approach-conclusions-and-recommendations-of-teeb/

Thorarinsdottir, T. L., Guttorp, P., Drews, M., Kaspersen, P. S., & de Bruin, K. (2017). Sea level adaptation decisions under uncertainty. *Water Resources Research, 53*, 8147–8163.

Toimil, A., Losada, I. J., Diaz-Simal, P., Izaguirre, C., & Camus, P. (2017). Multi-sectoral, high-resolution assessment of climate change consequences of coastal flooding. *Climatic Change, 145*, 431–444.

Toimil, A., Losada, I. J., Nicholls, R. J., Dalrymple, R. A., & Stive, M. J. (2020). Addressing the challenges of climate change risks and adaptation in coastal areas: A review. *Coastal Engineering, 156*, 103611.

Tomasicchio, G. R., Lusito, L., D'Alessandro, F., Frega, F., Francone, A., & De Bartolo, S. (2018). A direct scaling analysis for the sea level rise. *Stochastic Environmental Research and Risk Assessment, 32*(12), 3397–3408.

Torresan, S., Critto, A., Dalla Valle, M., Harvey, N., & Marcomini, A. (2008). Assessing coastal vulnerability to climate change: Comparing segmentation at global and regional scales. *Sustainability Science, 3*, 45–65.

UK NCC—United Kingdom Natural Capital Committee. (2013). *Natural Capital Committee's first state of natural capital report*. https://www.gov.uk/government/publications/natural-capital-committees-first-state-of-natural-capital-report

Visser, H., Dangendorf, S., & Petersen, A. C. (2015). A review of trend models applied to sea level data with reference to the acceleration-deceleration debate. *Journal of Geophysical Research, Oceans, 120*(6), 3873–3895.

Watson, P. J. (2011). Is there evidence yet of acceleration in mean sea level rise around mainland Australia? *Journal of Coastal Research, 27*(2), 368–377.

Wong, P. P., Losada, I. J., Gattuso, J.-P., Hinkel, J., Khattabi, A., McInnes, K. L., Saito, Y., & Sallenger, A. (2014). Coastal systems and low-lying areas. In C. B. Field et al. (Eds.), *Climate change 2014: Impacts, adaptation, and vulnerability. Part a: Global and sectoral aspects. Contribution of working group II to the fifth assessment report of the intergovernmental panel on climate change* (pp. 361–409). Cambridge University Press.

Zscheischler, J., Westra, S., van den Hurk, B. J. J. M., Seneviratne, S. I., Ward, P. J., Pitman, A., AghaKouchak, A., Bresch, D. N., Leonard, M., Wahl, T., & Zhang, X. (2018). Future climate risk from compound events. *Nature Climate Change, 8*, 469–477.

Felice D'Alessandro graduated in civil engineering in 2002 and achieved his Ph.D. degree in hydraulic and environmental engineering in 2006. Currently, he is assistant professor at the University of Milan, Department of Environmental Science and Policy. He was visiting research scientist at "Laboratorio Nacional de Engenharia Civil" (LNEC), Lisbon, Portugal, and at Coastal and Hydraulics Laboratory, U.S. Army ERDC, Vicksburg, Mississippi, USA. He was/is partner of national and EU projects (e.g., FP6-Hydralab III, FP7-Hydralab IV, Guideport). He carried out physical model tests in different international laboratories (e.g., LSTF, Vicksburg; CIEM-LIM, Barcelona; DHI, Hørsholm). His scientific research has been mainly related to the fields of coastal engineering and environmental hydraulics and shows a high degree of eclecticism covering: wave breaking modelling in Boussinesq-type equations (BTE), wave–dune interaction and beach resilience, longshore transport, multivariate design and structural risk in coastal and off-shore engineering, wave and wind energy converters, and coastal ecosystem restoration. Investigations combine numerical modelling with both controlled laboratory experiments and field observations. He is author/coauthor of more than 100 publications in international peer-reviewed journals and conference proceedings. He acts as referee for high-qualified international journals.

Chapter 7
Climate Change: From Science to Policies, Backward and Forward

Sara Valaguzza

Abstract The chapter investigates the relationship between science and politics, focusing on the characteristics of this complex relationship in the debate regarding climate change.

The thesis presented in the chapter aims to reassess the role of science concerning politics, given that the relationship between the two has often been misunderstood.

Moving from the analysis of the different nature of politics and science, the author points out that it is an exclusive responsibility of politics to balance both the public and private interests of a specific community and that the principle of democracy prevents science from controlling politics.

Moreover, the author highlights how politics has a duty towards the society to engage with the scientific community and to motivate, in terms of public interest, the decisions taken: to succeed, policies require a continuous dialogue with science and technology—meaning by "science" not only hard sciences but also social, economic, and legal sciences.

In other words, the author's thesis is that science should support the goals of politics but, at the same time, should not erase political negotiations and rules.

7.1 Introduction: The Convergence of Politics and Science

Politics, which embodies the principle of democracy, cannot be determined exclusively by science. But it does have a duty to enter into dialogue with scientific research and evidence.

S. Valaguzza (✉)
University of Milan, Milan, Italy
e-mail: sara.valaguzza@unimi.it

There is no doubt that scientific issues and climate policies are interconnected (Boehmer-Christiansen, 1994; Forsyth, 2003; Grundmann, 2007; Adger et al., 2009; Allan, 2017; Dessler and Parson, 2020) and that the former inspire and inform the latter. There is also no doubt that politics must take a stand anytime—there exists the possibility of an event that could cause irreparable harm to human beings or the environment.

However, even if environmental problems require a highly technical approach, the resulting policies are not established by scientific considerations alone. The outcome, in fact, is the result of political mediation. As such, the policies are drafted according to public-interest assessments that are rooted in scientific evidence.

Politics is often described as a "consensus factory," meaning a system that makes decisions, with the aim of maintaining peace and social justice, and facilitates the development of society. Politics and power are inseparable. Power, by definition, is free to make determinations of merit, within the constraints of the rule of law.

This freedom would be enough to emancipate politics from science, except that in the environmental field, it's the scientific community that has issued warnings to governments about the risks rooted in a lack of strategies to reduce consumption.

This "sensitivity" of science and its warning calls make it necessary to have an open dialogue with politics, albeit not one inspired by the consequentiality between the discoveries of the former and the strategies of the latter.

Increasingly, in the exchange between science and politics, science is used to support political evaluations that are rooted in economic and social necessities, which are external to the scientific debate. It is fundamentally important, therefore, that science retains its independence and the traditional objective knowledge brought about by the scientific method. And, in fact, "Perhaps the most striking observation is that this policy issue, clearly on the agenda as a result of science and permeated with scientifically-based questions, is not now a scientific issue" (Skolnikoff, 1999).

But what is then the link between scientific evidence and political determinations? This chapter addresses this question, investigating the relationship between science and politics and focusing on the characteristics that this complex relationship plays in the climate debate.

In the past, the most problematic—and at the same time, most compelling—aspect in the relationship between science and politics regarded the (troubled) connection with power.

In 1949, David Bohm, a physicist at Princeton University, was charged with anti-American activity because of his research and was later arrested for alleged contempt before Congress. After a few months, the trial ended with his full acquittal. Princeton University, however, decided not to renew his contract.

Equally serious incidents still occur in Europe and the rest of the world, especially in countries ruled by totalitarian regimes.

It is an undeniable fact that scientific research has political implications, since it has the power to strengthen or discredit governments' political choices.

In a large part of the free world, the greatest concern is not about quelling the opinions of scientists—even though the risk of censoring scientists cannot be considered entirely overcome—but about the credibility of politics, on the one hand, and the practical value of science, on the other.

The thesis presented in this chapter reassesses the role of science with regard to politics, given that the relationship between the two has often been misunderstood. This re-assessment will lead to the consideration that politics has a social responsibility toward the community for engaging with the scientific community, and motivates, in terms of public interest, the decisions made.

It would be a mistake for politicians to adopt policies to mitigate emissions without meeting with the scientific community, which collects data and forecasts the consequences of various decisions.

On the other hand, scientists also have a social responsibility to embrace a practical approach for determining the most useful aspects to pursue political goals, in terms of tools and techniques.

The thesis presented in this chapter concludes with a further observation: experience shows a certain difficulty in combining different disciplines in building effective policies to protect the environment. Still, we contend that not only hard sciences can contribute in making the adaptation process more effective but also social, economic, and legal sciences. These disciplines can help identify regulatory tools, economic and contractual models, nudging strategies, and *ad hoc* regulations.

7.2 Preliminary Concepts About Science and Politics

Science is an ideal type of knowledge, based on conventional rules and principles subjected to regular validation and development. It is also characterized by the consensus and knowledge of various groups of experts who bear the burden of the field's advancement and growth. The methods and internal processes of validation for scientific discoveries depend on these experts.

Politics is the activity of governing. Its purpose is the constitution, organization, administration, and direction of public life. In a democratic context, politics introduces and proposes laws that are the result of negotiations with the electorate body. Politics, according to Aristotle, is at the mercy of "differences and fluctuations" (Barker, 1995).

Thus, while science elaborates hypotheses and tests them according to the rigor of the scientific method, politics creates our future scenarios, designing them according to the democratic link that brings together elected public officials and the populace that voted them into office.

This section addresses the need to understand correctly the link between scientific analysis and political decision-making and identify how a virtuous cycle can be established between the two disciplines.

With regard to environmental matters, in particular climate change, the relationship we are investigating is immediately tense—almost in conflict.

In some parts of the world, priorities—set by local living conditions—are different. These might include measures to combat poverty or improve the educational systems. In other regions, political actors simply do not want to be bound by scientific evidence, leaving scientists shouting into the wind while demanding to influence policy.

Ultimately, the friction between science and politics is inevitable, with the reasons lying in the fact that politics deals with community as a whole. Politics sets general priorities and therefore, by definition, has a wider scope than science. The role of science is limited and determined, while politics has an extensive and inclusive goal.

Scientists were the first to identify the serious risks facing the planet as a consequence of global warming and to link climate change to human activity.

From there, the gravity of the phenomenon and the difficulty of containing it—along with the possible unpopularity of measures adopted to confront it, given their potential impact on economic development as it's traditionally understood—have become part of political debates worldwide, entering into the global agenda. And yet, negotiators still have not succeeded in identifying effective and efficient policies to confront the problems of global warming and pollution.

The tension between science and politics arises when the canons and the relationship between the two disciplines are misunderstood.

If, on the other hand, it is understood that science and politics operate in different, albeit adjacent, territories, and that there is no hierarchy between the two, it creates the conditions needed to stress the importance of a fruitful dialogue. This allows science and politics to cooperate, like two communications channels, in an equal relationship where each operates with profoundly different aims.

The debate must therefore open by discussing questions of identity: What are science and politics, respectively, and what do they deal with?

As noted earlier, the purpose of science is to contribute to knowledge, while the goal of politics is to determine a development model that is consistent with the principles of democracy and freedom.

Scientific evidence serves to provide knowledge and broaden the information base available to politicians while setting goals. However, it cannot replace the evaluation of public interests and discretionary evaluations, which are entrusted to the institutions exercising public powers.

It follows that there should not be tension between science and politics.

The error in perspective arises when politics seeks to be accredited as a scientific authority and when science discredits politics because it does not act according to the scientific community's wishes. As it has been recently affirmed, "Political questions cannot be 'solved' in the same way as scientific ones because they are not purely analytical, they require normative trade-offs; science can only answer analytical questions about how the world 'is' not normative ones about how it 'ought' to be. 'Evidence-informed policy' is more accurate to 'evidence-based policy' as it makes clear that evidence is an input to the political process and not the ultimate authority" (Mair et al., 2019, 8).

Politics may seek confirmation in scientific evidence, but science should not demand a political response.

Presupposing that science and politics are both free from one another, the two can engage in a fruitful and collaborative dialogue. With respect to their reciprocal traits, politics could become more aware of the impact of certain phenomena and the consequences of certain choices, and science could acquire realism with respect to the needs of society.

Consider, for example, climate change. Scientists, neglected for decades, have highlighted since the 1990s the urgent need to take action and mitigate the effects of rising global temperatures.

Today, that evidence is objective and recognized by the vast majority of politicians, as well as researchers.

The fact that the politics neglected for decades the voices of scientists means that the problem is worse than it would have been if action were taken at the time.

What politics is guilty of was failing to carry out an adequate investigation: meaning, in less formal terms, not engaging in sufficient dialogue with the scientific community about the available data on the imminent risks for human beings and the environment.

This occurred despite countries around the world appointed the Intergovernmental Panel on Climate Change (IPCC), established in 1988, as a United Nations body (Agrawala, 1998; Beck and Mahony, 2018), "to provide policymakers with regular scientific assessments on the current state of knowledge about climate change" (https://www.ipcc.ch/).

The IPCC's analytical function with respect to political assessments is clearly stated in the Panel's description: "The reports are neutral, policy-relevant but not policy-prescriptive. The assessment reports are a key input into the international negotiations to tackle climate change" (https://www.ipcc.ch/).

The mistake leading politics to shut down a dialogue with science, with the former flexing its power, is to treat the scientific community like any other stakeholder, when in fact, science is able to offer contextual analysis and foresee different scenarios in a special and privileged way. Science makes it possible to carry out an initial check on the reliability of the starting data and to analyze the implications of the different scenarios under examination in politics.

Those in power therefore have the burden of keeping the debate up-to-date and ongoing with the scientific community, which can foresee consequences and discern between risks and opportunities.

When it comes to risks, political inquiries must be timely. For instance, if researchers investigate tectonic movements, and politicians do not hold hearings when scientists foresee that a particular area is exposed to an imminent risk of earthquake, the responsibility of politicians (by omission) is clear.

In terms of opportunities, meanwhile, political inquiries are broader and more ideological. If, for instance, some biologists identify treatments for infertility, politics must determine whether or not to make available these treatments, and for who. It can debate on the right to be a parent, but the urgency with which it does so depends on the agenda and the priorities of the electorate.

In the environmental sector, both in the assessment of risks and in the creation of opportunities, the role of science is crucial. Science not only provides data to politics in order for the latter to take action but also undertakes the role of an operational arm supporting innovation and resilience goals. This holds true for both mitigation and adaptation policies.

We address the issue by considering three contact points between science and politics, for which the analysis will reveal commonalities and functional relationships that characterize the aforementioned dialogue. These are: method, implementation of goals, and measurement of phenomena.

7.3 The Method: Relativism and Complexity

As described above, the establishment of a relationship between science and politics is inevitable. In order to make decisions that impact society, those in power have a duty to carry out a preliminary survey providing an accurate understanding of the facts and consequences. An absence of such preliminary surveys exposes governments and administrations to complaints about superficiality.

Scientific investigation has different levels of opinion. It can certainly be said that scientific surveys belong to the field of data, making them "mathematical" so to speak. Verifying, for instance, the presence of polluting substances in water, or examining the toxicity in humans of certain polymers, means objectively ascertaining a situation within evaluation constraints.

In the sector at issue, measuring temperature falls within the category of scientific verification, albeit with a particularly high level of complexity. When, however, we move from surveying to analyzing data and formulating hypotheses about the causal relationship, mathematics no longer applies. We move into an environment where that which we seek to demonstrate inevitably becomes subjective. This is a good thing, as it allows science to progress (Beatty and Moore, 2010).

It makes sense, therefore, that scientists express opinions, analyze, and develop technical proposals. Their role, however, is instrumental with respect to politics and remains within the bounds of investigations aimed at acquiring useful elements for weighing each case. As it has been recently stated, "The desire to inform public policy by scientific evidence has not generally been seen as a highly political topic but rather a more technocratic one. The insights and findings of this report show that the argument that public policy is best informed by evidence can no longer be taken for granted. In the new complex information environment, where bad faith actors are taking advantage of the pressures on human behaviour, whether through disinformation, targeted political advertising or fake news, the case for evidence and expertise has to be argued on political as well as scientific grounds" (Mair et al., 2019, 65).

Despite this, reports published by international panels of researchers who investigate climate change—above all, IPCC reports—often become an underlying justification for various types of decisions. In these cases, scientific analysis implicitly becomes a sort of direct, consequential path to political evaluations—and sometimes even judicial decisions.

It is not a coincidence that, as the chapter dedicated to strategic litigation will show, judgments that challenge the inertia or ineffectiveness of public interventions are decided according to scientific evidence. Plaintiffs' reasons for bringing legal actions are rooted, in fact, in alarming scientific data, which show that we have reached a point of no return. States can therefore no longer tolerate delays, such that neglecting these issues is regarded as a serious infringement that solicits, at least in Europe, major convictions.

These judgments override the principle of separation of powers because, as a Dutch judge noted in the Urgenda case, this principle cannot hinder judicial intervention in defense of a current and imminent danger. Thus, the more public action is faced with a serious and potentially harmful factual situation and danger, the more public powers—including legislatures—must be made accountable. Consequently, the pervasiveness of judicial control must increase.

Science, however, has a hidden lever that should be defended against. The field is sometimes used as if it were invested with a type of veto power. It cannot be ruled out that scientific evaluation and the resulting risk assessment are oriented toward the definition of supranational policy, and are therefore linked to economic and geopolitical factors. This creates positions of power and allows strong economies to exploit weak economies—such as in the case of food safety or temperature-controlled supply chains—and to erect monopolies.

Therefore, if it is true that science's importance reduces the awkwardness and subjectivity of judicial decisions, and at least appears to remove discretion from political decisions, in reality it is not always neutral. The matter is more complex than it might appear based on a superficial analysis.

In fact, science and even mathematics distrust absolute truths and indisputable authorities, as they are continually subjected to a process of verification and authentication by trial and error (Jasanoff, 1990).

Even the scientific community is divided on the severity of upcoming risks facing the planet (Swart et al., 2009; Yohe and Oppenheimer, 2011). Some scientists think that the No. 1 political priority should be the oceans, which by 2040 will be completely devoid of fish and will no longer absorb carbon dioxide, if nothing is done. For others, the most important problem is atmospheric pollution, requiring a focus on investing in renewable energy, alternative mobility, and reducing consumption.

Thus, science—even climate science—is subject to the mechanism of relativity that has characterized the scientific method from its beginnings.

The truths held by politics and those of science belong to two different worlds. Scientific debate takes place within a group of experts, while politics dialogues with the community of voters. Yet the democratic and the scientific methods have many similarities.

The democratic system and the scientific method are both conducted according to certain protocols; the need for urgency generally does not allow for exceptions to the guiding principles. Neither science nor democracy advance based on guarantees; they are part of a slow loop and are continually subject to verification and choices. Both also fall prey to the formal language of the rules of the communities to which they belong. They are not subjected to a higher authority, and they grow through taking responsibility (Jasanoff, 2004).

The extent to which science can be a matter of opinion, however, is different and in some ways "inferior" to that of politics. The difference is rooted in their subject matters. In the former, specialist knowledge searches for objective connections and does not seek consensus; it ignores the acceptability of the results it obtains, as well as their approval by the community.

Conversely, politics cannot ignore the relationship with voters and grows through consensus. Therefore, political theses may not be rationally connected to a cause-and-effect relationship and could simply represent desired ideals and "visions" of the future.

It is therefore clear in what sense science and politics have rules in common. In both cases, they do not deal with absolute truths, but points of view, which are affected by the subjectivity of those who formulate the hypothesis or thesis. The point of view of both the scientist and the politician serves a purpose: the former seeks to demonstrate the validity of a theory, while the later responds to the interests of the majority.

It has been observed that point of view is more interesting than truth, because it has a body, a time, and occupies a space (Valerio, 2020) and implies interchangeability. It is accompanied by acceptance of diversity, the relativization of differences, and the demystification of ideologies—and even of science.

Therefore, if science and politics both express points of view, this means that there are no predefined hierarchies between things, other than in terms of probability and reasonableness with respect to a given context. In other words, there exists no absolute "right" choice, but rather a choice—scientific or political—that is appropriate to a specific context.

In the methodological introduction on complexity theory, we learned that even data collection—as well as its illustration and chosen analytical approach—involves ideological considerations. This depends on the purposes for which the data is collected and then analyzed.

Thus we understand that science, in its path toward the search for (relative) truths, is not endowed with objectivity. The scientific method, through which progress and knowledge are explored, through feedback mechanisms, implies an inevitable level of subjectivity. This subjectivity depends on what we want to find and what we choose as the main object of observation.

Failure is part of science, as is ideological opposition, diversity of views, and the parochialism of the various fringes of the scientific communities. So-called "climate deniers," who have adopted for themselves the more astute term of "realists," offer proof of this. Just as there are those who warn about the catastrophic consequences of the climate emergency, there exist those who argue that the grave scenarios and imminent danger are a false representation of reality (Helvarg, 2004; Hoggan and Littlemore, 2009; Anderegg et al., 2010; Dunlap and McCright, 2011; Washington and Cool, 2011).

It is also for this reason that it would be harmful for principle of democracy for the executive or judicial branches to make determinations based on scientific "certainty" that does not exist. Those who administer public affairs, including the justice system, should instead approach their decision-making on the basis of scientific reasoning, taken for what it is: a reading of reality based on trial and error.

Ultimately, the democratic principle also guarantees that political evaluations include an assumption of responsibility on the part of the ruling authority, with the public interest at the center of the evaluation.

Denialism belongs to the physiological dynamics of the scientific debate, in which opposing reconstructions are found to explain the same phenomena. The liturgy of opposing any thesis that is contrary to the dominant narrative is a recurring theme in history. This paradox is called "tobacco strategy." The expression refers to strategies adopted in the past to deny other environmental emergencies, such as asbestos, passive smoking, acid rain, and ozone depletion. There are well-known cases in which some scientists have bowed to the needs of politics or industry to keep alive doubts about the cause-and-effect relationship, and thus mitigate compensation that would otherwise be due for the inflicted damage.

With regard to climate change, the attacks experienced by Ben Santer, an American physicist and atmospheric scientist, are notorious. Santer was among the first to highlight, through scientific data and publications, that human activity is one of the primary causes of global warming. Santer was also criticized in 1995 when he wrote Chap. 8 of the first volume of the second report issued by the IPCC, entitled "Detection of Climate Change and Attribution Causes." There, he presented evidence that global warming is in fact caused by greenhouse gases.

The political community could have followed Santer's concerns by examining their reliability based on scientific reasoning and principles, or not. Above all, it could have listened to the public interest needs by combining the concerns underlying the principles of sustainable development.

Based on this, it becomes clear why the two fields—science and politics—do not represent linear logic or the cause-and-effect relationship but are different and detached—albeit interrelated.

Politics is not a consequence of science, and science does not advance with a political push. Yet, as we've seen, democracy and the advancement of technical knowledge have much in common, as far as methodology is concerned. Both can therefore benefit from a reciprocal communication.

The key to understanding the relationship under investigation is purely dialogic: a process of communication, comparison, experimentation, and mutual adaptation that is able to qualify the juxtaposition between science and politics, benefiting both fields.

The lack of consequentiality between science and politics does not, in fact, negate the fact that government action, in order to be responsible and aware, should be mature and cognizant with regard to the area it seeks to regulate.

This is an obvious consequence of the underlying principles of government action, which can be impartial, transparent, and effective only if based on suitable information.

It is therefore science—relativity and all—that allows politics to acquire knowledge and information. It further allows decision-makers to carry out their responsibilities based on a developing awareness of the facts, and with tools capable of implementing the development strategies elaborated by the world's governments.

One model that, to a certain extent, exemplifies the relationship between science and politics, and the virtuous dynamic under examination here, is found in the Treaty on the Functioning of the European Union: "1. The Union shall have the objective of strengthening its scientific and technological bases by achieving a European research area in which researchers, scientific knowledge and technology circulate freely, and encouraging it to become more competitive, including in its industry, while promoting all the research activities deemed necessary by virtue of other Chapters of the Treaties" (cfr. Art. 179 TFUE).

7.4 The Identification and Implementation of Goals

The relationship dynamics between goals and instruments is different from that of the hierarchical relationship between science and politics that was repudiated in the previous paragraphs.

The government takes its strength from knowledge. The recognition of actual phenomena—their manifestation, action, and reaction—allows decision makers to intervene by adequately balancing interests and idealized scenarios with awareness of requirements and consequences.

Starting in 2020, the spread of COVID-19 exemplified this fact.

Scientists identified the severity of the virus and studied the speed with which it spread, the means of contagion, the burden it put on the health system, the effectiveness of various preventive measures, and potential vaccines.

These determinations have not taken away the government's role in deciding whether and how to limit contact between people, block economic activity, close educational institutions, stop public transportation, and assist the needy. Nevertheless, the decisions made were based on scientific evidence and were corroborated by confrontations with scientists. In this case, science has become a tool, providing knowledge and labor, to achieve goals.

A similar dynamic occurs when governments engage in creating mitigation and adaptation policies.

Mitigation and adaptation policies involve work for scientists, as well as for those who have to set up strategies for responding to the needs of the community and advancing economic and social development.

On the one hand, international agreements and the various negotiating tables at which the world's rulers meet require scientists to develop and propose solutions to shape the future in light of the need to reduce emissions and resource consumption. On the other hand, it is science itself that offers public policy makers ideas for emerging from the crisis.

In this regard, within the European Union, an unprecedented collaboration between scientists and governments is underway. Scientists are consulted to enable an informed decision-making process and to select the appropriate responses to achieve political goals. Thus, science dialogues with the government both upstream and downstream in defining and executing political goals.

The efficiency of various policies is, in fact, directly proportional to their adherence to reality and the accuracy of a precise technical analysis aimed at the underlying purpose.

In consolidating the fruitful dialogue between science and public policy makers, the European Union in 2010 initiated and financed the "CIRCLE-2" project: "Climate Impact Research and Response Coordination for a Larger Europe—2nd Generation ERA-NET—Science Meets Policy." The project supports a common agenda in the field of scientific research carried out in Europe and assists with decision-making processes. Its aim is to ensure that research on climate change is focused on real national and European political needs, with knowledge also being shared at the regional level. It's funded by a European network and has been joined by Austria, Belgium, Germany, Estonia, Greece, Spain, Finland, France, Hungary, Ireland, Israel, Italy, the Netherlands, Portugal, Sweden, and Turkey.

Moreover, within the European context, the need to acquire data at a local level and make it accessible for different processing purposes has led to the establishment of special agencies. Their role is to support public policies, with the primary body being the European Environmental Agency (EEA) tasked with providing independent and qualified information about the environment: "The European Environment Agency (EEA) works with the aim of fostering sustainable development and contributing to the achievement of significant and measurable improvements to the environment in Europe, by providing policy makers and the public with timely, targeted information, relevant and reliable."

The agency is a fundamental source of data collection and processing. It has a technical and scientific nature and was set up within the European Environment Information and Observation Network (Eionet) comprised of cooperating members states. The EEA brings together the environmental information of individual members and cooperating countries, engaging in the timely provision of high-quality, nationally validated data.

Within the EEA's work, scientific data constitutes a constant reference point for decision makers. It supports the policies of the European Union, whose pragmatism and consistency are highly valued.

After all, European climate strategy and, more specifically, the Green New Deal represent political actions promoted in close connection with technical feasibility studies and evaluation of socio-economic impacts.

In general, in order to succeed, policies structured through identification of targets require a continuous dialogue with science and technology; this dialogue must remain constant as the policies progress.

The relationship between science and politics becomes particularly close in the phase of implementing political goals. Not only through technology but also pure science becomes a tool that enables government action, as it becomes necessary to ensure the mitigation and adaptation policies goals are being met.

As we will see in Part III, which analyzes policies adopted to protect the environment, science and technology are crucial for enacting resilience strategies.

Nevertheless, it is precisely the importance attributed to the principles of precaution and prevention that have led to science's fundamental role. The precautionary principle (Beck, 1999; Morris, 2000; Sustein, 2005; Peel, 2005; Sadeleer, 2014), which is essential within European environmental policy, requires the adoption of precautionary behaviors, in order to protect common goods and unknown risks, in situations characterized by a lack of unambiguous scientific knowledge. The principle of prevention, meanwhile, prohibits those actions that could damage the environment.

In both cases, public authorities entrusted implementing the aforementioned principles cannot do so without the results of scientific analysis.

The need to define policies based on scientific data is expressly considered in the Treaty on the Functioning of the European Union, which also in this case represents a virtuous and noteworthy experience. Specifically, the Treaty provides for the following (emphasis mine):

- Art. 114 "3. The Commission, in its proposals envisaged in paragraph 1 concerning health, safety, environmental protection and consumer protection, will take as a base a high level of protection, taking account in particular of any new development based on scientific facts. Within their respective powers, the European Parliament and the Council will also seek to achieve this objective."
- Art. 114 "5. Moreover, without prejudice to paragraph 4, if, after the adoption of a harmonisation measure by the European Parliament and the Council, by the Council or by the Commission, a Member State deems it necessary to introduce national provisions based on new scientific evidence relating to the protection of the environment or the working environment on grounds of a problem specific to that Member State arising after the adoption of the harmonisation measure, it shall notify the Commission of the envisaged provisions as well as the grounds for introducing them."
- Art. 191. "3. In preparing its policy on the environment, the Union shall take account of: available scientific and technical data, environmental conditions in the various regions of the Union, the potential benefits and costs of action or lack of action, the economic and social development of the Union as a whole and the balanced development of its regions."

In particular, mitigation and adaptation policies require various technical and scientific contributions throughout implementation. How to protect a city from flooding, how to prevent concentrations of PM10 from exceeding certain thresholds, how to deal with desertification, and how to increase the use of alternative energies are all questions that only science can answer. This means that the implementation phase of environmental policies is characterized by a high level of technicality, as well as the need for a continuous scientific support.

It's important to remember that science isn't the only field that needs to collaborate with legislative bodies to implement such policies; law, economics, and statistics also play vital roles (Victor, 2015).

Only in this way is it possible to understand the relevant economic and social contexts, and also the changes that certain policies bring about. Policies are nothing more than norms, which become rules that guide the activities of citizens within different legal systems. Poorly drafted rules can sabotage even the most solid project, for example, anticipating the need to build barriers against floods, but not being able to implement their construction, due to an impeding regulatory framework, would sabotage the scientists' work.

7.5 Measurement of Phenomena, Knowledge, and Awareness

Our era is characterized by the constant attempt to interpret reality by reading and re-reading contextual scenarios. The analyses regarding rises in global temperature and projections about the future of the planet are precisely the result of measurements of natural phenomena conducted by science. In the same way, proposed solutions to solve the climate crisis are also the result of calculations, views, and scenarios based on the acquisition and re-processing of information.

Data collection, knowledge sharing, and pooling of scientific research results are all essential elements allowing the progress of science. These elements, however, acquire a political dimension when they become an asset for governments to plan the best course of action.

The effectiveness of political decisions and the ability of governments to solve upcoming issues require a metric approach. This means gaining full awareness of the problem's characteristics so as to intervene as precisely as possible; only through an accurate collection of the appropriate information, gathered before any political intervention, it is possible to measure the solution's effects.

Acquiring data before an intervention is a double-edged sword. On the one hand, it allows interventions to be applied in a circumstantial and rational manner; on the other, it makes evident the gradient of effectiveness of the action put in place. This is why reconstructive ambiguity or the secrecy of certain discoveries is often tolerated.

When it comes to environmental protection, the acquisition of technical-scientific data is unavoidable. This is, in fact, the only way in which opposing interests, be they environmental, economic, or social can be reasonably balanced.

There is no other way to show the reasonableness of a government action that positively affects some areas while negatively affecting others, other than demonstrating that the sacrifice on one side is justified based on the benefit to the whole system.

In order for the public to verify the action, whether approving or disapproving of the outcome, it is necessary to share, in a transparent manner, the information and data collected and re-elaborated by the scientific community, thereby generating knowledge and awareness.

The spread of so-called "science for all" leads to maximizing the sharing and processing of data. In fact, research results not only need to be shared; they also need to be made understandable to the average person. The dissemination of science is, in our times, essential for evaluating the decisions made by elected officials.

Reconstructing the context in which governments select lines of action is a prerequisite for forming political will and freedom from political power, which can be evaluated, removed, or confirmed.

In this sense, especially in the current period of mass media, there could be a potential risk for the spread of misinformation. In fact, going back to our original observation, it is noteworthy that all scientific data are a representation of a hypothesis. This is even more true when science dialogues with the community, because there is great uncertainty when it comes to environmental research and the evaluations that follow.

As a result, verification systems for relevant news are spreading, making it possible to corroborate the information's reliability and avoid, among other things, policy being conditioned by misleading scientific data. Thus, studying the necessary criteria for verifying the reliability of information—with reference to experts and, above all, meta-analysis, systematic reviews, and case studies—has become extremely timely.

It is clear that the risk of misleading information decreases when the interlocutor for governments is the scientific community itself, rather than relying on journalistic information with scientific content.

In fact, within each scientific field—understood as a community of experts that validates research within the sector—assumptions and discoveries are credited to the extent that they meet certain methodological and procedural standards designed to solidify their reliability. Each scientific publication is in fact subjected to control and peer review systems that provide preliminary scrutiny of their reliability and, consequently, their empirical nature.

Freedom of information, the difference between scientific publication and journalistic or public dissemination of information, and trustworthiness of texts are all aspects that cannot be neglected in the dialogue between science and politics.

7.6 Conclusion

Public policies are the tools through which governments decide how to shape public interests, where to focus their time and attention, where to invest, how to produce innovation, how to regulate competition, how to deal with the socio-educational aspects of society, and so on.

However, there is nothing automatic or obligatory about this process. Policies are the result of continuous choices and ideas to promote progress. They are the result of a selection—subjective and carried out within the democratic system—between one or more views of the future.

The reason why there is nothing automatic in this process lies with the principle of democracy. It is also the principle of democracy that prevents science from controlling politics. Scientists have no legitimacy to make decisions about the matters they are asked to evaluate or to balance the interests of the community.

Science supports the goals of politics, but it cannot erase political negotiations and rules. In other words, politics is fed with scientific data, but it cannot be bound by them.

Climate science brings to light an imminent and very serious danger for humanity and for the planet. The decisions made—or not made—in this regard are a political responsibility, and decision-makers must answer to electors. Science cannot save politics from its enormous responsibility nor can the gravity and imminence of the danger justify exceptions to the rule of law or the principle of democracy.

Science cannot insert any political determination, at least not through a direct consequentiality link. The nexus between government responsibility, the principle of democracy, decision-making autonomy, and ideological strategies sever the automatic link between science and politics.

Science, therefore, can also be placed upstream with respect to political evaluations, for example when it fulfills the task of alerting public authorities of an imminent risk (i.e., research into earthquakes formation) or an opportunity (i.e., the evolution within the biosciences for treatment for certain diseases). However, science should never be in a direct causal relationship with a political decision, because doing so would eliminate the ideological discretion of the inquiry that precedes policy decisions.

More often, in fact, science is placed within the political evaluation process, because it deepens the available tools for achieving certain objectives and identifies ways to accomplish the goals set by policy makers.

Only lawmakers are responsible for the policies they adopt, including their effectiveness, consistency, and adequacy. Thanks also to the phenomenon of "science for all," our judgment must be increasingly sharp, trained to find ideology and identify superficiality.

Upon acquiring the data produced and processed during scientific dialogue, it is up to those in power to propose, negotiate, and implement public policies to assess whether and how to intervene.

Consequently, when the relationship between science and politics is correctly identified, there is a continuous synergy of steps, both backward and forward, in which the dynamism of a network relationship and the narrative between two communicating sides prevails over the logic of causality.

References

Adger, N. W., Lorenzoni, I., & O'Brien, K. L. (2009). *Adapting to climate change: thresholds, values, governance*. Cambridge University Press.

Agrawala, S. (1998). Context and early origins of the Intergovernmental Panel on Climate Change. *Climatic Change, 39*(4), 605–620.

Allan, B. B. (2017). Producing the climate: States, scientists, and the constitution of global governance objects. *International Organization, 71*(1), 131–162.

Anderegg, W., Prall, J., Harold, J., & Schneider, S. (2010). Expert credibility in climate change. *Proceedings of the National Academy of Science, 107*(27), 12–109.

Barker, E. (1995). *The Politics of Aristotle*. Oxford University Press.

Beatty, J., & Moore, A. (2010). Should we aim for consensus? *Episteme, 7*(3), 198–214.

Beck, S., & Mahony, M. (2018). The IPCC and the new map of science and policy. *WIREs Climate Change, 9*, e547.

Beck, U. (1999). *World risk society*. Polity Press.

Boehmer-Christiansen, S. (1994). Global climate protection policy: The limits of scientific advice: Part 1. *Global Environmental Change, 4*(2), 140–159.

Dessler, A., & Parson, E. (Eds.). (2020). *The science and politics of climate change*. Cambridge University Press.

Dunlap, R. E., & McCright, A. M. (2011). Organized climate change denial. In J. S. Dryzek, R. B. Norgaard, & D. Schlosberg (Eds.), *The Oxford handbook of climate change and society* (pp. 144–160). Oxford University Press.

Forsyth, T. (2003). *Critical political ecology: The politics of environmental science*. Routledge.

Grundmann, R. (2007). Climate change and knowledge politics. *Environmental Politics, 16*(3), 414–432.

Helvarg, D. (2004). *The war against the greens*. Johnson Books.

Hoggan, J., & Littlemore, R. (2009). *Climate cover-up: The crusade to Dany global warming*. Greystone Books.

Jasanoff, S. (1990). *The fifth branch: Science advisers as policymakers*. Harvard University Press.

Jasanoff, S. (Ed.). (2004). *States of knowledge: The co-production of science and the social order*. Routledge.

Mair, D., Smillie, L., La Placa, G., Schwendinger, F., Raykovska, M., Pasztor, Z., & Van Bavel, R. (2019). *Understanding our Political Nature: How to put knowledge and reason at the heart of political decision-making*. Publications Office of the European Union.

Morris, J. (Ed.). (2000). *Rethinking risk and the precautionary principle*. Butterworth-Heinemann.

Peel, J. (2005). *The precautionary principle in practice: Environmental decision-making*. The Federation Press.

Sadeleer, N. (Ed.). (2014). *Implementing the precautionary principle*. Routledge.

Skolnikoff E. B. (1999) *From science to policy: The science-related politics of climate change policy in the U.S.* Retrieved 8 June, 2021, from http://hdl.handle.net/1721.1/3601.

Sustein, C. (2005). *Beyond the precautionary principle*. Cambridge University Press.

Swart, R., Bernstein, L., Ha-Duong, M., & Petersen, A. (2009). Agreeing to disagree: Uncertainty management in assessing climate change, impacts and responses by the IPCC. *Climatic Change, 92*, 1–29.

Valerio, C. (2020). *La matematica è politica*. Enaudi.

Victor, D. G. (2015). Embed the social sciences in climate policy. *Nature, 520*, 27–29.

Washington, H., & Cool, J. (2011). *Climate Change Denial*. Eartscan.

Yohe, G., & Oppenheimer, M. (2011). Evaluation, characterization, and communication of uncertainty by the intergovernmental panel on climate change: An introductory essay. *Climatic Change, 108*(4), 629–639.

Sara Valaguzza is a full professor of administrative law at the University of Milan, where she teaches administrative and environmental law, sustainable development, green procurement, and public–private partnership. She also leads a research group on environmental issues and policies.

Professor Valaguzza is an expert on public contracts and environmental law and policy. Her publications on governing by public contracts and collaborative agreements are an important reference for national and international scholarship. Her researches on climate litigations, on the juridical dimension of sustainability, and on the issues of resilience and government of the environment are often discussed in academic and institutional seats.

She is the promoter of the global partnership among universities, centers of research, public authorities, and private operators that in 2020 launched the first edition of the Interdisciplinary Approaches to Climate Change (IACC) in the University of Milan, which she coordinates.

She is also the founding president of the European Association of Public Private Partnership—EAPPP and the scientific director of the Italian Centre of Construction Law & Management.

She wrote six monographs, both in Italian and in English, and more than 70 essays on administrative law topics. She has written extensively on public–private partnership, public contracts policy, European multilevel governance, sustainable development, *res iudicata,* and judicial review.

She also practices as an attorney for Italian and international clients and leads a research group to promote collaborative contracts in the public and private sectors in Italy, to reach environmental and social targets, and improving public value.

Part III
Ethics and Policies

Chapter 8
The EU Perspective from Setbacks to Success: Tackling Climate Change from Copenhagen to the Green Deal and the Next-Generation EU

Agostino Inguscio

Abstract While the world will remember 2020 as the year of the COVID-19 pandemic, it was also an exceptional year in at least one other aspect. Scientists at Copernicus and at NASA have confirmed that, globally, 2020 tied with 2016 as the hottest year on record. It must be clear to all that the climate commitments we reach in 2021 will very likely shape the next decade. The stakes could not be higher for the climate action agenda.

We are facing a crisis unprecedented in magnitude and reach. We need to transform the COVID crisis into an opportunity to build back better and pave the way for a more sustainable and just future. Advancing new and ambitious nationally determined contributions, long-term strategies, and a re-alignment of global financial flows to tackle mitigation and adaptation must allow us to advance rapidly to meet the objectives of the Paris Climate Agreement.

While the scale of the challenge cannot be underestimated, we have many reasons to feel confident about the path ahead. One of the reasons for optimism is that the key objectives of the green transition are in line with the spirit of the European Green Deal and with the vision embodied in the Next Generation EU as discussed in this chapter.

8.1 Introduction

While the world will remember 2020 as the year of the COVID-19 pandemic, it was also an exceptional year in at least one other aspect.

Scientists at Copernicus and at NASA have confirmed that, globally, 2020 tied with 2016 as the hottest year on record.

A. Inguscio (✉)
Italian Prime Minister's Office—G7/G20 Office, Rome, Italy
e-mail: a.inguscio@governo.it

Coming—as it did—at the end of the warmest decade ever recorded, we might be tempted to pay little attention to this fact.

However, the global community now fully understands the gravity of the existential threat posed by climate change. It is telling—for instance—that the 2021 Global Risks Report, published by the World Economic Forum, identified how, despite the centrality of COVID-19, climate-related matters constitute the core of this year's risk list.

It must be clear to all that the climate commitments we reach in 2021 will very likely shape the next decade. The stakes could not be higher for the climate action agenda.

We are facing a crisis unprecedented in magnitude and reach. We need to transform the COVID crisis into an opportunity to build back better and pave the way for a more sustainable and just future. Advancing new and ambitious nationally determined contributions, long-term strategies, and a re-alignment of global financial flows to tackle mitigation and adaptation must allow us to advance rapidly to meet the objectives of the Paris Climate Agreement.

While the scale of the challenge cannot be underestimated, we have many reasons to feel confident about the path ahead. One of the reasons for optimism is that the key objectives of the green transition are in line with the spirit of the European Green Deal and with the vision embodied in the Next Generation EU.

Furthermore, the ambition of reaching climate neutrality by 2050—the destination set by the European Green Deal—is shared beyond the European Union by several major countries, including the United Kingdom, Japan, the Republic of Korea, and Canada, while China has pledged to get there before 2060. This is an extremely relevant consequence of a deeper acceptance of science and use of evidence-based policy-making. Indeed, the need to achieve climate neutrality by mid-century is one of the key recommendations of the 1.5° IPCC report, which highlights also the differences that reaching a 1.5 or a 2° increase wordlwide by the end of the century would make in terms of suffering and damages. Furthermore, it is crucial that the world collectively shows to businesses and investors that the global economy can and must recover through an acceleration along the path of climate neutrality. In this sense, it cannot be underestimated how important it is that President Biden has made the return of the United States to the Paris Agreement the centerpiece of his first day in office and the new targets set by the United States.

Therefore, in 2021, we have a unique opportunity to meet our responsibilities for the young generations—well represented here at the University of Milan by the first cohort of students to take part in the Interdisciplinary Approaches to Climate Change Master.

The crucial appointment of the year, the Conference of the Parties (COP) 26, sees —once again—the European Union (EU) as a strong advocate for more ambitious joint commitments to tackle climate change.

Therefore, it is an appropriate moment to reflect on the negotiating positions, strategic approaches, legislative behaviors, and common aspirations that have shaped the EU role in tackling climate change.

In this chapter, we will start by considering the role of the EU in shaping climate negotiations until the ratification of the Paris Agreement in 2015. Afterward, we will consider the road traveled so far by the continental bloc, summing up the main policies developed to tackle climate change and considering the role of the European Green Deal and the Next-Generation EU as accelerators of such effort. We will then discuss the crucial role that research and innovation will play in reaching climate neutrality by 2050. To conclude, we will sum up the position of the EU in view of COP26.

8.2 The EU Road to Paris

As we said, the challenge is unprecedented, but we must look at the future with optimism. Part of this optimism rests on the fact that the assumptions made at the start of the European Green Deal proposal still hold. The Green Deal can and must be a strategy for growth and reconstruction.

Of course, the internal dimension of the Green Deal and its perceived role as a way of relaunching the EU engine growth sustainably also depends—both for its ultimate objective of countering climate change and for its objective of strengthening European competitiveness in the new economy—on the capacity to share the climate and the green deal growth strategy with the world.

Therefore, it is important to reflect on the EU relative successes and failures in shaping the global climate agenda.

The EU is often defined as one of the pulling forces in the United Nations Framework Convention on Climate Change (UNFCCC).

In particular, in the period 1997–2005, during which the parties of the Convention negotiated the details on the implementation of the Kyoto Protocol (KP) —committing the most developed countries to reduce their emissions of greenhouse gases (GHG) —and its ratification.

The EU had a crucial role in this process, especially when the United States decided not to ratify the KP. This role did not mean that, as the negotiations unfolded, its objectives have always been met. During the 2019 COP15 in Copenhagen, for instance, the EU had to give up on many of its most ambitious objectives, such as the ultimate goal of ratifying a legally binding agreement where all the signing parties would have had to respect specific GHG reduction goals (Groen, 2015).

The Copenhagen negotiations have been described on the page of his recent—instant classic—A Promised Land. The description portrays how the U.S. proposal for an interim agreement broke the stalemate between the Europeans "holding out for a fully binding treaty," and China, India, and South Africa "appeared content to let the conference crash and burn and blame it on the Americans." The President's last-minute actions to convince the Chinese, South African, and Brasilian delegations —"Gentlemen! I have been looking everywhere for you. How about we see if we can do a deal?"—are described by one of his aides as "some real gangster shit." On the contrary, the EU's disappointment with the results of the conference—and its own supposed incapacity to broker a deal—is summarized by Angela Merkel's acknowledgement that "what Barack describes is not the option we hoped for…but it may be our only option today. So…" (Obama, 2020, 512–513).

After Copenhagen, the capacity of the EU to reach its objectives has improved according to most observers. The COPs of Cancun and Durban (2010 and 2011) saw an EU able to reach its negotiation mandate (more limited than those of Copenhagen). The result in Durban was a compromise—strongly desired by the EU—on a roadmap toward a new legally binding agreement on climate change to be reached in 2015.

Crucial for future developments was the capacity of the EU to build a broad coalition with ambitious partners, such as South Africa and Brazil. Before COP21 in Paris, therefore, the EU had the time to hone its position. The level of ambition before Paris can be understood through the conclusions of the EU Council of October 2014 on the "2030 Climate and Energy Framework." This document posed the basis for the INDC (Intended Nationally Determined Contribution) presented by the EU in March 2015 to prepare for COP21. This INDC stated that the EU committed to reducing GHG of at least 40% below the levels of 1990 by 2030.

The Council then adopted a mandate for the negotiations in September and conclusions on Climate Finance in November. Reflecting on the results obtained by the EU in the various COPs, it is important to reflect on the concept of the relative bargaining power of each member of the UNFCCC. Essentially, everything else being equal, the bargaining power of each party at the COP can be measured in terms of its global share of emissions. The first and foremost objective of the UNFCCC is to reduce emissions of GHG at the global level. The higher the percentage of emissions of a given party, the higher the other parties' need to include it in any agreement. Thus, at least in theory, countries with a higher percentage of emissions have a stronger negotiating space. Given this reflection, the EU was in a weak bargaining position at COP15, considering how China (22.2%) and the United States (15.6%) both had higher shares of global GHG emissions than the EU (11.1%) (Groen, 2015, 4–6).

Moreover, the share of global GHG emissions of the EU and of the United States was downward, while the Chinese one was growing. Therefore, it is important to consider how the European position at COP requires and will require an increasing capacity to mediate and to build solid coalitions.

As decided during COP17 in Durban, the objective in Paris was that of reaching a legally binding agreement for all parties in the convention. The success of the Paris Conference was considered decisive for the future of international climate action.

8.3 From Paris to the Green Deal and the Next-Generation EU

Since the signing of the Paris Agreement, the European Union has developed and updated a number of crucial policies to tackle climate change.

In December 2019, the European Commission took a historic step with the launch of the European Green Deal at a systemic level. The Green New Deal also guides the EU to review all these crucial policies to make them fit-for-purpose for the new objective of climate neutrality by 2050. After the formation of the European Coal and Steel Community, the creation of the Single Market and the Eurozone, German reunification, and the peaceful reunification of the continent in 2004, the European Green Deal could be considered the fifth great mission of the EU. The Green Deal requires an unprecedented level of engagement with the democratic system in a new way, ensuring a socially fair transition, working with citizens to imagine new destinations, and creating a fairer, greener, cleaner society. The Green Deal has the ambition of being a guiding star for the next 30 years, and the actions undertaken have the potential to reverberate, affecting future generations, so many of whom marched in the world's streets asking for, demanding, a better future.

8.4 The Impact of COVID-19

Of course, when the Green Deal was launched, COVID-19 had not yet upended the planet. The ambitious European Green Deal was devised to give the new Commission a propositive agenda to meet a clear objective—climate neutrality by 2050—that would ensure that we could collectively keep the 1.5° objective within reach—as clearly stated in the 1.5° IPCC report of 2018, which further underlined the urgency of the climatic crisis in the aftermath of the Paris Agreement.

In the intention of the Commission, the European Green Deal marked the passage from a Commission that had been perceived as operating in reaction to various crises (financial, Eurozone, migratory, etc.) to one that could set a new positive vision for the EU.

It was undoubtedly a dramatic coincidence that the first 100 days in office of Ursula von der Leyen coincided with both the adoption of the European Green Deal and with the spreading of the Covid-19 virus in Europe.

The European Green Deal and the entire agenda of the European Commission have found themselves in a particular situation—where we are all perhaps. They are much older than their years and they owe the measure of their past not to the time that has passed, but to the fact that they belong to a period that came before a great fracture "in life and conscience." In The Magic Mountain, Thomas Mann spoke so about events that were very recent but had happened before the Great War "the beginning of many things that perhaps have not ceased beginning yet" (Mann, 1924).

It is unquestionable that the ongoing pandemic represents a drastic break from the past and that every strategy of the EU, starting with the Green Deal, needs to be assessed again to factor this systemic shift.

At the beginning of the COVID-19 crisis, the main arguments used against considering the European Green Deal, the strategy to emerge from the aftermath of the pandemic relied on the consideration that its realization would be too costly and uncertain, given that it would depend to a certain extent on changes in human behaviors. These challenges have dissipated in the course of the crisis. The pandemic has shown that rapid public intervention is possible, behavioral change may happen faster than we think, and re-direction of financial resources happens relatively fast even in our rigid financial system (financial support packages put in place by numerous governments, the EU, and also international organizations such as the World Bank). Re-directing our financial and economic systems toward resilience and regeneration will be critical.

Indeed, in her State of the Union address in September, President von der Leyen emphasized that, despite the turmoil caused by COVID-19, this moment represents an "opportunity to make change happen by design."

Indeed, the Green Deal was initially presented as the EU's new growth strategy, and this is still true—it will be the motor and compass of our recovery.

The cost of dealing with the climate crisis, with biodiversity loss, with pollution, with all the challenges presented in the Green Deal and compounded by COVID-19, is huge, but the cost of non-action is truly immense and irreconcilable with the society that we want to nurture.

We cannot talk about competitiveness without also considering sustainability because they go hand-in-hand. It is now widely shared that investing in the Green Deal is the right thing to do—economically, environmentally, and societally. Indeed, the sooner we act, the sooner we, and future generations, will reap the benefits.

A green recovery from the pandemic will be essential in reaching European climate goals.

Indeed, the economic stimulus packages and recovery plans that governments are now putting in place have the potential to create a recovery that is green, inclusive, and just. This would allow governments to progress toward climate and environmental goals, such as the transition to a climate-neutral future with net-zero emissions, while also providing the only kind of growth that will remain meaningful in the future, a sustainable one.

A green recovery will significantly enhance the resilience of economies and societies in the face of accelerating climate challenges. In the midst of the pandemic, governments' responses need to focus on short-term liquidity and emergency health measures. However, this focus must quickly shift to enable a green, inclusive, and just recovery by, for instance, focusing on large-scale infrastructure, clean energy innovation, and public works that could play a major role in the economic recovery, as well as reducing global emissions.

Unless properly oriented, recovery measures may lock in existing industrial structures, perpetuate a business-as-usual approach, reduce global economic and social resilience, and move the Paris Agreement goals further beyond reach.

The European Union aims to put a green recovery at the center of its strategy to bounce forward. This commitment is demonstrated by the European recovery and resilience plan, Next-Generation EU. Indeed the European Union stressed that the member-states should consider reforms and investments to support the climate transition as a priority. The European Council committed to achieving a climate mainstreaming target of 30% for both the multiannual financial framework and Next-Generation EU. This is why each recovery and resilience plan will have to include a minimum of 37% of expenditure related to climate.

Furthermore, within the framework of the European Green Deal, all major climate change policies at the EU level are to be revised to become more ambitious. Such policies include the following ones, which will be briefly described: the EU Emissions Trading System; Effort Sharing; Land Use and Forestry; Adaptation; Energy; Research and Innovation.

8.5 EU Emissions Trading System

The EU Emissions Trading System (EU ETS) was devised to reduce greenhouse gas emissions from the power sector, industry, and flights within the EU. The EU ETS is the world's first international emissions trading system, having started in 2005. It played an important example in the development of emissions trading in other countries and regions. The ETS has proven to be an effective tool in driving emissions reductions cost-effectively. Emissions from installations, covered by the ETS, declined by about 35% between 2005 and 2019. Under the European Green Deal, the Commission presented in September 2020 an impact-assessed plan to increase the EU's greenhouse gas emission reduction target to at least 55% by 2030. By June 2021, the Commission will present legislative proposals to implement the new target, including revising and possibly expanding the EU ETS.

8.6 Effort-Sharing

Most sectors that are not covered by the EU ETS are covered by the Effort-Sharing legislation, which establishes binding annual greenhouse gas emission targets for member-states for the periods 2013–2020 and 2021–2030. These targets concern emissions from most sectors not included in the EU Emissions Trading System (EU ETS), such as transport, buildings, agriculture and waste. The Effort-Sharing legislation forms part of a set of policies and measures on climate change and energy to help move Europe toward a low-carbon economy and increase its energy security. Greenhouse gas emissions from sectors falling outside the scope of the EU emissions trading system are regulated by the so-called effort sharing regulation, which sets binding targets for annual greenhouse gas emission reductions for member-states for 2021–2030. The regulation aims to ensure that these sectors contribute to reducing greenhouse gas emissions. The sectors include buildings, agriculture (non-CO_2 emissions), waste management, and transport (excluding aviation and international shipping). To achieve a climate-neutral EU by 2050 and the intermediate target of at least 55% net reduction of greenhouse gas emissions by 2030, the Commission proposes revising effort-sharing regulations. The Commission has published an inception impact assessment and launched an open public consultation on the revision.

8.7 Land Use and Forestry

In May 2018, a new regulation to improve the protection and management of land and forests was adopted. Through this regulation, greenhouse gas emissions from land use, land-use change, and forestry (LULUCF) have been incorporated into the 2030 climate and energy framework.

8.8 Adaptation

Adapting to climate change means taking action to prepare for and adjust to both the current effects of climate change the predicted impacts in the future.

Global emissions of greenhouse gases are still on the rise. Even with our commitment to cut net global emissions to zero by 2050, the concentration of greenhouse gases in the atmosphere will continue to increase for the coming decades, and average global temperatures will climb. As the climate heats up, it will bring with it all kinds of risks. From more frequent extreme weather events like heat waves, droughts or floods, to coastal erosion from rising sea levels, the impacts will affect everyone.

That is why the Commission adopted a new EU Adaptation Strategy on February 24, 2021. The strategy is a key part of the European Green Deal, and it aims to increase and accelerate the EU's efforts to protect nature, people, and livelihoods against the unavoidable impacts of climate change. In March 2021, EU environment ministers held an exchange of views on a Commission communication on the new EU strategy on adaptation to climate change. The strategy outlines a long-term vision for the EU becoming a climate-resilient society that is fully adapted to the unavoidable impacts of climate change by 2050.

8.9 Energy

Three-quarters of the EU's greenhouse gas emissions are caused by the production and consumption of energy. The EU is working toward the decarbonization of the energy sector—a central element of the green transition. In December 2020, the Council adopted conclusions on the strategies on offshore renewable energy (December 2020) and hydrogen (July 2020), which had been proposed by the Commission. Thus, they can be understood as integral parts of the overarching European Green Deal Strategy. Furthermore, within the context of the Energy Union, EU member-states are required to present and regularly update their national energy and climate plans (NECPs) to report on their contributions to energy efficiency and renewable targets and also to emissions reduction targets. The NECPs were introduced within the energy union strategy and the first ones cover the period 2021–2030.

The plans for the period 2021–2030 were submitted to the Commission by the end of 2019, and the progress reports must be submitted every 2 years. The Commission will, as part of the state of the energy union reports, monitor EU progress as a whole toward achieving these targets.

The plans must specify how each country expects to deal with the following issues: energy efficiency; renewables; GHG reductions; interconnections; research, and innovation.

8.10 Research and Innovation

Research and innovation, in particular, is a crucial area to deliver on the EU ambition to fight climate change.

COVID-19 represents an opportunity to increase public trust in science, research, and innovation. When navigating unchartered waters, the voice of experts is a valuable currency. The new coronavirus comes together with many uncertainties, half-truths, and even myths. To that end, expertise and the voice of science becomes a strong weapon in fighting the crisis and providing guidance to citizens on best practices, as well as hope for long-term solutions. Evidence-informed policies become a prerequisite for effective action.

With COVID-19, we witnessed the massive and fast mobilization of public and private funds with a very clear and undisputed directionality (due to the clear challenge ahead). Perhaps for the first time in recent history, the entire interconnected world focuses on solving a single problem. With remarkable speed, scientists have uncovered fundamental insights about the novel coronavirus and the interventions that might best address the disease it causes, leading to the vaccines that are now being deployed. The current pandemic has seen scientists working together on common platforms using new tools and technologies such as machine learning and AI for testing through the use of large amounts of data from multiple sources. This outbreak has demonstrated in real-time how scientific understanding can indeed be a global public good. In the situation of a global pandemic, the value of open collaboration is obvious. This can serve as an example to replicate in other global challenges centered around human interconnectedness, a sense of common fate, and shared responsibility, such as climate change. Europe has a unique opportunity to lead efforts and become a global champion on open science and innovation.

Science is, of course, at the basis of the European Green Deal and of the EU commitment to reach net-zero by 2050. This target is not arbitrary—it is evidence-informed, based on a detailed impact assessment, and underpinned by scientific modeling.

Science, knowledge, and evidence—not short-term crisis contingencies—provide us with the means to chart a path toward a more resilient, stronger, greener economy.

Of course, research and innovation contributes much more than just data and the scientific base for achieving the Green Deal objectives. Research and innovation is an enabler of the Green Deal transitions across all the major work streams—climate, energy, industry and circular economy, buildings, mobility, food and farming, biodiversity and ecosystems, and pollution. A forward-looking, mission-oriented, and impact-focused research and innovation agenda can power the Green Deal transformation. By developing, de-risking, demonstrating, and deploying breakthrough innovation at scale, we can help drive long-term systemic shifts, beyond the normal policy, electoral, institutional, and investment cycles.

Horizon Europe is integral to all of this, supported by the European Innovation Council and collaboration in the European Research Area. Under Horizon Europe, different Green Deal Missions will be launched to foster ambitious, long-term research and innovation in key areas: healthy oceans, seas, coastal and inland waters; climate-neutral and smart cities; soil health and food; and climate adaptation, including societal transformation.

These missions aim to produce European public goods on a grand scale, acting as excellent platforms for citizen participation and instilling confidence in the green transition. This will help to bring European research and innovation closer to people, demonstrating the concrete solutions that we can provide to face up to major societal challenges.

Beyond European efforts such as the ones described, research and innovation offers the perfect tool to develop international alliances to address common challenges. This is the case of the initiative Mission Innovation. Mission Innovation is a global initiative—launched during COP 21 in Paris—of 24 countries and the European Commission (on behalf of the European Union). These 25 members—from five continents—have committed to seek to double public investment in clean energy R&D and are engaging with the private sector, fostering international collaboration, and celebrating innovators.

8.11 European Green Deal Call

When considering the role of research and innovation, it is telling that the last call under Horizon 2020—adopted last year—was the European Green Deal Call.

The Call addresses the urgent challenge of aiding Europe's recovery in the wake of the COVID-19 crisis, directly contributing to the EU's Recovery Plan, with the Green Deal at the center of the EU strategy. The Call wants to underline the massive contribution that research and innovation can make to the objectives of the Green Deal and the transitions ahead of the EU, across all sectors of the economy and society. With a budget of €1 billion, it is a major cross-cutting Call, structured to reflect the priorities of the European Green Deal with 10 areas, encompassing 20 topics in total. The crucial role of this approach is also to demonstrate feasibility. Citizens need to witness that the transitions are beneficial, for their health and well-being, for the economy, for the planet, and, ultimately, for their children and grandchildren and the society they will inherit.

Research and innovation, as a policy for opportunity and the long-term, is ideally placed to play a major role, and the Green Deal Call encapsulates the EU ambition to be at the forefront of the transformation.

8.12 Road to Glasgow

The same ambition to lead the transformation toward a sustainable future is enshrined in the European Council conclusions adopted in January 2021 on "Climate and Energy Diplomacy: Delivering on the external dimension of the European Green Deal." These conclusions are a good place to start to reflect on the latest positions of the EU on the need to scale-up ambitions to tackle climate change at the global level.

In its conclusions, the Council recognizes that climate change is an existential threat to humanity. It notes that global climate action still falls short of what is required to achieve the long-term goals of the Paris Agreement and the 2030 Agenda for Sustainable Development.

The Council confirms that Europe is showing leadership and setting an example by stepping up its domestic commitments—we will shortly return to this leadership role—but also stressed that there is an urgent need for collective and decisive global action. The EU sees this coherent pursuit of external policy goals as crucial for the success of the European Green Deal. It is easy to see why such external dimensions are deemed critical, and we will return to this point later in the chapter.

Therefore, the Council calls on all parties to enhance the ambition of the nationally determined contributions and to present long-term low emissions development strategies well ahead of the 26th Conference of the Parties in Glasgow in November 2021, whilst welcoming recent mid-century climate neutrality, as well as carbon neutrality commitments, in particular, those recently taken by major economies.

After the broad reflection on the need to step up ambition and the acknowledgment of the efforts done at the global level, the conclusions address the pathways that are deemed necessary to achieve the EU's objectives.

The first of these means is finance. This is why we read of the confirmation of the EU's continuous commitment to further scale up the mobilization of international climate finance, including sustainable finance practices, as a contribution to the transition toward climate neutrality. The Council notes, in this context, that the EU is the largest contributor to public climate finance, having doubled its contribution from the 2013 figure to EUR 23.2 billion in 2019.

The second leg of the strategy proposed by the conclusions addresses the crucial nexus between climate change and energy. The Council stresses that EU energy diplomacy will aim, as its primary goal, to accelerate the global energy transition, promoting energy efficiency and renewable technologies, among other things. At the same time, the EU's energy diplomacy will discourage further investments into fossil-fuel-based infrastructure projects in third countries unless they are aligned with an ambitious climate neutrality pathway and will support international efforts to reduce the environmental and greenhouse gas impact of existing fossil fuel infrastructure.

In line with this strategy, the Council also calls for a worldwide phase-out of unabated coal in energy productions and plans to launch or support international initiatives to reduce methane emissions.

It is also crucial to underline that while the main focus is on the opportunities created by a clean energy transition, an important consideration is also given to the social and inclusiveness part of the transition.

In line with this, the Council notes that while the energy transition is central to the path toward climate neutrality, it will have a significant impact on societies, economies, and geopolitics globally. Therefore, the strategy is for the EU to enhance its energy diplomacy to continue playing a key role in maintaining and strengthening the energy security and resilience of the EU and its partners.

Finally, acknowledging the plurality of multilateral fora, the conclusions highlight the importance of effective multilateral structures and deepening international cooperation in relevant international fora, whilst identifying the Paris Agreement as the indispensable multilateral framework governing global climate action.

8.13 Conclusion

The European Green Deal is the compass to emerge from the COVID-19 crisis.

The impacts and response to the COVID-19 crisis provide a real-time global experimentation on what major disruption entails and what type of (and how fast) solutions need to be taken and implemented on the ground. To that end, the current crisis serves as a dress rehearsal exercise for the (more profound) structural changes and arrangements that we will need to put in place in our efforts to fix our failing relationship between humanity and nature. As the impacts of runaway climate change and environmental degradation will most likely be more devastating and lasting (as there are no solutions able to reverse climate trends within months/years, in a similar way that vaccines reverse infection trends within weeks/few months), the importance of the European Green Deal (EGD) is paramount in this context. It should re-boot sustainable economic development by reinforcing the links and accelerate the diffusion of solutions required from the various tipping points in human health, climate change, biodiversity, food systems, while promoting just transition and increasing societal resilience. The urgent action required for combating climate change as designed in the EGD comes together with improving living standards and regenerating our environment, unlike the action on pandemics which occur under crisis conditions and involves lockdowns of the economy and society. To that end, viewing the COVID-19 crisis as a separate (or even competitive) policy will be a mistake.

References

Groen L (2015). *On the road to Paris: How can the EU avoid failure at the UN Climate Change Conference (COP 21)?* IAI Working Papers 15/33. September 2015. Retrieved 7 June, 2021 from, https://www.iai.it/sites/default/files/iaiwp1533.pdf.

Mann, T. (1924). *Der Zauberberg*. Fischer Verlag. English edition: Mann T (2018) The Magic Mountine (trans: Lowe-Porter HT). Vintage books, London.

Obama, B. (2020). *A promised land*. Penguin.

Agostino Inguscio is the policy lead on Climate Change and Green Finance in the G7/G20 Sherpa Office of the Italian Presidency of the Council of Ministers.

He joined the Presidency from the European Commission, where he was a policy officer (2017–2020) working on ecological transitions in the Climate Action task force of the Healthy Planet Directorate.

Before joining the Commission, Agostino held a British Academy Advanced Newton Fellowship (2015–2016), he was a lecturer in economic history at the University of Cape Town (2014–2015), he worked as post doc at the Economic Growth Center of Yale University (2012–2014), and he taught European business environment at the ESCP Business School (2010–2011).

He is a visiting lecturer for the LLM—International Sustainable Development at the University of Milan, where he teaches the course "Tackling Climate Change," and he is a visiting lecturer at LUMSA University in Rome, where he teaches the course "Africa and Europe."

He holds a master's (St John's College—2007/2008) and a doctorate (Hertford College—Senior Scholar—2008/2012) in economic and social history from the University of Oxford.

Chapter 9
Carbon Pricing from the Origin to the European Green Deal

Isabella Alloisio and Marzio Galeotti

Abstract Economists have long advocated the widespread use of carbon pricing as the chief policy to combat climate change. Consensus is growing also among governments and businesses on the fundamental role of carbon pricing in the decarbonization of the economy. For governments, carbon pricing can be a source of revenue, which is particularly important in an economic environment of budgetary constraints. Businesses increasingly evaluate the impact of mandatory carbon prices on their operations and use them as a tool to identify potential climate risks and revenue opportunities. Finally, long-term investors use carbon pricing to analyze the potential impact of climate change policies on their investment portfolios, allowing them to reassess investment strategies and reallocate capital toward low-carbon or climate-resilient activities. By and large, carbon pricing takes two basic forms: a carbon tax and an emission trading system. In this chapter, we briefly review these two instruments from a conceptual standpoint. We then review the actual experience with these instruments with reference to European countries and the European Union (EU), the region where those policies have been introduced and implemented first.

I. Alloisio (✉)
European University Institute—Florence School of Regulation Energy and Climate, Florence, Italy
e-mail: isabella.alloisio@eui.eu

M. Galeotti
University of Milan, Milan, Italy
e-mail: marzio.galeotti@unimi.it

© The Author(s), under exclusive license to Springer Nature Switzerland AG 2022
S. Valaguzza, M. A. Hughes (eds.), *Interdisciplinary Approaches to Climate Change for Sustainable Growth*, Natural Resource Management and Policy 47,
https://doi.org/10.1007/978-3-030-87564-0_9

9.1 Introduction

Carbon dioxide (CO_2) is an important greenhouse gas (GHG) that helps to trap heat in our atmosphere, driving global warming and climate change. Emissions of CO_2 accumulate in the atmosphere, increasing GHG concentrations, leading to a series of changes in the climate system, of which the increase in global average temperature is the most important one. The adverse impacts of climate change are multiple and vary from sector-to-sector of socio-economic activity (agriculture and food security, health, biodiversity, terrestrial and freshwater ecosystems, marine ecosystems and coastal areas, human settlements, insurance industries, financial services, energy production, and distribution, tourist flows), and also vary in space and over time. These adverse effects are already being felt today, but it is well known that they will unfold all their negative potential in the more or less distant future and to the detriment of generations to come. It is our scientific knowledge on the damage of climate change that is increasingly consolidating and becoming more precise, to motivate the need to adopt policies and reduce GHG emissions.

Around three-quarters of global CO_2 emissions are related to fossil fuels, which in turn account for about 85% of global energy consumption. Despite the recent growing penetration of renewable sources and the traditional presence of nuclear power, world economic growth is still heavily dependent on coal, oil, and gas. These energy sources are necessary inputs of production activities: their increasing use is, therefore, at the basis of economic growth and improvements in social wellbeing.

The economic theory considers pollution (air, water, soil, etc.) to be a negative externality, a side effect of economic activities with adverse consequences for the community. In modern market economies, those carrying out production or consumption activities that generate pollution—CO_2 emissions in this case—do not bear the social cost of the environmental damage caused. In homage to the principle according to which "the polluter pays," carbon pricing has precisely the purpose of "internalizing" the cost of externality, adding it to the private costs that the producer has to bear (or to the price that the consumer has to pay). Externalities are, as economists say, a cause of market failure since no price reflecting the damage caused by GHG emissions is paid by the polluter so that the amount of generated pollution exceeds the socially acceptable level. Thus, a price on carbon shifts the burden for the damage from GHG emissions back to those who are responsible for it and who can avoid it. Instead of dictating who should reduce emissions, where, and how, as would be the case under "command and control" policies, a carbon price provides an economic signal to emitters and allows them to decide to either transform their activities and lower their emissions or continue emitting and pay for their emissions. Environmental economists prefer carbon pricing instruments over mandatory instruments because the overall goal of emission reduction is achieved in the most flexible and least-cost way to society. In addition to cost efficiency, carbon pricing instruments provide a dynamic incentive to help mobilize the financial investments required to stimulate clean technology and market innovation, fueling new low-carbon drivers of economic growth.

Economists have long advocated the widespread use of carbon pricing as the chief policy to combat climate change. Consensus is growing also among governments and businesses on the fundamental role of carbon pricing in the decarbonization of the economy. For governments, carbon pricing can be a source of revenue, which is particularly important in an economic environment of budgetary constraints. Businesses increasingly evaluate the impact of carbon prices on their operations and use them as a tool to identify potential climate risks and revenue opportunities. Finally, long-term investors use carbon pricing to analyze the potential impact of climate change policies on their investment portfolios, allowing them to reassess investment strategies and reallocate capital toward low-carbon or climate-resilient activities.

By and large, carbon pricing takes two basic forms: a carbon tax and an emission trading system. In this chapter, we briefly review these two instruments from a conceptual standpoint. In addition, we review the actual experience with these carbon pricing instruments with reference to European countries and the European Union (EU), the region where those policies have been introduced and implemented first.

9.2 Carbon Taxes

The carbon tax is a tax on CO_2 emissions, the most important greenhouse gas. Since CO_2 is released into the atmosphere by the combustion of fossil energy sources—coal, oil, natural gas—the instrument is a tax on the carbon content of the currently most used energy sources. The use of so-called clean sources—wind, sun, hydro—and nuclear power do not generate CO_2: as such, they are not subject to this form of taxation. The tax rate is the carbon price.

At the international and institutional level, carbon taxation became a debatable topic in the early 1990s, in particular with the establishment of the IPCC (1988), the publication of its first report (1990) and the establishment of the UNFCCC during the Earth Summit of Rio de Janeiro (1992). The topic was also carefully considered by the European Commission until 1991.

Although the motivation and the advantages of this environmental policy tool are sound and known to economists for some time, carbon taxation has so far experienced limited national applications, particularly in Europe, although the scope for it is increasing everywhere. When carbon taxes were initially implemented, rates were set at relatively low levels and exempted for many energy-intensive industries. More generally, several OECD countries have adopted forms of taxation of energy products and motor vehicles that are only indirectly linked to carbon content and CO_2 emissions.

9.2.1 Economic Rationale and Optimal Structure of the Carbon Tax

The carbon tax is a specific form of taxation of pollution. Economist Arthur Pigou was the first to propose an economic solution to the problem of negative environmental externalities. In 1932, he suggested a tax on goods whose production generated such externalities—in this case, the use of fossil fuels—whose rate was to be equal to the cost of the marginal damage caused at the optimum level of pollution. Two important implications follow from this fundamental result: (1) the tax proportional to polluting units represents an increase in costs that causes the producer to contain the level of pollution/emissions compared to the unregulated case; (2) if the tax is set as described, this level of pollution is the socially optimal one (it is not zero!) as it can optimally reconcile, with a given technology, the need to produce energy services that create value with that of containing polluting emissions.

9.2.2 Tax Base and Carbon Tax Rate

If Pigou's recipe for an optimal tax is straightforward in theory, its translation into practice is very difficult. For each level of pollution/emission, it is very difficult to know or be able to estimate the cost of environmental damage, especially when it comes to a global phenomenon such as climate change, whose impacts—as mentioned above—spread over space and time. Climate economists typically resort to simulations of climate-economy models that, under different assumptions of stabilization of GHG concentrations or temperature reduction targets, produce optimal paths of tax rate values. As these impacts are bound to become more severe with time, tax rates gradually increase over time (in the case of stabilization at 450ppm, according to climate economist Richard Tol, the Carbon tax should on average rise from \$102/tCO_2 in 2020 to \$4004/tCO_2 in 2100).

National governments define the tax base and rates in practice. The first decision is which fuels or energy sources to tax. Carbon taxes are usually imposed on fuels and other petroleum products, coal, and natural gas. Certain industries, as we will see below, are often exempted or charged with reduced tax rates. The other decision governments need to make is whether to introduce upstream or downstream taxation. Taxing upstream sources can provide an administratively efficient way to collect tax revenues, while taxing downstream sources, such as electricity and gas consumption for heating, provides a more direct signal to consumers.

Tax rates vary from jurisdiction to jurisdiction, in part according to their function. Because of lower emission content, natural gas is typically taxed more lightly than petroleum derivatives and, in turn, coal. However, the choice of the rate and their structure is partly dictated, rather than by "Pigouvian" reasons, by less noble budgetary reasons. As such, they can be considered simply indirect taxes, rather than carbon taxes. However, some of the highest rates have been imposed in Europe:

in Sweden, the standard rate is equal to $105/tCO_2$ (for industry, it drops to $23), the petrol tax in Norway is $62/tCO_2$, while in Finland the rate is $30/tCO_2$.

Since the carbon tax attributes a cost to the carbon emitted, it is a price instrument (as opposed to standards and "command and control" measures that are quantity instruments), and a reference can be given by the price of the emission permits in the cap-and-trade systems, such as the European Union Emission Trading System (EU ETS). In the French proposal regarding the initial price of the EU ETS permits, the rate was around $25/tCO_2$. By contrast, some of the lowest rates apply in California. Other countries that unilaterally introduced different forms of carbon taxation include British Columbia, Chile, Iceland, Japan, Mexico, South Africa, Switzerland.

9.2.3 Costs and Benefits of the Carbon Tax

The carbon tax is an indirect tax, a tax on transactions. Although its aim is to reduce harmful emissions, the policy objectives may vary: not only must the tax base be determined but also the sectors and the rate are to be set. The tax revenues that the tax generates and the use of proceeds must also be considered. Similarly, the impact on consumers and the economy must be considered, not only the effectiveness of reducing emissions to the desired level.

When addressing effectiveness in an ex-ante evaluation, it should be remembered that the tax is a price instrument. As such, the price of carbon is certain, while the amount of emissions is uncertain. If priority is given to the achievement of specific emission reduction targets, as typically happens in policies to combat climate change, a quantity instrument offers greater guarantees. An emission permit trading scheme, such as the ETS offers this advantage, even if the market can provide a weak or insufficient price signal to induce operators to adopt choices consistent with the objectives set by the policymaker. From the point of view of the ex-post evaluation of the effectiveness of the carbon tax, the task is typically complicated by the presence of many confounding factors that affect emissions, including the growth of the economy and other environmental policy programs. Rarely, governments that have enacted a carbon tax also include assessments of its effectiveness, not to mention that it would be politically non-trivial to increase the rate as a result of insufficient effectiveness (unless automatic increase mechanisms are in place). There is no shortage of studies and research that, with the help of econometric methods or simulation models, have tried to quantify the reduction in emissions in the presence of a carbon tax relative to the level in the absence of a tax.

As it is an indirect tax, the carbon tax is regressive, a feature shared by all energy taxes. The higher burden on low-income consumers compared to the more affluent classes is an important consideration in the design of these forms of taxation and is also a typical criticism. Since equity issues are always very important in tax matters, environmental taxation has often been accompanied by compensation measures in favor of the most vulnerable social groups, such as pensioners, the unemployed, and low-income households. This is the case with taxes on heating fuels or electricity that tend to be regressive. Overall, however, it has been observed that value-added taxes tend to be three times more regressive than environmental taxes as a whole. Moreover, fuel and vehicle taxes are progressive, as high-income taxpayers tend to drive more expensive and more fuel-hungry vehicles than low-income people. There are several ways to mitigate regressivity. Carbon tax revenues can in fact be substantially recycled by financing: (1) emission mitigation programs (renewable sources, energy efficiency), (2) income tax reduction programs, (3) public deficit or burden reductions on public debt. The choice depends on the overall objectives that the government pursues and its political acceptability. For example, the government of British Columbia provides for a "climate action tax credit," a 5% reduction in the personal income tax rates of the first two brackets. In France, President Sarkozy planned to return all revenues to households and businesses by lowering income taxes or by assigning a green check.

Economists call "double dividend" the possibility of catching two birds with the one stone of environmental taxation: improving the environment and contributing to increasing employment levels through appropriate use of tax revenues. The environmental and economic literature has devoted a growing interest to revenue-neutral environmental taxation, particularly relevant for European countries traditionally characterized by scarcely flexible labor markets. Theoretical studies and the empirical results make this perspective attractive under dual environmental and employment objectives. Interventions to reduce social security contributions weighing on labor costs due to the introduction of different forms of energy and CO_2 taxation were carried out in Denmark (1994), Netherlands (1996), the United Kingdom (1996), Finland (1997), Germany (1999), and in the Italian proposal (1999). These articulated interventions usually are the backbones of the proposals for "environmental tax reform" pushed forth especially by the European Commission.

If, in the short-term, the carbon tax offered the double advantage of inducing a more efficient use of energy and of redirecting the choice of consumption toward clean sources by altering the relative prices of the various sources, in a longer-term perspective, it would provide the incentive to invest in technologies which are associated with lower or even zero emissions. This action can be strengthened by using tax proceeds in favor of incentives and subsidies for research and the adoption of new clean technologies and energy efficiency programs.

In addition to consumers, carbon taxes also have an impact on businesses. Businesses may show hostility toward this form of taxation by complaining about an increase in energy costs. The opposition may come in particular from the so-called energy-intensive industries (production and processing of metals, glass, plastic and paper, petrochemicals, electricity generation) and those industries exposed to international competition, especially from countries where environmental policy is less stringent or altogether absent. In this case, the risk, or often the threat, is the so-called carbon leakage, that is, the relocation abroad of production plants due to the higher carbon costs at home. In this case, the solution envisaged by policymakers consists in the total exemption or, more often, in the granting of reduced tax rates. In Sweden, as already mentioned above, the tax rate for the industry is reduced to a quarter from $105/tCO_2$ to $23/tCO_2$, while in Denmark, companies that sign savings agreements with the Ministry of Transport and Energy can pay reduced fees.

The discussion of measures that could be taken to induce countries that do not have carbon taxes has intensified in recent years. Above all, they are so-called border tax adjustments, import tariffs, export discounts, or an obligation on importers to return carbon quotas for the amount of CO_2 emitted as a result of the production of a good. While, on the one hand, there is still no consensus among economists about the effectiveness of such a measure, on the other hand, the World Trade Organization (WTO) has not yet issued specific legislation on forms of taxation linked to climate that may be of relevance to the freedom of trade.

9.2.4 The European experience

Finland (1990), Sweden (1990), Norway (1991), and Denmark (1992) were the first countries to introduce taxes aimed at reducing carbon dioxide emissions. At the same time, income-tax reduction packages have been launched to make carbon taxes overall neutral to the public budget. This trend was followed by the Netherlands (1996), Slovenia (1997), Germany (1998), and the United Kingdom (2000). Switzerland and Ireland joined in 2008 and 2010, respectively.

A case worth mentioning is the 1998 Italian Carbon Tax project, which has never seen concrete implementation. As part of the 1999 Budget Law to reduce polluting emissions according to the commitments undertaken by Italy in Kyoto in 1997, it provided for heavier rates on coal than fuel oil and diesel and therefore natural gas, initially low in 1999 and decidedly higher in 2005. Of the expected revenue of over 2,000 billion lire for 1999, over 60% was to be used to reduce labor costs by reducing social security contributions. While 31% would have been allocated to compensation measures for energy-intensive industries, the remaining 8.4% was aimed at financing energy-efficiency improvement projects.

In many countries, energy taxation has been widely introduced, typically on fuel, but with a budgetary goal and rarely with explicit emissions targeting. One important example is energy taxation at EU level according to Directive 2003/96/EC. Energy products are only taxed when used as fuel and for heating. Taxation applies to electricity although the member-states may provide various exemptions. The uses of energy products as raw materials in chemical, electrolytic, metallurgical, and mineralogical reduction processes are excluded. The level of taxation cannot be lower than the minimum rates set by the Directive (the minimum rates are higher in transport than in heating), biofuels can be exempted while the tax base are the volumes in the case of mineral oils and energy content in the case of coal, gas, and electricity. In light of the introduction of the EU ETS in 2005, and of the new CO_2 reduction targets of the 20-20 climate and energy package in 2008, which include differentiated national reduction obligations in sectors not covered by the EU ETS (transport, buildings, agriculture, and waste), in 2011, the Commission put forward a much-needed revision of the 2003 Directive. The tax should be based partly on energy content and partly on CO_2 content. The minimum rates would be introduced in stages until 2018: the part on carbon dioxide equal to €20/tCO_2 in 2013 while the tax on the energy content would grow to €9.6/GJ by 2018. In this way, the tax would be "technology-neutral" and would provide an incentive to save energy regardless of source. The draft directive would introduce a European carbon taxation scheme and control a sufficient amount of emissions outside those pertaining to the EU ETS. Taxation would be complementary to the EU ETS to avoid a double burden on firms and cover sectors and businesses that are excluded from the EU ETS by size. After a long process and despite the determination of the European Commission, there has been intense opposition from some member-states, particularly from Eastern Europe such as Poland and Romania. Since unanimous voting is required in tax matters, the European Commission eventually withdrew its proposed reform in 2015.

Within the European Green Deal (European Commission, 2019), the revision of the Energy Taxation Directive forms part of a group of policy reforms to deliver on the EU increased climate ambition. The main objectives of revising the Directive are: (1) aligning taxation of energy products and electricity with EU energy and climate policies to contribute to the EU 2030 energy targets and climate neutrality by 2050; (2) preserving the EU single market by updating the scope and the structure of tax rates, and (3) rationalizing the use of optional tax exemptions and reductions. Adoption of the new Directive is planned for 2021 and in July 2021 a proposal for a Directive's revision has been adopted.

9.3 Emission Trading System

The second main instrument to price carbon is the Emission Trading System (ETS), where emitters can trade emission units to meet their emission targets. To comply with their emission targets at least cost, regulated entities can either implement internal abatement measures or acquire emission units in the carbon market, depending on the relative costs of these options. By creating supply and demand for emissions units, an ETS establishes a market price for carbon emissions.

9.3.1 Tradable Permits: From Principle to Practice

In the standard theory of externalities, developed in the first half of the last century, a negative externality can be approached by allocating a full set of property rights among those causing the externality and/or those affected by the externality, which can uphold trading.

One of the main theorems of environmental economics holds that a well-designed tradable permit system can maximize the value received from the resource, under specific conditions, given the sustainability constraint. According to Montgomery (1972), this theorem is true regardless of how the permits are initially allocated, meaning regardless of whether permits are auctioned or are allocated for free. The logic behind this corollary is that whatever the initial allocation, the trading of the permits allows them to ultimately flow to their highest-valued uses. So, even when permits are allocated free-of-charge, they can still support a cost-effective allocation

However, some preconditions must hold. Tradable permit systems may not maximize the value of the resource in the presence of imperfect market conditions. These include the possibility for market power, in the presence of high transaction costs, insufficient monitoring and enforcement, and the absence of large uninternalized externalities. The latter would imply that "maximizing the net benefits of permit holders would not necessarily maximize net benefits for society as a whole even with a fixed environmental target" (Tietenberg, 2002).

If externalities are internalized, an approach fixing a quantity cap in the volume of emissions, the so-called cap-and-trade system, is the second most important policy instruments for climate mitigation control after carbon tax. The cap-and-trade system involves an absolute baseline, an upper aggregate limit on emission, and the trading of emission allowances or tradable pollution permits. As already observed, once the aggregate number of allowances is set, they can be allocated in an infinite number of ways, either auctioned or allocated for free.

In this framework, the Kyoto Protocol has implemented three tradable emission permits mechanisms: Emission Trading, Joint Implementation, and the Climate Development Mechanism (CDM). In the interest of this chapter, we will focus on Emission Trading, and more specifically on the EU emission trading system (EU ETS).

9.3.2 ETS International Experience

Since the launch of the EU ETS in 2005, the diffusion of large-scale emissions trading systems for climate change mitigation has been limited, with few exceptions, to developed economies. Along with the EU ETS, which currently remains the world's largest ETS in operation, other well-established systems include New Zealand, South Korea, Switzerland, California, Quebec, and the Regional Greenhouse Gas Initiative (RGGI), which includes 10 northeastern U.S. Among emerging economies, Mexico's pilot ETS started in January 2020, whereas in China, a pilot ETS started in 2013 at the local level in cities and provinces, and a national ETS has become operational at the end of 2020. Emission trading systems are under consideration in Brazil, Chile, Indonesia, Pakistan, Thailand, Turkey, and Vietnam (ICAP 2020).

9.3.3 EU ETS: Origins and Reforms

The EU ETS is considered the cornerstone of EU climate policy, and it is the oldest and the largest ETS operating system worldwide. It covers 45% of EU GHG emissions and 5% of global emissions. Since its inception in January 2005, the EU ETS was meant to be the main instrument to reach its first Kyoto Protocol target of reducing GHG emissions by 8% below 1990 levels over the period 2008–2012.

The EU ETS has undergone major reforms over the past 15 years. The first main reform was undertaken in 2009, which shaped Phase III of the EU ETS (2013–2020). The last major reform was passed in 2018, setting the new rules for Phase IV (2021–2030).

The most important reform of Phase III concerns the total volume of emission allowances determined at the EU level (as opposed to being the sum of caps determined at the national level) and a single set of rules that governs their allocation. Free allowances are distributed by applying emission-efficiency benchmarks and auctioning has replaced free allocation as the default allocation method, first and foremost in the electricity sector. 10 years later, the reform for Phase IV pursued three main objectives: (a) strengthening the price signal by further tightening the cap whose linear reduction factor increases to 2.2% from the previous 1.74%, (b) supporting low-carbon innovation and modernization of the energy sector (in the lower-income member states) through funding mechanisms based on auction revenues, and (c) better addressing the free allocation of European Union Allowances (EUAs) and revising the sectors at risk of carbon leakage.

9.3.4 EU ETS: Scope and Coverage

The EU ETS regulates emission trading in 31 states, including 27 EU member-states and the three European Economic Areas - European Free Trade Association States (Iceland, Liechtenstein, and Norway). It covers carbon dioxide (CO_2), nitrous oxide (N_2O), and perfluorocarbon (PFC) emissions from about 11,000 heavy energy-using installations, including power stations and industrial plants (oil refineries, steelworks, and production of iron, aluminum, metals, cement, lime, glass, ceramics, pulp, paper, cardboard, acids, and bulk organic chemicals), as well as from air flights. To limit administrative costs and to avoid disproportionate burdens for small firms, only installations above certain production capacity thresholds are subject to the EU ETS in most sectors. As regards aviation, only flights within the European Economic Area are currently subject to the EU ETS. The future regulation of this sector will depend on the specific design of the Carbon Offsetting and Reduction Scheme for International Aviation (CORSIA) by the International Civil Aviation Organization (ICAO).

9.3.5   EU ETS: *Allowance Allocation*

In the EU ETS, free allowance allocation is used to protect the competitiveness of the regulated industries and avoid carbon leakage. Carbon leakage happens when a firm decides to move its operation to a country or region where climate mitigation policy and regulations are less stringent. This poses severe consequences to the competitiveness of European emission-intensive sectors and requires careful consideration by EU policymakers to safeguard the competitiveness of EU firms, especially in the sectors at the highest risk of carbon leakage.

In the first two phases of the EU ETS, allowances were mainly distributed for free. With the start of Phase III, auctioning became the default method for allowance allocation, whereas in the electricity sector power generators must buy all their allowances, with derogations for eight lower-income member states. By contrast, industrial installations in sectors deemed at significant risk of carbon leakage are given free allowances covering 100% of their benchmarked emissions. Here, the level of benchmarked emissions is determined by multiplying the relevant benchmark by the installation's recent output level. As a rule, the benchmark corresponds to the average performance of the 10% most efficient installations. The European Commission developed 52 product-specific benchmarks and two fallback approaches based on heat and fuel consumption. As to the installations that are not in sectors at risk of carbon leakage, free allocation is less generous. For them, free allowances cover progressively smaller shares of benchmarked emissions: from 80% in 2013 to 30% in 2020 (Verde et al., 2018).

Because the aggregate amount of preliminary free allocation calculated by the member states exceeded the maximum allocation available, a uniform cross-sectoral correction factor has been applied to all installations. Moreover, special allocation rules were set for the aviation sector, with 82% of allowances freely allocated, 15% auctioned, and 3% withheld for new entrants and fast-growing companies. At the end of Phase III, 43% of total allowances were freely allocated, while the remaining (57%) were auctioned by member states.

In Phase IV, the European Commission introduced some modifications to the allowance allocation rules. The EC proposed to keep the share of EUAs to be auctioned at 57%, and it designed a more efficient and stringent criterion for identifying the sectors at risk of carbon leakage, as well as a set of new rules for updating the benchmark values. A sector would be considered at risk of carbon leakage only if the product of its trade and carbon intensities exceeds 0.2. The benchmarks should be updated to take into account the technological progress made since 2008. The benchmark values would therefore be reduced by a standard rate of 1% yearly since 2008. According to the new benchmark values, the installations in the sector at risk of carbon leakage would continue to receive free allowances covering 100% of their efficient emissions level, whereas this level would be lower at 30% for the other sectors.

Within this framework, the first list of sectors at risk of carbon leakage was defined in 2009, for the years 2013–2014. Out of 258 sectors, the "carbon leakage list" classified 165 sectors as being at risk. The second carbon leakage list was defined in 2014 for the years 2015–2019. As of Phase IV, a more stringent rule is applied to identify the sectors at risk of carbon leakage, and the resulting carbon leakage list includes 63 sectors, 100 sectors less than in 2009.

Among the reforms for Phase IV, one of the most relevant is the decrease of the yearly reduction factor of the EU ETS cap from 1.7% to 2.2%. Given that the cap decreases over time, fewer free allowances will be available. Therefore, it is fundamental that those industries that have limited potential for emission abatement with current technologies invest in more advanced low-carbon technologies, which is one of the overarching objectives of the EU ETS.

9.3.6 EU ETS: Prices, Oversupply, and Price Containment

The EUAs, which cover one ton of CO_2-eq emissions each, have been characterized since the beginning and across Phases I, II, III by a fluctuating and quite low price. This is due to different and self-reinforcing factors. First, a systemic factor brings a persistent excess supply of allowances (oversupply) of EUAs in the market. Second, a contingent factor was the result of the 2007–2008 financial crisis, which reduced the demand for EUAs, and consequently depressed their prices. Third, the system of free allocation of allowances. Fourth, the increasing import of international emission credits since 2012 contributed to the low EUAs prices.

By the start of Phase III in 2013, the EU ETS had accumulated a surplus of about two billion allowances (more than the total volume of annual emissions under the EU ETS). The Great Recession was the main cause of the initial fall in allowance demand. The persisting surplus of EUAs originated from this exceptional event, combined with the perfect rigidity of allowance supply, severely depressed EUAs prices. In 2012, the European Commission started tackling the problem by postponing the auctioning of 900 million allowances from 2014–2015 to 2019–2020, a measure known as "backloading."

However, backloading did not bring the expected results, and in order to address a persistent excess of allowances, the Market Stability Reserve (MSR) was established in 2015 and reinforced under the Phase IV reform. The MSR consists of a systematic and long-term solution to cope with possible shocks to allowance demand. By introducing some flexibility in allowance supply, the MSR is intended to mitigate the impacts on EUAs prices of any shock affecting allowance demand. With the MSR, an automatic system is implemented and the number of allowances to be auctioned partly depends on the market surplus: if surplus exceeds 833 million EUAs, allowances equal to 12% (24% in the period 2019–2023) of the surplus are withheld from auctions and added to the reserve; if the surplus is lower than 400 million, 100 million allowances are taken from the reserve and injected into the market through auction; if the surplus is anywhere between 400 and 833 million allowances, no intervention is triggered. Furthermore, from 2023 onward, the number of allowances held in reserve will be limited to the auction volume of the previous year via the invalidation of those in excess. The European Commission monitors the operation of the MSR, which is formally reviewed every five years. However, the advantage of the MSR is an automatic adjustment of the system without the need for political intervention. A revision of the MSR is expected in 2021.

9.3.7 EU ETS and the EU Green Deal

As part of the European Green Deal, EU climate policy is undergoing an unprecedented relaunch in its scope and ambition. A cornerstone of the Green Deal, which is a top priority for the current European Commission, is climate neutrality for the continent by 2050. The most closely relevant proposals of the EU Green Deal impacting on the EU ETS are found under the heading "Achieving climate neutrality." To achieve climate neutrality by 2050, the intermediate 2030 target for overall GHG emissions has been set by the European Council at 55% relative to 1990 levels, which is much more ambitious if compared with the 40% previous target. For the EU ETS, this implies a steeper reduction of the cap over Phase IV. If a yearly linear reduction factor was set at 2.2% from 2021 to achieve an emission reduction target of 40% relative to 1990 by 2030, the 55% target would require a more ambitious yearly linear reduction factor.

Another critical aspect of the EU Green Deal interacting with the EU ETS is the possible extension of the EU ETS to other sectors, notably shipping, road transport, and buildings, by 2030. Moreover, in the aviation sector, which is already partially regulated for intra-EU flights, the EU Green Deal would push toward a progressive reduction of free allowance allocation.

Furthermore, another relevant proposal from the EU Green Deal is the introduction of a Carbon Border Tax. Depending on other countries' level of ambition in climate mitigation, the European Commission proposed introducing a Border Carbon Adjustment (BCA) Mechanism (compatible with WTO's rules), leveling the playing field for European products. The heterogeneity in ambition in climate policy with non-comparable carbon constraints has reopened discussions over the BCA as an alternative approach to free allocation for addressing carbon leakage in the European Union. By leveling carbon costs on embodied emissions, a BCA aims to avoid carbon leakage from vulnerable sectors while strengthening incentives for abatement across industrial value chains, both domestically and abroad (Mehling et al., 2019).

The mechanism is quite straightforward: imports from countries that do not have adequate emissions pricing requirements should be penalized by higher tariffs. However, putting this idea into practice is not as simple. Indeed, given the complicated international accounting rules, a BCA is difficult to be calculated since it is virtually impossible to determine what percentage of value was added in a given country.

The inception impact assessment report for the EU Green Deal (European Commission, 2020) considers a few policy options to reduce the risk of carbon leakage: first, a carbon tax on selected products, both imported and domestic. Second, a new carbon customs duty or tax on imports. Third, an extension of the EU ETS to imports so that the type of instrument and the level of carbon constraints applied to imports would be aligned with the EU domestic approach. The first two options would require revising the EU Energy Taxation Directive with the constraint of unanimous vote under article 113 of the Treaty on the Functioning of the European Union. Despite requiring a review of the ETS regulations, the latter option seems to be the most technically, politically, and legally feasible. However, two main barriers exist: first, the EU ETS only applies to specific regulated sectors making the scope of the BCA too limited with respect to the whole EU economy; second, intra-EU competition issues since countries that already have a domestic carbon tax in addition to the EU ETS should be allowed to adjust the BCA upward to include both the domestic carbon tax and the EU-wide EU ETS.

In conclusion, there are many challenges in the design and implementation of a BCA, both in theory and in practice. The complementarity of the mechanism with internal carbon pricing, particularly the EU ETS, and the BCA relation with the measures already undertaken to avoid the risk of carbon leakage will have to be carefully assessed.

Other measures under the European Green Deal with significant implications for the EU ETS include (a) the phase-out of coal, (b) the abolition of any remaining fossil fuel subsidies, and (c) the review of the EU Energy Taxation Directive setting minimum tax rates on energy goods.

9.4 "Price-based" vs "Quantity-based" Regulation

Carbon pricing is one of the most diffused and efficient policy instruments to mitigate climate change. As already explained, it takes two basic forms. The carbon tax is a tax on CO_2 emissions, whose tax rate is the carbon price, and the emissions trading system, which by creating supply and demand for emissions units establishes a market price for carbon emissions.

The main difference between the two carbon-pricing instruments is that carbon taxation is a price instrument that sets a price (for a certain environmental externality) and lets the market determine its supply. In contrast, emission trading is a quota instrument which specifies a particular supply level and lets the market set the price.

Carbon pricing, whether in the form of carbon taxation or emissions trading, reduces greenhouse gases emissions at a minimum cost for society. This is explained by the fact that carbon pricing equalizes marginal abatement costs across polluters, thus minimizing the cost of achieving a given abatement level. This equi-marginal principle holds in perfectly competitive markets, with perfect information and no transaction costs, where both carbon taxation or emissions trading can be used to set a common price signal across countries and sectors. However, this principle does not hold in the real world where uncertainty exists on the timing and scale of impacts of climate change and on the costs of abatement and where other concomitant energy and climate policies are implemented. According to Weitzman (1974), carbon taxation is preferable where the benefits of further GHG emissions reduction vary less with the level of emission than do the costs of these reductions (in other words, where the marginal social cost of carbon is flat compared with the marginal abatement cost curve, as GHG emissions rise). On the other hand, quota-control systems are preferable where the benefits of further GHG emissions reductions increase more with the level of emissions than do the costs of delivering these reductions, in other words where costs associated with surpassing a given level of emission are potentially large and rising.

In practice, the fact that carbon taxation is less politically acceptable and that with a carbon trading the amount of emissions allowed is certain, are the main considerations that prompted the EU to adopt the ETS, together with the political and institutional difficulty of going ahead with the principle of a common European carbon tax. However, from an administrative standpoint, introducing and managing a carbon tax is probably simpler, while a permit exchange scheme is a more insider's mechanism, and it does not contain the word "tax" while pursuing the same purpose.

Furthermore, it is observed that the tax has the advantage, compared to the other instrument, of providing a price signal to companies that allow them to plan strategies and investments in low-carbon technologies, given that energy technologies are long-lived or very long-lived. Conversely, these same firms would suffer from the volatility of carbon prices that might occur in a cap-and-trade mechanism. It should also be added that the tax always generates tax revenues to be used, while permits trade generates revenues only when their initial distribution takes place by auction, a practice that has not been fully utilized so far.

In any case, the carbon tax and the emission trading system should not be seen as alternative instruments to reduce emissions, but complementary, as in many realities is already happening, or will soon happen.

9.5 Conclusions

Climate change is a global public bad whose effective contrast can be achieved only through coordinated international action. Many climate economists, such as Nobel laureate William Nordhaus (2019), have spoken out in favor of a global carbon tax that has been uniformly introduced among the participating countries, owing to the superior advantages it would have over other forms of intervention. In January 2019, few economists published a statement in *The Wall Street Journal* calling for a carbon tax, describing it as "the most cost-effective lever to reduce carbon emissions at the scale and speed that is necessary." By February 2019, the statement had been signed by more than 3,000 U.S. economists, including 27 Nobel Laureate Economists. Reality may, however, take another direction, namely that of a diffusion of national or regional cap-and-trade systems that ideally could subsequently be linked together, perhaps within an agreement under the aegis of the United Nations, to end up simulating a global permit market.

According to the World Bank (2020), carbon pricing initiatives worldwide are expanding across national and state lines, with increased cooperation among jurisdictions to align their carbon markets. In Europe, the Swiss ETS and the EU ETS became linked on January 1, 2020. In the United States, the Regional Greenhouse Gas Initiative (RGGI), a collection of Northeastern states with a regional carbon market for the power sector, has expanded to include New Jersey and Virginia, with Pennsylvania interested in joining. There are now 61 carbon pricing initiatives in place or scheduled for implementation, consisting of 31 ETSs and 30 carbon taxes, covering 12 gigatons of carbon dioxide equivalent ($GtCO_2e$) or about 22% of global GHG emissions (up from 20% in 2019). Governments raised more than $45 billion from carbon pricing in 2019. While increasing in many jurisdictions, carbon prices remain substantially lower than those needed to be consistent with the Paris Agreement. The High-Level Commission on Carbon Prices estimated that carbon prices of at least $40–80/$tCO_2$ by 2020 and $50–100/$tCO_2$ by 2030 are required to cost-effectively reduce emissions in line with the temperature goals of the Paris Agreement. As of today, less than 5% of GHG emissions currently covered by a carbon price are within this range, with about half of covered emissions priced at less than $10/$tCO_2$, and the International Monetary Fund calculates the global average carbon price to be only $2/$tCO_2$.

Under the Paris Agreement, the 196 parties to the United Nations Framework Convention on Climate Change (UNFCCC) committed to shifting the world's course toward sustainable development and "holding the increase in the global average temperature to well below 2 °C above pre-industrial levels and pursuing efforts to limit the temperature increase to 1.5 °C above pre-industrial levels." In order to achieve these goals, all participating countries are required to set national GHG emissions reduction targets—nationally determined contributions (NDCs). The review of the NDCs is taking place and will continue throughout 2021, and countries must increase their ambition to meet the goals of the Paris Agreement, recognizing that this would significantly reduce the risks and impacts of climate change (Alloisio et al., 2020).

One of the keys to this increased ambition lies in the implementation of Article 6 of the Paris Agreement, which could establish a policy foundation for an emissions trading system, helping to lead to a global price on carbon. Specifically, according to Art. 6.2, parties can cooperate directly with one another. This makes it possible for emission reduction measures to be implemented in one country and the resulting emission reductions to be transferred to another and counted towards its NDC through the Internationally Transferred Mitigation Outcomes (ITMOs). ITMOs aim to provide a basis for facilitating international recognition of cross-border applications of subnational, national, regional, and international carbon pricing initiatives. This requires a transparent process and accurate accounting of the emission reductions achieved to avoid double counting of emission reductions, for example, in the emissions inventory of the country in which the reduction activities are conducted, as well as in the country to which the resulting emission reductions are transferred. The issue of international cooperation related to Article 6 of the Paris Agreement is the only left out of the Rule Book at the COP24 in Katowice in December 2018. The operationalization of the new mechanisms under Article 6 is one of the challenges which need to be overcome to enable carbon pricing to deliver on its potential for cost-effective decarbonization and adaptation.

There is substantial pressure to move rapidly toward consensus on climate change issues and expectations are directed at COP26 negotiations' results in Glasgow. In this respect, the EU will likely lead the implementation of ambitious climate policy and will reaffirm its global "climate leadership."

References

Alloisio, I., Nicolli, F., & Borghesi, S. (2020). Increasing the ambition of the EU Nationally Determined Contribution: Lessons from a survey of experts and students. *Economia Politica.* https://doi.org/10.1007/s40888-020-00193-6

European Commission. (2019). *Communication from the Commission to the European Parliament, the European Council, the Council, the European Economic and Social Committee and the Committee of the Regions, the European Green Deal, COM(2019) 640 final.* Retrieved 7 June, 2021, from https://eur-lex.europa.eu/legal-content/EN/TXT/?uri=COM%3A2019%3A640%3AFIN.

European Commission. (2020). Inception Impact Assessment, Ref. Ares(2020)1350037 - 04/03/2020. Retrieved 7 June, 2021, from https://eur-lex.europa.eu/legal-content/EN/ALL/?uri=PI_COM%3AAres%282020%291350037.

ICAP. (2020). *Emissions trading worldwide: Status report 2020.* Retrieved 7 June, 2021, from https://icapcarbonaction.com/en/icap-status-report-2020.

Mehling, M., Van Asselt, H., Das, K., Droege, S., & Verkuijl, C. (2019). Designing border carbon adjustments for enhanced climate action. *American Journal of International Law, 113*(3), 433–481.

Montgomery, W. D. (1972). Markets in licenses and efficient pollution control programs. *Journal of Economic Theory, 5*(3), 395–418.

Nordhaus, W. D. (2019). Climate change: The ultimate challenge for economics. *American Economic Review, 109*(6), 1991–2014.

Tietenberg, T. (2002). The tradable permits approach to protecting the commons: What have we learned? Retrieved 7 June 2021, from doi: https://doi.org/10.2139/ssrn.315500.

Verde, S. F., Teixidó, J., Marcantonini, C., & Labandeira, X. (2018). Free allocation rules in the EU emissions trading system: What does the empirical literature show? *Climate Policy, 19*(4), 439–452.

Weitzman, M. L. (1974). Prices versus quantities. *Review of Economic Studies, 41*(4), 477–491.

World Bank. (2020). *State and trends of carbon pricing 2020.* Retrieved 7 June, 2021, from http://hdl.handle.net/10986/33809.

Isabella Alloisio is senior manager at PwC Italy where she leads a team on ESG (Environmental, Social, Governance). She graduated from the University of Milan and she holds a Ph.D. in international law and economics from Bocconi University and a M.Phil. in international relations from the Geneva Graduate Institute. She is former visiting scholar at the University of California at Berkeley. She is advisor of the Florence School of Regulation Energy and Climate of the European University Institute, and a visiting fellow of the Centre for Finance and Development at the Geneva Graduate Institute. She is associate lecturer at the Ph.D. in Science and Management of Climate Change and at the Master in Global Economics and Social Affairs of Ca' Foscari University of Venice.

She was finance and fiscal lead at ECCOClimate, research associate at Florence School of Regulation Climate, senior researcher at Fondazione Eni Enrico Mattei (FEEM) where she coordinated the sustainable development and sustainable finance research programs, scientist at Fondazione Centro Euro-Mediterraneo sui Cambiamenti Climatici (CMCC), and scientific coordinator at the International Center for Climate Governance (ICCG).

Marzio Galeotti is professor of environmental and energy economics at the Università degli Studi di Milano. He graduated from Università Bocconi in Milan and holds a M.Phil. and a Ph.D. in economics from New York University. He is director of scientific research at Fondazione Eni Enrico Mattei (FEEM) in Milan, where previously he was the coordinator of the Climate Change Modelling and Policy research program. He is fellow of the Centre for Research on Geography, Resources, Environment, Energy & Networks (GREEN) at Università Bocconi of Milan and a visiting fellow at King Abdullah Petroleum Studies and Research Center (KAPSARC). He is review editor for Chap. 4 ("Mitigation and Development Pathways in the Near- to Mid-Term") of the Sixth Assessment Report (AR6), IPCC WGIII, 2021. Founder and first president of the Italian Association of Environmental and Resource Economists (IAERE), he is a member of the editorial board of lavoce.info, of the scientific committee of Centro per un Futuro Sostenibile, and of Fondazione Lombardia per l'Ambiente. He has published extensively in scholarly journals and actively participates to the policy debate through media interviews, comments and articles in newspapers and magazines and speeches and presentations in non-academic public events.

Chapter 10
Technology Innovation in the Energy Sector and Climate Change: The Role of Governments and Policies

Francesco Ciaccia

Abstract This chapter focuses on the importance of technology innovation in the energy sector that has been responsible for more than two-thirds of the global CO_2 emissions in 2020, highlighting the clear link between energy and climate. Climate change is undoubtedly the main challenge of the twenty-first century. For this reason, the energy sector will face a dual challenge: tackling the impact of rising temperatures, on one hand, which will affect the entire energy value chain, and changing its mode of operation, on the other, investing in mitigation. The global energy system will need to undergo a clean energy transition, whereby sources of energy that emit GHG are replaced by increasingly cleaner sources. The next decade is all that is left to effectively limit the climate crisis, and technology innovation will play a crucial role in this context.

Specifically, the chapter describes the role of governments in technology innovation, the impacts of the Covid-19 pandemic on the energy transition policies, and the importance of the international cooperation on innovation in the energy sector, with a particular focus on the G7/G8 and G20 experiences.

10.1 Introduction. The Importance of Technology Innovation in the Energy Sector

All economic revolutions have happened when new technologies and innovations have generated a significant change in economic and social structures. The first Industrial Revolution was triggered using steam in mechanized production, the second one using technologies based on electric power for mass industrial processes, and the third one using electronics and information to automate human activities.

F. Ciaccia (✉)
Italian Prime Minister's Office: G7/G20 Sherpa Office—Eni: Public Affairs Department, Rome, Italy
e-mail: f.ciaccia@governo.it; francesco.ciaccia@eni.com

© The Author(s), under exclusive license to Springer Nature Switzerland AG 2022
S. Valaguzza, M. A. Hughes (eds.), *Interdisciplinary Approaches to Climate Change for Sustainable Growth*, Natural Resource Management and Policy 47,
https://doi.org/10.1007/978-3-030-87564-0_10

Now, the world is living the fourth industrial revolution, where computers and digitalization are changing many aspects of business, economy, and society (Schwab, 2016).

The driving force of technological innovation is particularly evident in the energy sector, where policies have historically tried to combine and achieve different objectives at the same time: meeting the demand for energy security, energy access to all at a minimum cost, environmental sustainability. This task has always been difficult to achieve because national strategies, and their implementation in policies, often have not been able to match the various goals altogether. It is the historical and objective conditions that determine from time-to-time the prevalence of one priority over the others. In this context, in many countries, the policies of a large part of the twentieth century have focused on the need to guarantee the security of supply, while those starting from the late 1990s have focused on promoting the competition and liberalization of the markets. In both cases, technological progress has played a primary role in favoring a more effective pursuit of the purposes.

Climate change is undoubtedly the main challenge of the twenty-first century. According to the "Special Report on the impacts of global warming of 1.5 °C above pre-industrial levels" of the Intergovernmental Panel on Climate Change (IPCC) (IPCC, 2018), the near totality of the scientific community recognizes that anthropogenic activity is the main cause of global warming. Anthropogenic greenhouse gas emissions (GHG) have increased since the pre-industrial era, largely driven by economic and population growth and are now higher than ever. This has led to unprecedented atmospheric concentrations of carbon dioxide, methane, and nitrous oxide. At the international level, CO_2 emissions are identified as one of the main drivers responsible for climate change. The energy sector produced more than two-thirds of the global CO_2 emissions in 2020, highlighting the clear link between energy and climate.

For this reason, the energy sector will face a dual challenge: tackling the impact of rising temperatures, on the one hand, which will affect the entire energy value chain, and changing its mode of operation, on the other, investing in mitigation. The global energy system will need to undergo a clean energy transition, whereby sources that emit GHG would be replaced by increasingly cleaner sources. The next decade is all that is left to limit the climate crisis effectively, and technology innovation will play a crucial role in this context.

According to the International Energy Agency (IEA), technological innovation means "the process of improving the means of performing tasks through the practical application of science and knowledge, usually resulting in higher-performing equipment as measured by, for example, energy efficiency, user-friendliness or cost. This process includes learning-by-researching (R&D) and learning-by-doing, and their interaction with the technology innovation systems to which they contribute." Technological innovation, therefore, consists of a series of phases necessary to implement improvements or develop a new production process, product, or service. Each stage (concept/prototype; demonstration; early adoption; maturity) is associated with uncertainty and complexity, as well as with funding, technical and market risks, which are influenced by various social and political factors. For this reason, in order to have successful innovation systems, it is vital to achieve an alignment in the interests of all the actors in the sector, including researchers, private companies, investors community, civil society, consumers/citizens, and governments.

Private companies are at the core of technological innovation in the energy sector, conducting market-oriented experiments necessary to establish radical changes and turn ideas into business opportunities. They use corporate venture capital to fund laboratories, universities, and research centers, allowing new technologies to move from lab to market. Moreover, private companies can help create the space and political legitimacy that enables making new policies and spreading knowledge through professional societies, international investment, and standardization. The contribution of the investors' community and financial institutions is just as important, providing venture capital funds and growth equity that bring technologies to markets and help small start-ups and entrepreneurs. Further value can be provided by the civil society that can advocate policies to support new technologies, including consumer incentives and public procurement and provide funds through foundations and donations to universities. Finally, if appropriately engaged and empowered (Mazzucato, 2019), consumers, citizens, and local communities can support the energy transition through participatory and behavioral change, acting as a bridge between innovation, technological progress and nationally driven measures. In order to enable innovation to be deployed at scale, the support provided by governments could be of vital importance to foster an ecosystem in which technological innovations can develop and mature, mitigating the pressures and obstacles related to social, political, and economic variables that characterize the energy markets.

10.2 The Role of Governments in Technology Innovation

The role of governments is particularly crucial in the energy sector, setting the overall legislative and regulatory framework for markets and finance in which the prerequisites and essential conditions for developing technologies are identified. Governments can leverage environmental standards, pricing policies, tax exemptions, CO_2 taxes, feed-in rates, etc., in order to support breakthrough technologies with low market value but with a significant social and environmental benefit. The public leverage is important to offset the need of private companies to increase their market share or keep their position in offering products and services to the customers. Competition is one of the main drivers of technological innovation in the energy sector, but incumbents can hinder new technologies. Novel technologies often cannot guarantee secure and immediate profits and may not be competitive with incumbent technologies. For this reason, governments can be the game-changer in terms of attractiveness and openness of markets to new entrants by preventing the abuse of monopoly power.

Public actors could be involved outside the framework of "market failure," toward the co-creation and the formation of markets. Governments can create artificial markets to stimulate innovation through regulations and information sharing.

For instance, public policies have promoted renewable energy technologies through subsidies, tax incentives, regulated feed-in tariffs, procurement policies, minimum production quotas, and exemptions from regulation, especially at the beginning in their early commercialization (Raven, 2007, 2390–2400). Another interesting example is currently provided by the technologies enabled to produce hydrogen. In "Hydrogen strategy for a climate-neutral Europe (European Commission 2020a)," the European Union (EU) pointed out hydrogen as one of the key priorities to achieve the European Green Deal and Europe's clean energy transition. To unlock investment and accelerate the strong deployment of hydrogen, the EU highlighted the need for all the actors, public and private, to work together across the entire value chain to build a dynamic hydrogen ecosystem in Europe.

Public intervention is also fundamental for the knowledge development that is at the basis of any innovation process. Policymakers can launch extensive public-resource funding programs to support R&D projects and activities, laboratory experiments, patents, research plans, and other initiatives focused on learning by researching, prioritizing studies, and ideas able to reach the ambitious targets of the energy transition. At the same time, they can promote and support knowledge diffusion through conferences, workshops, and alliances to stimulate knowledge exchange, partnership, joint projects, and intergovernmental collaborations (Planko et al., 2017, 614–625). The public sector can also provide and boost the development of networks that facilitate the exchange of information between various actors. Knowledge is a public good, and it is difficult to control or restrict its use. For this reason, governments must implement an adequate intellectual property rights protection system.

The guarantee offered by the public sector represents a significant incentive for investors. Governments can establish appropriate fiscal mechanisms to stimulate private spending on research and provide useful information that allows investors to accurately evaluate the risk-return profile of new technologies. At the same time, they can leverage on investments of state-owned enterprises, fund projects, and start-ups with grants, loans, and equity financing and use different types of finance to help small-and-medium companies scale up and face gaps in capital markets. Public actors can also support entrepreneurs by funding incubators and investor networks or reducing investment risks for third-party capital.

Public policies can also support innovation through demonstration projects, advertising, providing information, and raising public expectations and confidence in technological change. Policies focused on the education of employees, citizens, and local communities can be strategic to promote a behavioral change and provide and upgrade skills for new occupations and sectors in which innovative technology is likely to emerge. Moreover, governments can use public procurement, portfolio standards, and purchase incentives to create successive "niche" markets. The technology-push and demand-pull policies are complements rather than substitutes.

The creation of legitimacy is another essential element for convincing the main or potential actors to apply new technologies, including formal and informal lobbying activities. Public institutions may contribute to this process, as in planning agencies advising regional or national governments to develop supporting policies for emerging technologies (Gallagher et al., 2012, 137–162). Public actors must act to set regulatory frameworks that represent interests transparent and non-discriminatory, preventing powerful lobbies with interests in consolidated technologies from hindering the development of competing novel technologies.

10.3 Innovative Technologies, Energy Transition, and Covid-19

The energy transition depends on the development and diffusion of new technological, economic, social, behavioral, and business model innovations. Technological change is key, but diffusion and scale-up need to be supported, even in the case of existing solutions. However, policymakers' strategies to reach climate goals are different between countries, and it is difficult to have a shared definition of clean energy, as well as a common set of technologies able to contribute to the transition. Nevertheless, the energy transition relies mainly on three broad types of technologies to reduce CO_2 emission: renewable energy, energy efficiency, and energy storage (OECD, 2019).

Renewable energy can be applied both in power generation (e.g., solar PVs, hydrogen) and in the transport sector (e.g., fuel cells, electric vehicles, biofuels). Some of these technologies are already gaining competitiveness, as is the case of offshore wind, where technology improvements, supply chain efficiencies, and logistical synergies in closely linked markets in Europe have caused rapid cost reductions. Other technologies, such as ocean energy, the largest unused renewable energy source on the planet, are still mainly in the research and development stages, although projects at small scales are starting to be deployed and are increasing in size. Some technologies, such as those focused on scaling up low or zero-carbon hydrogen production, can strengthen implementation and delivery of climate goals in the short-medium term. Green hydrogen has also started gaining momentum as a source of low-cost renewable electricity, thanks to the ongoing technological improvements and the benefits of greater power-system flexibility as an energy carrier able to contribute to achieving net-zero emissions from hard-to-abate sectors, such as heavy industry and transport.

Energy efficiency is another important leverage to contain CO_2 emissions while stimulating economic growth. Technologies focused on improving efficiency are essential to managing energy performance, especially in buildings (i.e., heating, cooling, cooking, and powering appliances), transportation, and industrial activities. Efforts in this field can produce positive effects in terms of energy and economic savings, environmental benefits, security in energy supply—also in relation to energy increased demand due to climate change impacts—and competitiveness in the business sector.

Energy storage technologies are also critical for the transition. Energy storage is necessary for electromobility and, in the electricity production sector, it eases the demand for peak loads and increases the flexibility of the energy system. For example, the electrification of consumption, mainly driven by renewable sources, is accelerating the need to have energy storage capacity. The intermittency of the "renewables," their strong geographical concentration, and the relative hourly peaks of the demand can dramatically stress the transmission networks. Innovative technologies such as high-energy-density storage can have a big potential for the future.

A crucial contribution to the energy transition could be provided by carbon capture, utilization, and storage (CCUS) technologies that can help reduce emissions directly in key sectors and remove CO_2 to balance emissions that cannot be avoided (CCUS). This solution is one of the most cost-effective ways to reduce carbon emissions in the power sector, with successful examples showing the technology viability.

Beyond the technologies previously described, there are other solutions that are not strictly defined as green, but that will be necessary to achieve the energy transition. These options include digital technologies such as Artificial Intelligence, the internet of things, and blockchain. For example, the development of smart grids and the contribution of digitalization can improve the management of billions of daily connections of energy flows and data in grids and between grids. Furthermore, a reduction in costs, better efficiency, and lower GHG-related emissions will be achieved in the management/maintenance of the networks through self-repair processes and other services via Artificial Intelligence systems, learning machines, complex algorithms, and forecasting models, able to guarantee energy security to the consumers and increasingly accessible costs.

In order to be on target to reach net-zero globally by 2050, the world must decrease CO_2 emissions by nearly 40% relative to pre-pandemic levels by 2030 (IEA, 2020a, 2020b, 2020c, 2020d). This requires rapid action within this decade to decarbonize the energy sector by enhancing sustainable mobility, renovating large numbers of existing buildings, and bringing nascent renewable technologies to full commercialization. In addition, part of the investments must be oriented to enhance new business models, such as circularity within the energy sector, that can contribute to reaching green goals.

There are already large public investment plans around the world for technological innovation and R&D aimed at encouraging the energy transition. For example, one of the world's largest funding programs for energy technology demonstration is China's National Major Science and Technology Projects program, where the bulk of funding from the Ministry of Science and Technology goes to "National Key R&D Projects." Another example is provided by the European Commission that is finalizing its next multiannual R&D funding program, Horizon Europe (European Commission 2020b), which will run from 2021 to 2027. Finally, the 17 National Laboratory system, which is the core of the United States' energy research and development architecture, addresses large-scale, complex research and development challenges with a multidisciplinary approach that places emphasis on translating basic science to innovation. The U.S. Advanced Research Projects

Agency–Energy (ARPA–E) program receives funds to nurture new strategic energy technologies to achieve rapid deployment of radical technologies with high market potential, including by combining expertise across disciplines to seek spillovers (IEA, 2020a, 2020b, 2020c, 2020d).

The post-pandemic reconstruction offers an unprecedented opportunity to accelerate the shift of global financial flows toward climate neutrality and enhance resilience to climate change. The economic stimulus packages and recovery plans that governments are now putting in place have the potential to create a recovery that is green, inclusive, and just. This will allow governments to progress toward climate and environmental objectives, such as the transition to net-zero emissions, while also providing the only kind of growth that will remain meaningful in the future, a sustainable one. Green stimuli can lead to strong economic growth shortly after their implementation, with effects on quality job creation. The energy sector must play a central role in bouncing forward from the current crisis, but investments and policies need to be coordinated to align the countries' efforts in achieving climate goals. For this reason, international cooperation is essential, and it has to be strengthened.

10.4 The International Cooperation on Technology Innovation

Global warming and climate change risk becoming irreversible. This threat has no geographical or political boundaries, and its effects have worldwide consequences that need quick and consistent reactions. For this reason, government coordination in implementing policies to contain and limit the climate crisis is becoming more and more crucial. Considering the inextricable nexus between climate sustainability and energy transitions, it is necessary to find tools that ensure, or at least facilitate, the alignment between the measures that countries can put in place. The energy transition is the key pathway toward the decarbonization of the whole sector, and it requires urgent actions on a global scale, coordinating domestic and international interventions. In this context, the challenge of global energy governance is focused on the effective control of this matter at an international level, establishing principles, rules, norms, and institutions.

In terms of the definition of common rules, the convergence of the policymaker's activities would boost incentive frameworks for the wider diffusion of new energy technologies. The portfolio of technologies needed to decarbonize the world energy system needs to be supported along all the stages of their development by strong and consistent policies focused on ensuring certainty for private companies, investors, and researchers. These technologies are at widely varying stages of development, and they span all sectors of the economy in various combinations and applications. Governments' coordination could support the development of a shared regulatory and legislative framework and guarantee a more profitable market in terms of goods, services, and competitiveness. Moreover, the cooperation between countries could strengthen intellectual property rights protection and management and improve capacity-building activities through appropriate financial support for developing countries. Important results could also be provided by the standard harmonization, the best practices, and information sharing.

An interesting example of international collaboration in this field is Mission Innovation (MI), a global initiative of 24 countries and the European Commission (on behalf of the EU), launched at COP21 in November 2015. The MI works toward reinvigorating and accelerating global clean energy innovation, to make clean energy widely affordable. The members are committed to double their investments in R&D over 5 years, working closely with the private sector, sharing information, and promoting collaboration and cooperation. Their commitment was renewed at the fifth MI Ministerial meeting in September 2020, where the members agreed to work together to develop an ambitious second phase of the initiative. The EU offered another example of coordination on innovation policies. The Regulation on the governance of the energy union and climate action (EU/2018/1999) rules that the EU countries need to establish a 10-year integrated national energy and climate plan (NECP) for the period from 2021 to 2030. One area of focus of the NECPs is on research and innovation, which is considered one of the five dimensions where the countries need to be coordinated to reach their common climate goals.

Different models can be implemented for the governance at the global level of the energy transition (Rosenau, 1995, 13–43). The first one is through international governmental organizations, which are founded and sponsored by national governments. International organizations such as the International Energy Agency (IEA) and the International Renewable Energy Agency (IRENA) are trying to create a shared vision of how the energy market is changing, as well as to provide guidelines and white papers relevant to industrial actors and policymakers, stressing the technology innovation as a key factor. For example, the IEA established "The Technology Collaboration Programme" (TCP), a multilateral mechanism created to advance common research and applications of specific energy technologies between policymakers and experts. Although these international institutions can formulate forecasts on future scenarios and provide suggestions and best practices to policymakers, they do not have the power to cooperatively determine joint action at the global level and set the long-term changes needed to be in line with the Paris Agreement. Today the problem is not a lack of institutions in global energy governance, as many have been created since the 2000s, but rather the fragmentation of the governance structure, where coordination and coherence are often missing (Roehrkasten & Westphal, 2016, 12–18).

Another option of governance is through multilateral financial institutions, such as the development banks. They offer government's economic and technological support, as well as loans, in energy-related issues. A third model is through regional organizations where countries of one specific region join their efforts in reaching the same goal. In all these models, countries have so far failed to find an adequate response in formulating the necessary cross-cutting policies because there are opposite interests linked to specific energy sources, and they are reluctant to transfer substantial authority over the matter to formal multilateral settings.

The global forum model has given the most satisfactory results by allowing governments to discuss different subjects and make commitments on a voluntary basis. For example, the Clean Energy Ministerial (CEM) is a high-level global forum focused on improving energy efficiency worldwide, enhancing clean energy supply, and expanding clean energy access. The CEM leverages on the promotion of policies and programs that advance a global energy transition away from carbon-intensive technologies and infrastructure toward technologies for "clean energy." It was launched at the 15th Conference of the Parties (COP) to the UNFCCC in 2009, and it is now composed of 26 members (Australia, Brazil, Canada, Chile, China, Denmark, the European Commission, Finland, France, Germany, India, Indonesia, Italy, Japan, Korea, Mexico, Norway, Russia, Saudi Arabia, South Africa, Spain, Sweden, the Netherlands, the United Arab Emirates, the United Kingdom, and the United States), owing its success to the development and implementation of many concrete policy measures (Yu, 2019).

Other two informal and high-level forums that are developing an interesting activity are the Group of Seven/Eight (G7/G8) Summit and the Group of Twenty (G20), which fulfill a key function to ensure continuous dialogue and deliberation on highly strategic and complex policy issues and can provide "leadership" in global energy governance. These summits have no fixed regulations or memberships and can create a relatively flexible approach to address multilateral issues of common urgency.

10.5 The Energy Transition Policies in the G7/G8 Experience

The G7 has a consolidated experience with energy matters. This forum is an informal group of industrialized economies (France, Germany, Italy, Japan, the United Kingdom, the United States, and Canada) established to respond jointly to the oil shock and economic crisis of the early 1970s. This group of countries, plus Russia (G8), was the first candidate global governor in energy policy. Although during most of the 1990s, the G7 remained silent or divided on energy, in 2005, it turned its attention to this matter again. The activities carried out on energy throughout the 2005 Gleneagles Summit are considered a milestone, and the main outcome was the Gleneagles Plan of Action on Climate Change, Clean Energy, and Sustainable Development that contained 63 non-binding commitments related to climate change and energy. On that occasion, the Presidency of the G8 invited the IEA to investigate on energy efficiency, cleaner fossil fuels, CCUS and renewables, and the World Bank to create a framework for investment and financing on clean energy initiatives.

The following summits were less focused on climate change. The St. Petersburg Summit in 2006 had shifted attention to energy security, and the Heiligendamm Summit in 2007 put energy efficiency in the spotlight, presenting it as the solution to both climate change and energy security concerns. The approach of Gleneagles was revitalized at the Hokkaido/Toyako Summit in 2008, which emphasized the importance of climate change (G8 Research Group, 2008), leading the G8 to pledge to accelerate research on the second-generation of biofuels. The G8 also committed to supporting the launching of 20 large-scale carbon capture and storage demonstration projects globally by 2010 (G8 Hokkaido Toyako Summit Leaders' Declaration, July 8, 2008) and welcomed the establishment of the International Partnership on Energy Efficiency Cooperation (IPEEC), a forum to promote international cooperation for the improvement of energy efficiency (The G8 and Climate Change since Heiligendamm, Final Compliance Report for the G8 and Outreach Five Countries, July 20, 2008). The IPEEC was officially launched at L'Aquila Summit in 2009, which undoubtedly can be considered a very successful meeting because, in this context, G8 countries agreed for the first time on the fundamental principle that "main economies" need to control their emissions and that the world's temperature must never rise more than 2° above pre-industrial levels and set the goal to cut their GHG emissions by 80% by 2050 (G8 Research Group, 2009). In 2009, the G8 also highlighted that it was considering to erecting a "low-carbon energy technology global platform" managed by the IEA. At the Muskoka Summit in 2010, energy was not treated as a key topic. However, the G8 leaders encouraged eliminating or reducing tariff and non-tariff barriers to trade in environmental goods and services to promote cleaner low-carbon energy technologies and associated services worldwide. In this context, they endorsed the CCUS as an important technology in transitioning to a low-carbon emitting economy (Muskoka Declaration: Recovery and New Beginnings, Muskoka, Canada, June 26, 2010).

In 2011, at the Deauville Summit, leaders reaffirmed that green growth is an essential element to ensure sustainable global growth and fight climate change and promoted R&D for clean technologies and energy efficiency as leverage to encourage investments. At the same time, they confirmed their support for international initiatives launched by the G8, such as the IPEEC and the IEA International Low Carbon Energy Technology Platform (G8 Declaration: Renewed Commitment for Freedom and Democracy, Deauville, May 26–27, 2011). Throughout the Camp David Summit in 2012, several substantial steps forward were taken to produce a sound success on energy and climate change (G8 Research Group, 2012). The role of the innovative technology was highlighted as crucial, and leaders recognized that increasing energy efficiency and reliance on renewables and other clean energy technologies were able to contribute significantly to energy security, savings, fight to climate change, promotion of sustainable economic growth, and innovation (Camp David Declaration, Camp David, Maryland, United States, May 19, 2012).

In Lough Erne's summit, in 2013, the G8 leaders did not highlight innovation in technology as a driver of emission cutting. However, they committed to reach "ambitious and transparent action" on climate change through various international forums and organizations, including the Major Economies Forum (MEF), the

International Civil Aviation Organization (ICAO), and the International Maritime Organization (IMO). In 2014, France, Germany, Italy, Japan, the United Kingdom, the United States, and Canada canceled the planned G8 summit that was to be held in the Russian city of Sochi and suspended Russia's membership of the group due to Russia's annexation of Crimea. The summit was replaced with a meeting of the remaining G7 nations in Brussels, where the leaders stressed the need to promote clean and sustainable energy technologies and continue investments in research and innovation. They also recognized the link between energy security and climate change, noting that reductions in GHG emissions and a move to a low-carbon economy were both necessary for energy security.

The G7 obtained substantial success on climate change at the Elmau Summit in 2015 (G7 Research Group, 2015). It repeatedly affirmed the core principle that all major emitters must control their carbon emissions, and the leaders committed to doing their part to achieve a low-carbon global economy in the long-term, including developing and deploying innovative technologies, striving for a transformation of the energy sectors by 2050. To this end, they also committed to developing long-term national low-carbon strategies (Leaders' Declaration: G7 Summit Schloss Elmau, Germany, June 8, 2015). After the Paris Agreement of December 2015, the G7 summit of Ise Shima of 2016 tried to lead the international community's efforts to address the challenge of climate change. In this regard, the leaders recognized the important role that the energy system has to play in implementing the Paris Agreement. To do this, they were determined to accelerate the work toward the transition to an energy system that enables decarbonization of the global economy, committing to invest further in supporting innovation in energy technologies and encouraging clean energy and energy efficiency to ensure economic growth with reduced GHG emissions (G7 Ise-Shima Leaders' Declaration, Ise-Shima, Japan, May 27, 2016).

With the G7 Taormina Summit, a new negative season of this forum was inaugurated because it was characterized by the new Donald Trump's administration divisive position. The 2017 summit failed to reach an agreement on climate change. Six world leaders reaffirmed their commitment to the Paris Agreement to reduce their GHG emissions, but the United States stated that it was in the process of reviewing its policies on climate change and that it was not in the position to join the consensus on this topic. However, the other leaders were determined to harness the significant economic opportunities, in terms of growth and job creation, offered by the transformation of the energy sector and clean technology. In the Charlevoix Summit of 2018, Canada, France, Germany, Italy, Japan, the United Kingdom, and the EU reaffirmed their strong commitment to implement the Paris Agreement through ambitious climate action. Nevertheless, all G7 leaders (the United States included) agreed on the key role of energy transitions by developing market-based clean energy technologies and the importance of carbon pricing, technology collaboration, and innovation. The 2019 G7 leaders meeting in Biarritz failed again on climate change. The summit produced no commitments on climate change, confirming the negative trend of Donald Trump's U.S. on these multilateral contexts. The G7 summit of 2020, which was meant to take place at Camp David in the United States, was definitively canceled due to the Covid-19 pandemic after being postponed several times.

In 2021, The United Kingdom took on the Presidency of the G7, organizing its summit in Cornwall. After Biden's election as U.S. President, the ambition of the United Kingdom has been to revitalize the forum after years of divisive points of view, especially in terms of the fight against climate change. In this regard, one of the key goals of the U.K. Presidency is to support the transition to a low-carbon economy through increased investment and cooperation in green technologies. This ambitious objective should be achieved through a common consensus to leverage green recovery interventions after the pandemic, fostering investments and better regulations focused on reducing emissions and creating opportunities in prosperity and new jobs. The United Kingdom is also aiming to reach an energy transition commitment to net-zero power, phasing out all unabated coal power generation in the G7 by 2030 or soon after that. Moreover, the United Kingdom is working toward an accelerated transition to zero-emission vehicles, aiming at all new light-duty vehicle sales being zero-emission by 2040 or earlier and scaling up investments for battery development and production. These goals are supported by increasing clean energy innovation investments through a partnership with Mission Innovation and by strengthening institutions to support effective international collaboration within each of the emitting sectors.

10.6 The Energy Transition Policies in the G20 Experience

Although the G7 has been dealing with energy since the 1970s to respond jointly to the oil shocks, in recent years, its role on the matter has been called into question by the emergence of the G20 leaders' summit. The G20 was founded in 1999 as an informal international forum of the Finance Ministers and Central Bank Governors from 19 countries and the EU to discuss policy focusing on the promotion of international financial stability. The G20 has expanded its agenda since 2008 to include heads of government or heads of state after the global recession. In addition to its "Finance track," the G20 currently works on a wide range of issues in its "Sherpa track," such as sustainable development, energy, anti-corruption, climate change, employment, and food security. The composition of the membership is very diversified and involves all the main energy-consuming economies. The members are the G7 countries (Canada, France, Germany, Italy, Japan, the United Kingdom, and the United States) plus the EU, Argentina, Australia, Brazil, China, India, Indonesia, Mexico, Russia, Saudi Arabia, South Africa, South Korea, and Turkey. The Group plays a central role in the global energy landscape since it accounts for over 80% of global primary energy consumption, almost all of the world's renewable power generation, and it counts the four biggest oil producers among its members. The affirmation of the G20 has demonstrated the importance of considering a growing multipolarity and the need to involve other key players in the discussions for the world economy. Its added value is the result of two important leverages.

On the one hand, some members have key roles in other energy institutions such as the IEA and the IRENA, facilitating the definition of consistent actions and messages with other international institutions. On the other hand, the G20 can also be powered by the B20 (Business 20) platform, which can foster a continuous exchange between the private and the public sectors. Compared to the G7, the G20 membership is more widespread at a territorial level and is not concentrated among the more industrialized Western countries. This makes the G20 an increasingly important forum as it allows consensus or agreement on a topic to be projected on a global scale.

The growing importance of the G20 on the energy matter was confirmed in 2009 at the Pittsburgh Summit when the members pledged to rationalize and phase out the medium-term inefficient fossil fuel subsidies that encourage wasteful consumption (Van de Graaf & Westphal, 2011). On that occasion, energy was one of the key matters discussed to react to the economic recession. G20 countries highlighted their support for clean energy, renewables, and technologies that can help to reduce GHG emissions. They committed to facilitating the diffusion or transfer of clean energy technology, including by conducting joint research and building capacity (G20 Leaders Statement: The Pittsburgh Summit, September 24–25, 2009, Pittsburgh). In 2010, the G20 organized two summits. In June, leaders met in Toronto, where the members reiterated their commitments to inefficient fossil fuel subsidies. In November, they met again in Seoul, where they committed to creating enabling environments for the development and deployment of energy efficiency and clean energy technologies and stimulating investments in these sectors to address the threat of global climate change. To reach these objectives, G20 members agreed to promote cross-border collaboration and coordination of national legislative approaches to mobilize finance, establish clear and consistent standards, develop long-term energy policies, and support education and R&D (The Seoul Summit Document).

In 2011, at the Cannes Summit, leaders directly linked climate and energy for the first time. The country members committed to promoting low-carbon development strategies to optimize the potential for green growth and to ensure sustainable development. On that occasion, they committed to encouraging effective policies to overcome barriers to efficiency and to spur innovation and deployment of clean and efficient-energy technologies (Cannes Summit Final Declaration – Building Our Common Future: Renewed Collective Action for the Benefit of All, Cannes, November 4, 2011). The following year, an energy working group was established at the summit of Los Cabos, and the members acknowledged the G20 countries' efforts to foster investment in clean energy and energy-efficiency technologies through the sharing of national experiences on challenges for technology deployment (G20 Leaders Declaration Los Cabos, Mexico, June 19, 2012).

Since 2013, the G20 has been addressing energy issues more comprehensively, and the energy-working group created the previous year changed its title to "Energy Sustainability Working Group." At the G20 summit in St. Petersburg, the members welcomed the efforts aimed at promoting sustainable development, energy efficiency, inclusive green growth, clean energy technologies, and energy security, and they are committed to continue cooperation with international organizations sharing national experiences and case studies (G20 Leaders' Declaration, September 6, 2013, St Petersburg).

At the 2014 summit in Brisbane, the G20 leaders dedicated a whole session to discuss global energy issues for the first time and endorsed the "G20 Principles on Energy Collaboration." On that occasion, leaders agreed on the principles underlying the group's energy policies: strengthen market principles, increase the security of supply, abolish fossil fuel subsidies, and support sustainable growth and development. It is worth highlighting that the members included among the principles the commitment in encouraging and facilitating the design, development, demonstration, and widespread deployment of innovative energy technologies, including clean energy technologies (G20 Principles on Energy Collaboration). At the Brisbane Summit, the G20 members also endorsed the "Action Plan for Voluntary Collaboration on Energy Efficiency," a document focused on efficiency, emissions performance of vehicles, industrial processes, and electricity generation (G20 Leaders' Communiqué, Brisbane, November 16, 2014). Energy efficiency has been prominent in G20 action plans, as it represents an area in which the group's members can easily agree on expanding their activities.

The first G20 energy ministers meeting was held in 2015 at the Istanbul Summit. The opportunity for the members to have a specific moment focused on energy enabled them to address issues related to the Brisbane principles in greater depth. The ministers stressed the need to take strong and effective actions to tackle climate change, and they highlighted the key role of energy efficiency and increasing investments in clean energy technologies. Ministers confirmed their support to the international coordination on clean energy research and welcomed the private and public sectors in making the investments and developing the technologies needed to enhance productivity, efficiency, and sustainable development (Communiqué: G20 Energy Ministers Meeting, Istanbul, October 2, 2015). In this context, it is worth mentioning the adoption of the "G20 Toolkit of Voluntary Options on Renewable Energy Deployment," a background paper that identified action areas for G20 policymakers for the development of roadmaps for the renewable energy deployment (G20 Toolkit of Voluntary Options on Renewable Energy Deployment – document annexed to the G20 Leaders' Communiqué G20 Leaders' Communiqué, Antalya, Turkey, November 16, 2015).

The following year, the Chinese Presidency carried out the "G20 Principles on Energy Collaboration" to make energy institutions more inclusive and effective under the title "Global Energy Architecture." The ministerial meeting of Beijing confirmed the members' commitment to promoting the deployment of renewable technologies and their integration in the power system, smart grids, energy storage, electric vehicles, and modern bioenergy, including advanced biofuels. Moreover, the G20 countries highlighted the key role of governments in supporting cooperation on knowledge-sharing, capacity-building, technology transfer, financial innovation, and pilot projects. In Beijing, the G20 members adopted a "Voluntary Action Plan on Renewable Energy," following the works of the "G20 Toolkit" and the "G20 Energy Efficiency Leading Programme" (EELP) in order to take the lead in promoting efficiency and to provide the basis for a comprehensive long-term framework on the matter (G20 Energy Ministerial Meeting Beijing Communiqué, Beijing, Final Draft, June 29, 2016). China also promoted the adoption of a "G20 Innovation Action Plan" that was mainly focused on a general commitment to stimulate dialogue and best practice sharing, aiming at encouraging innovation through practical actions (G20 Innovation Action Plan 2016 Hangzhou Summit, Hangzhou, September 5, 2016).

Energy ministers did not meet in 2017, under the German Presidency, but the matter was discussed at the leader summit in Hamburg, where the United States stated its decision to withdraw from the Paris Agreement. Germany established the Climate Sustainability Working Group, which together with the Environment Sustainability Working Group developed a" G20 Climate and Energy Action Plan for Growth" to promote clean energy, align financial flows, and support countries to implement their Paris Agreement commitments. The G20 countries remained collectively committed to mitigate GHG emissions through increased innovation on sustainable and clean energies and energy efficiency. Recalling the G20 Principles on Energy Collaboration, they confirmed the commitment to work on international cooperation to develop, deploy, and commercialize sustainable and clean energy technologies (G20 Leaders' Declaration Shaping an Interconnected World July 8, 2017, Hamburg).

The year 2018 was marked by unprecedented natural disasters and record temperatures that boosted a collective and coordinated action to tackle climate change. At the 2018 Buenos Aires Summit, G20 leaders launched the first-ever Climate Sustainability and Energy Transitions steering committees, bringing together several topics such as adaptation, climate financing, sustainable consumption, flexibility, transparency, and digitalization of energy grids. This format enabled more inclusive collaboration by addressing numerous energy principles, including those tied to better access, renewables, transparency, clean energy technologies, and the phase-out of inefficient fossil fuel subsidies. In Argentina, the United States reiterated its decision to withdraw from the Paris Agreement. Nevertheless, G20 members encouraged energy transitions combining growth with decreasing GHG emissions toward cleaner, more flexible, and transparent energy systems (G20 Leaders' Declaration: Building Consensus for Fair and Sustainable Development, June 15, 2018, Bariloche, Argentina). It is worth highlighting that the Argentinian Presidency introduced the concept of the "diverse energy transitions in G20 countries," leveraging the assumption that national resources, GDP growth, per capita energy use, and emissions vary from country to country. In this context, leaders acknowledged the possibility to leave individual G20 members to choose their national pathways to transform their respective energy sectors (IEA, 2018).

In 2019, the G20 Energy and Environment Ministers met in Karuizawa Town (Japan). The concept of the energy transitions was at the core of the meeting's discussion, which focused on the importance of improving the "3E + S" approach (Energy Security, Economic Efficiency, and Environment + Safety) and on global issues, such as climate change. The Japanese Presidency promoted innovation as the crucial leverage to boost the transition. The G20 countries adopted the "G20 Karuizawa Innovation Action Plan on Energy Transitions and Global Environment for Sustainable Growth" to accelerate the virtuous cycle as a collaborative endeavor to facilitate voluntary actions. The ministers stressed the importance of international cooperation and private finance in strengthening research, development, and deployment of innovative technologies and approaches for a clean energy transition and confirmed, in line with the Argentinian Presidency, the existence of different possible national paths to achieve this goal. In addition, the G20 Energy Ministers stressed the need to successfully transform energy systems by increasing investments in cleaner technologies, cooperation in energy efficiency, and deployment of renewables. While past G20 works on the topic were mainly focused on research and development finance and best practice sharing, Japan elected disruptive innovation for climate action as a key priority of its presidency. Hydrogen was the focus of the G20 energy innovation commitment, reflecting Japan's strong interest in the matter as a key solution to its energy security concern, as well as the nation's desire to lead the world in realizing a hydrogen economy. Throughout the G20 energy ministerial meeting, the IEA released "the Future of Hydrogen," a report on the status of hydrogen development also guiding future efforts. The G20 Energy Ministers also recognized the potential of other technologies, such as sustainable biofuels and bioenergy, innovative technologies for sector coupling, and the developing and deploying of CCUS technologies (Communiqué, G20 Ministerial Meeting on Energy Transitions and Global Environment for Sustainable Growth, June 15–16, 2019, Karuizawa, JAPAN).

In 2020, the energy-related G20 agenda was drastically influenced by the effects of the COVID-19 pandemic. The discussion has been dominated by two main challenges: the destabilization of energy markets caused by the health emergency and the need to transition toward cleaner energy systems for a sustainable future. In order to address the issue related to the energy markets' stability, the G20 Saudi Presidency organized an Extraordinary G20 Energy Ministers' virtual meeting in April, just the day before an Extraordinary OPEC and non-OPEC Ministerial meeting that laid the basis for collective actions to rebalance the oil market. On that occasion, the G20 members established a short-term Focus Group to monitor the situation and the effectiveness of the response measures (G20 Extraordinary Energy Ministers Meeting Statement, Extraordinary G20 Energy Ministers Meeting By videoconference, April 10, 2020). In September, the issue was also discussed at the Energy Ministerial meeting where the G20 members faced the topic of the energy transition. Despite the high initial ambition, the G20 left to individual countries the possibility to choose the technologies and solutions to adopt in line with their national strategies, missing the opportunity to set common principles to coordinate the policymakers' action (G20 Research Group, 2020). The G20 Energy Ministers endorsed the Circular Carbon Economy (CCE) framework, a holistic, integrated, inclusive, and pragmatic approach to managing emissions that can be applied reflecting the country's priorities and circumstances. Such an approach results from a deal between the EU and Saudi Arabia, whereby the EU endorsed Saudi Arabia's vision for a circular carbon economy in exchange for a "joint commitment" on the "phasing-out of inefficient fossil fuel subsidies that encourage wasteful consumption." The CCE framework is built on the principles of circular economy and applies them to managing carbon emissions: reduce the carbon that must be managed in the first place; reuse carbon as an input to create feedstocks and fuels; recycle carbon through the natural carbon cycle with bioenergy; remove carbon emission and store it (4Rs framework) (G20 Energy Ministers Communiqué, September 28, 2020). All this considered, in the G20's 2020 communiqué, the collective actions aiming at emissions reductions were characterized by a strong focus on the carbon circular economy. Leveraging on the CCE, the G20 members recognized the importance of accelerating the development and deployment of innovative, scalable, and efficient technologies to advance energy for all. Moreover, they acknowledged the potential of hydrogen as a clean energy carrier, committed to strengthening international collaboration, to advance its development, usage, and dissemination, and highlighted the cross-cutting role of bioenergy and biofuels.

On December 1, 2020, Italy took over the G20 Presidency for the first time, focusing on three main pillars of wide and interconnected action: people, planet, and prosperity. A hard challenge for the G20 in a time when the world economy needs to find a way to rise from the pandemic crisis generated by Covid-19. Italy decided to make the Climate Sustainability Working Group (CSWG) and the Energy Transition Working Group (ETWG) work together toward a joint-ministerial meeting, to capitalize on the clear synergies existing between energy and climate matters. First of all, the Italian Presidency decided to focus the G20 works on fostering the role of sustainable and smart cities that can be considered strategic laboratories of sustainable growth. The aim is to leverage the role of the cities in the energy transition process through a wide range of innovative solutions, such as buildings efficiency and renewable energy projects with small-scale batteries, promoting sustainable mobility at public and private levels, developing digital-energy technologies and tools enabling local communities to be at the center of their urban energy systems. Moreover, urban areas provide an ideal scale and context to test and deploy circular economy approaches.

In addition, the Italian Presidency focused the G20 activities on advancing toward a sustainable and green recovery, seizing the opportunities offered by innovative energy technological solutions. In this regard, the aim is to investigate the potential impacts of a "green" approach in the recovery funds of the G20 countries, analyzing the national recovery plans, the related effects on the review of the Nationally Determined Contributions, and the long-term low-emission strategies. It is worth highlighting that the core of the discussion was focused on examining the potential of new energy technologies (such as smart digital grids, solutions for energy communities, energy efficiency in the construction sector) and the contribution of research and innovation in new fields (i.e., offshore renewables and multiple forms of ocean energy) and towards a decarbonized energy system beside the use of decarbonized hydrogen and its concrete industrial applications.

The G20 is committed to discussing the opportunities offered by the COVID-19 crisis to accelerate the alignment of global capital flows toward a green and sustainable transition in a public and private synergy. Sustainable finance has a key role in boosting the energy transition, strengthening capacities for climate change adaptation, and achieving an emission path in line with the objectives of the Paris

Agreement. For this reason, the post-Covid-19 reconstruction offers an unprecedented opportunity to accelerate the shift of the global financial flows in the direction of climate neutrality and resilience to climate change, and the Italian Presidency of the G20 intends to promote it. Energy, particularly clean technologies, represents an enabling factor for economic development, and it is an essential tool to promote inclusion in line with the objectives of the 2030 Agenda.

10.7 Conclusions

Energy transition and the strictly connected development of innovative technologies will expose the global energy system to face new crucial challenges that governments should be able and ready to face. First of all, the progressive digitalization of electricity grids and the diffusion of platforms focused on the exchange and management of information and physical flows have increased the resilience of the system by favoring efficiency, operational management, and the widespread exploitation of the sources, however, they expose to growing risks from a cybersecurity point of view (Cyber challenges to the energy transition, World Energy Council, 2019). Technology complexity, data sharing, and interconnectivity increase vulnerabilities to malfunction or sabotage in a digital global economy (Advancing Cyber Risk Management: From Security to Resilience. FireEye and Marsh & McLennan Insights, 2019). For this reason, governments will have to find ways, at the national and international level, to cooperate in identifying tools for the prevention, management, and mitigation of digital disruption and cyber threats. In this context, important results can be achieved at the multilateral level through adequate investments in cybersecurity that will allow the entire system to react quickly and with adequate responses.

Another challenge posed by digitalization is data protection and privacy. In the absence of an international rules-based framework, big tech companies that collect data from millions of consumers can manage and use such information not just to analyze behaviors but also to predict them and, more worrying, to change them (Zuboff, 2019). Therefore, governments should join their efforts to clearly rule data management, defining the regulatory and legislative context, and boost transparency.

The transition can produce inequality and social disruption if not correctly managed. The advent of new technologies is including an irreversible change of jobs' categories, requiring a deep awareness by governments that must provide for the relocation of part of the workforce in the energy sectors, through retraining and professional updating programs, as well as the creation of specific training courses.

Energy transformation, driven by the growing deployment of renewables, could have severe geopolitical implications. Rare earth elements are often considered to be critical components of renewable energy hardware. They are found in many countries, including China, Russia, Australia, the United States, Brazil, India, Malaysia, and Thailand. However, two countries—China and Russia—have together most global reserves (O'Sullivan et al., 2017). For this reason, controlling rare earth elements is strategic for the whole cleantech sector, and it is a destabilizing factor because it is disputed between the main international players, in particular China and the United States. Therefore, the rapid growth of renewable energy is likely to alter the power and influence of some states and regions relative to others and redraw the geopolitical map in the twenty-first century (IRENA, 2019). In this context, a possible future scenario will be characterized by the competition between rivaling tech blocs that will be focused on facing concerns linked to a new and wider concept of energy security. An increase in political instability is also a possible threat for all the fossil-fuel-producing countries that will not be able to convert their economies quickly and adequately. These countries will face major economic and social imbalances linked to the failure in reconfiguring their economic models.

All the challenges related to the energy transition require coordinated intervention by the policymakers to mitigate or prevent some unavoidable global risks. As analyzed in the previous paragraphs, the international scene is still looking for adequate governance of the energy transition process to manage all its variables. The G20, with its model of discussion based on the principle of voluntariness and on "soft" modes of steering that do not involve the adoption of binding commitments, is trying to give some answers. The G20 can leverage agenda-setting, coordination among the members, knowledge sharing, and involvement of international organizations, offering an opportunity to discuss the main energy issues. However, it is still too early to judge the real impact of this deliberative and delegating process. Concrete action to steer investments in the right direction is still needed, as well as a shared definition of taxonomy on the investments and clear tools that can help countries prevent, manage, and mitigate crises promptly and in a coordinated manner. In addition, mechanisms that allow greater dialogue between G20 members and other states must be identified to produce the effects of the agreements reached within the forum also in other areas, especially in developing countries.

References

European Commission. (2020a). *Communication from the Commission to the European Parliament, the Council, the European Economic and Social Committee and the Committee of the Regions, Hydrogen strategy for a climate-neutral Europe*. Retrieved 4 June, 2021, from https://op.europa.eu/en/publication-detail/-/publication/5602f358-c136-11ea-b3a4-01aa75ed71a1/language-en.
European Commission. (2020b). *Horizon Europe (2021–2027)*. Retrieved 4 June, 2021, from https://ec.europa.eu/info/sites/default/files/research_and_innovation/funding/presentations/ec_rtd_he-investing-to-shape-our-future.
G20 Research Group. (2020) *Ambitious Words, Constrained Commitments: G20 Energy Performance at the 2020 Riyadh Summit*. Retrieved 8 June, 2021 from http://www.g20.utoronto.ca/analysis/201127-kokotsis.html.
G7 Research Group. (2015). *A summit of significant success: The G7 at Elmau*. Retrieved 8 June, 2021 from http://www.g7.utoronto.ca/evaluations/2015elmau/kirton-performance.html.
G8 Research Group. (2008). *Assessment of the 2008 G8's Climate Change Performance*. Retrieved 8 June, 2021 from http://www.g8.utoronto.ca/evaluations/2008hokkaido/2008-kirton-climate.html.

G8 Research Group (2009). *The Performance of the G8 at L'Aquila 2009: A summit of sound success*. Retrieved 8 June, 2021 from http://www.g8.utoronto.ca/evaluations/2009laquila/2009prospects090706.html.

G8 Research Group (2012). *Several substantial steps forward: The 2012 G8 Camp David's performance on energy and climate change*. Retrieved 8 June, 2021 from http://www.g7.utoronto.ca/evaluations/2012campdavid/kirton-energy.html.

Gallagher, K. S., Grübler, A., Kuhl, L., Nemet, G., & Wilson, G. (2012). The energy technology innovation system. *Annual Review of Environment and Resources, 37*, 137–162.

IEA. (2018). *Energy Transitions in G20 Countries*. Retrieved 8 June, 2021 from https://www.iea.org/reports/energy-transitions-in-g20-countries.

IEA. (2020a). *Energy technologies perspectives*. Special Report on Carbon Capture Utilisation and Storage, in clean energy transitions. Retrieved 4 June, 2021, from https://www.iea.org/reports/ccus-in-clean-energy-transitions..

IEA. (2020b). *Global energy-related CO_2 emissions, 1900–2020*. Retrieved 8 June, 2021, from https://www.iea.org/data-and-statistics/charts/global-energy-related-co2-emissions-1900-2020.

IEA. (2020c). *World Energy Outlook 2020*. Retrieved 8 June, 2021, from https://www.iea.org/reports/world-energy-outlook-2020.

IEA. (2020d). *Energy Technology Perspectives 2020*. Retrieved 8 June, 2021 from https://www.iea.org/reports/energy-technology-perspectives-2020.

IPCC. (2018). *Global warming of 1.5°C. an IPCC special report on the impacts of global warming of 1.5°C above pre-industrial levels and related global greenhouse gas emission pathways, in the context of strengthening the global response to the threat of climate change, sustainable development, and efforts to eradicate poverty*. Retrieved 8 June, 2021 from https://www.ipcc.ch/site/assets/uploads/sites/2/2019/06/SR15_Full_Report_Low_Res.pdf.

IRENA. (2019). *A new world: The geopolitics of the energy transformation*. Retrieved 8 June, 2021 from http://geopoliticsofrenewables.org/assets/geopolitics/Reports/wp-content/uploads/2019/01/Global_commission_renewable_energy_2019.pdf.

Mazzucato, M. (2019). Governing missions in the European Union. In *Directorate-general for research and innovation*. Publications Office of the European Union. Retrieved 8 June, 2021 from https://ec.europa.eu/info/sites/info/files/research_and_innovation/contact/documents/ec_rtd_mazzucato-report-issue2_072019.pdf.

O'Sullivan, M., Overland, I., & Sandalow, D. (2017). *The geopolitics of renewable energy*. HKS Working Paper No. RWP17–027. https://doi.org/10.2139/ssrn.2998305

OECD. (2019). *Innovation and business/market opportunities associated with energy transitions and a cleaner global environment*. Retrieved 8 June, 2021 from https://www.oecd.org/g20/summits/osaka/OECD-G20-Paper-Innovation-and-Green-Transition.pdf.

Planko, J., Cramer, J., Hekkert, M. P., & Chappin, M. M. H. (2017). Combining the technological innovation systems framework with the entrepreneurs' perspective on innovation. *Technology Analysis & Strategic Management, 29*(6), 614–625. https://doi.org/10.1080/09537325.2016.1220515

Raven, R. (2007). Niche accumulation and hybridisation strategies in transition processes towards a sustainable energy system: An assessment of differences and pitfalls. *Energy Policy, 35*(4), 2390–2400. https://doi.org/10.1016/j.enpol.2006.09.003

Roehrkasten, S., & Westphal, K. (2016). The G20 and its role in global energy governance. In S. Röhrkasten, S. Thielges, & R. Quitzow (Eds.), *Sustainable energy in the G20: Prospects for a global energy transition* (pp. 12–18). IASS Study.

Rosenau, J. N. (1995). Governance in the twenty-first century. *Global Governance, 1*(1), 13–14. https://doi.org/10.1163/19426720-001-01-90000004

Schwab, K. (2016). *The fourth industrial revolution*. World Economic Forum.

Van de Graaf, T., & Westphal, K. (2011). The G8 and G20 as global steering committees for energy: Opportunities and constraints. *Global Policy, 2*(1), 19–30. https://doi.org/10.1111/j.1758-5899.2011.00121.X

Yu, B. (2019). *Climate clubs and global decarbonization: A comparison of the APP and the CEM, EGL*. Working Paper 7, Munk School of Global Affairs and Public Policy, Retrieved 8 June, 2021, from https://munkschool.utoronto.ca/egl/files/2019/01/Yu_EGL_Working_Paper_2019-1.pdf.

Zuboff, S. (2019). *The age of surveillance capitalism: The fight for a human future at the new frontier of power*. PublicAffairs.

Francesco Ciaccia has been working at Eni since 2007 and he currently manages the relations with Ministry of Foreign Affairs and Diplomatic Representations for the Public Affairs Department. He is qualified in law and has more than 15 years of experience in the field of energy. Between 2013 and 2014, he has been the deputy head of the Eni International Relation Office in London. He worked on national, International and EU competition and regulatory topics and scenarios for the gas and power chain. Francesco is an alumnus of the Oxford Saïd Business School's Corporate Affairs Academy, and he is the Energy Senior Expert of the G20 Sherpa Office of the Italian Prime Minister since July 2020.

Chapter 11
How Emerging Technologies Are Finally Matching the Policy Leverage of Cities with Their Political Ambitions

Mark Alan Hughes, Angela Pachon, Oscar Serpell, and Cornelia Colijn

Abstract U.S. cities are typically "policy-takers," operating within a narrow range of possibilities for housing, transportation, education, health, and energy policy as defined by states and the federal government. Cities' energy use is bound to a large network of interlocking infrastructure, market, and policy systems—all of these cities have limited capacity and ability to directly influence. Despite these governance limitations on local energy control, the past decade has brought remarkable technological innovations that are set to disrupt the status quo, creating opportunities for cities to become powerful platforms for emerging clean energy systems. The generation, movement, and sale of energy is becoming increasingly decentralized. Whereas cities are already playing a powerful role in shaping local adaptation to climate impacts, the decentralization of energy will increasingly allow cities to play a pivotal role in emission mitigation as well. Thanks to distributed generation, battery storage, smart metering, microgrids, load aggregation, and electrification, cities are beginning to look like "prosumers"—both producing and consuming electricity depending on the time and day. This new autonomy will allow cities to leverage local policies targeting improved efficiency and circularity, decarbonization, renewable fuels, and offsets, thereby furthering a just and efficient energy transition.

Cities around the world recognize that the energy system produces most of the emissions that drive climate change. In response, they are setting goals—often ambitious ones—and leveraging the policies and programs tied to their own local energy systems. For cities, energy supply and demand largely determine the outcome of their CO_2 reduction goals.

In previous work, we have argued that city efforts are grossly insufficient to confront climate change (Hughes & Colijn, 2018; Serpell, 2017). This is not for lack of

---

M. A. Hughes (✉) · A. Pachon · O. Serpell · C. Colijn
Kleinman Center for Energy Policy, University of Pennsylvania, Philadelphia, PA, USA
e-mail: mahughes@upenn.edu; apacho@upenn.edu; serpello@upenn.edu; ccolijn@upenn.edu

commitment but for lack of jurisdiction. In this chapter, however, we modify this critique. Ongoing changes in the energy system are disrupting the balances of power in the governance of the energy system. These changes in technology and resulting market structure create new opportunities for cities to realize their aspirations.

11.1 Introduction

In the United States, cities traditionally exert jurisdiction over a range of issues that are formally granted to them by higher levels of government. Typical examples include land and building regulation, waste management, or public safety. Often these issues create benefits beyond a city's boundaries, and these spillovers are sometimes reflected in intergovernmental transfers, user fees, and regional authorities. But on a much larger range of issues, U.S. cities are "policy-takers," playing by rules and operating under constraints defined by higher levels of government and with little, if any, autonomy. States and the federal government largely define a narrow range of possibilities for housing, transportation, education, health, and environmental policy in cities.

With the century-long development of a global energy system defined by initial endowments of, and international trade in, fossil fuels, cities have been, until very recently, "policy-takers" on energy as well. State and federal law and regulations govern energy production and distribution. Cities import their energy through much larger policy and market structures, over which they exert little influence. While cities do have some leverage over energy consumption inside city boundaries, traditional pricing structures (especially tariffs and contracts that do not fully reward the system-level benefits of efficiency and conservation) have limited the impact of city policy making.

Consider building codes. In the United States, building codes are mandated and enforced by state legislatures, but many cities are granted an option to pass codes more stringent than their state requirements. For example, in 2017, the Pennsylvania legislature passed House Bill 409, bringing statewide building codes up to the 2015 International Energy Conservation Code (IECC). This legislation also allowed Philadelphia a one-time opportunity to update to the IECC 2018 code for commercial construction, surpassing state mandates and creating a real opportunity to improve efficiency in a city where 60% of all carbon emissions are associated with the construction and operation of the building stock (Philadelphia Office of Sustainability, 2018). Philadelphia estimates that new commercial construction will be as much as 30% more efficient under the adopted code. But, even this local span of control is not complete and faces constraints that are significant and often underappreciated. Perhaps the most significant example is that building codes almost exclusively limit their requirements to new construction, leaving huge stocks of existing buildings operating with levels of energy performance from decades, or even generations, ago.

U.S. cities operate within the overlapping footprint of many larger, interlocking infrastructure, market, and policy systems—all of these cities have limited capacity and ability to directly influence. To continue with Philadelphia as an example, its wholesale electricity market is managed by PJM Interconnection LLC, a regional transmission organization (RTO) that serves 65 million customers in 13 states and numerous big cities, including Chicago, Philadelphia, and Washington D.C. PJM also monitors system load and coordinates and directs electricity generation across its geography. Grid transmission and distribution infrastructure (such as substations) in the Philadelphia region are maintained by the regional electric utility PECO, a subsidiary of Exelon. These powerful actors are regulated by the Federal Energy Regulatory Commission (FERC), which oversees the wholesale transmission and sale of electricity that crosses state boundaries, and also by the Commonwealth of Pennsylvania, which oversees intrastate energy matters (Hughes & Colijn, 2018).

Despite these governance limitations on local energy control, the past decade has brought remarkable innovations to the energy system that is starting to reveal a near future of transformative change. After decades of being largely captive to state and federal energy regulations and policy, emerging technologies being deployed by disruptive companies are creating opportunities for cities to become powerful platforms for emerging clean energy systems. This new role for cities has been triggered by technological breakthroughs in renewable energy that are weakening *both* the market actors that convert fossil reserves *into* useful energy and deliver them as fuels or electricity to consumers and the governing structures that regulate these actors. These technological breakthroughs are transforming the energy system from a centralized, concentrated series of producer conversions (extraction to combustion to consumer demand) into decentralized, distributed networks of prosumers managing ubiquitous energy supply and demand at small-scale. These distributed networks are much smaller than previously possible: households rather than utility service territories; cities rather than states. Disruptions remain small in the context of a U.S. energy system still dominated by fossil fuels powering multi-state wholesale electricity grids and over 250 million gasoline vehicles (Hedges and Company, 2020). However, on the margin, these new technologies are exerting enormous pressures on incumbents. It is increasingly clear that a fundamental energy transition is underway.

Climate change has become the single most important driver of energy policy in the last decade. Climate change and climate policy have forced a global reconsideration of the true costs of carbon-based energy and have forged a global commitment to limiting additional greenhouse gas emissions in accordance with a science-based "carbon budget." Cities are playing a powerful role in shaping local adaptation to climate impacts (at least to the extent that cities have the financial capacity without assistance from higher levels of government). But the role for cities in mitigating climate change itself has always been less clear and more limited. Emissions reduction can be achieved by competing energy policy goals, as demonstrated by the use of the phrase "net-zero emissions," but alone is insufficient to provide cities with an impactful role in the energy transition. Without an alternative to the existing carbon energy system, the world would be focused on lowering the carbon intensity of the existing energy system, such as switching from coal to gas, capturing new emissions from combustion and removing existing emissions from the atmosphere, improving the efficiency of buildings and processes, and so on. Under these conditions, cities would still be policy-takers in a centralized and concentrated system of producers and consumers. They would have little ability to divert any public funding away from local climate adaptation and toward climate mitigation through local energy policy.

11.2 Technology Drivers Empowering Cities

Technology innovation is the driver of energy policy that makes it possible for cities to play the impactful role that they have sought for decades. Cities now have the leverage they need to make significant progress toward their state policy goals, expanding the role for cities in shaping energy policy and energy systems. We now turn to a quick review of the most important of these technological breakthroughs.

Regional transmission grids are a marvel of modern engineering. However, the electricity grid has, historically, significantly limited cities' autonomy when it comes to energy governance. By geographically separating production and consumption, a unidirectional grid is a hindrance to any local effort to decarbonize energy, by placing the levers of change far beyond city boundaries.

Emerging technologies for generating, coordinating, pricing, metering, and consuming electricity are transforming this historically centralized system. Thanks to distributed solar generation, battery storage, smart metering, microgrids, load aggregation, and electrification, regional grids are quickly transitioning from unidirectional delivery systems to multidirectional networks of shifting supply and demand. In this emerging electricity network, cities need no longer be confined to the role of consumer. Instead, cities begin to look like "prosumers"—both producing and consuming electricity depending on the time and day. This finally allows cities to leverage the unique advantage of density to not only improve energy efficiency as consumers but to directly produce low-to-zero carbon energy at the point of consumption and profoundly impact regional energy systems.

11.2.1 Smart Grids

The internet has transformed how we share information in nearly every aspect of our lives. Now the "internet of things" enabled by 5G and 6G—the latest generations of cellular networking—will continue to expand the possibilities for distributed services on the electricity grid, allowing grid operators and utilities to monitor consumption and generation in real time and allowing for rapid changes and adjustments in pricing and load. The internet allows electricity markets to track sales in a far more complex and multi-nodal network, rather than as a unidirectional energy highway.

11.2.2 Energy Storage

A second technological breakthrough reshaping the electricity system is energy storage. Smartphones and laptops have become ubiquitous over the last three decades, and every year the power demand of these machines grows. Today, an iPhone has many times the computational power of a supercomputer from the 1980s, yet we require it to last a full day using only the electrical energy stored within it (Routley, 2017). This wireless existence would not have been possible without the development and improvement of advanced, high-density, rechargeable batteries. Thanks to these growing demands for storage of electricity, we now have lithium-ion battery packs capable of storing enough electricity to power a car or serve as a back-up power source for a home. Energy storage is a key technology in stabilizing the capacity for renewable energy sources like solar and wind to provide reliable and continuous power. Energy storage manages the intermittency of these energy sources over time and facilitates the flow of energy among uses, such as buildings and vehicles.

11.2.3 Distributed Generation

Together, a more ubiquitous energy supply via solar and wind and a more powerful storage system via batteries and other means have produced a third technological breakthrough known as distributed generation. Fossil fuels exhibit enormous economies of scale. Extracting, refining, and transporting fossil fuels require big, expensive capital investment that is most productively managed by huge companies and utilities. Solar and wind power, on the other hand, can be generated at the point of demand wherever the sun shines or the wind blows, and local storage can help manage times when it is dark or calm. Distributed generation carries many benefits to owners, beyond a drop in individual carbon emissions. In fact, in a country like the United States that has never placed a price on the social costs of carbon emissions, one could argue that nearly all individual benefits to the owner of distributed energy systems fall outside of the shared climate benefits of reduced emissions. Distributed generation and storage not only allow intermittent energy sources to meet a greater share of overall demand, these technologies also empower individual energy consumers, by allowing them greater flexibility to decide when and how they use grid electricity.

As areas of high-density, high-demand, intensive capital investment, and continuous innovation, cities also account for more than 75% of energy use and 60% of greenhouse emissions (United Nations 2021). The emerging energy transition to renewable, distributed, and decentralized prosumer networks gives cities an important role to play. In the remainder of this chapter, we present some of the significant ways in which cities around the world, and especially in the United States, are leveraging emerging technology to reinvent energy and mitigate climate change.

11.3 City Policies that Leverage Technology Drivers

The leading strategy for a just and efficient energy transition consists of four related components:

- Improve energy efficiency and circularity where possible
- Decarbonize major sectors of the economy (electricity, transportation, industrial processes, and buildings)
- Shift activities from difficult sectors into sectors that have made the greatest progress toward low or zero-carbon energy
- Replace any remaining fossil fuel use with renewable fuels (such as hydrogen) or offset remaining emissions with carbon capture investments

In the following brief review, we organize major policy examples from cities around the world in terms of energy efficiency in buildings and other energy sectors, followed by electricity decarbonization through ownership and other instruments, then electrification of sectors that currently rely on combusting fossil fuels, and finally on capturing or otherwise avoiding the carbon emissions that remain in industry and other sectors.

11.3.1 Energy Efficiency and Circularity

Owing to the density of cities, their high carbon emissions are primarily attributable to residential and commercial buildings. When considering the steps cities can take to improve urban energy efficiency, a pragmatic place to start is with residential and commercial buildings—employing policies and actions that reduce energy consumption and carbon emissions. To achieve progress in this area, city officials need to work closely with building owners, real-estate developers, and building occupants in both residential and commercial properties.

11.3.1.1 Energy Building Codes

Energy building codes are powerful tools that cities and states can adopt to ensure minimum energy efficiency standards for new construction. Recent codes have provided 30% more energy savings than codes from a decade ago (U.S. Department of Energy, 2016a, 2016b), and for the period of 2010–2040, residential and commercial building energy codes are projected to save $126 billion in energy costs, 841 million metric tons (MMT) of avoided CO_2 emissions, and 12.82 quads of primary energy (U.S. Department of Energy, 2016a, 2016b). The stringency of these codes is partly a reflection of the technological progress made over the years in lighting, insulation, and heating. Without technological progress in these areas, new building codes would be prohibitively costly.

The codes are developed by two private organizations (ASHRAE and the International Code Council) and updated every three years. The most recent codes, approved for implementation in 2021, failed to include electrification provisions, such as installing electric vehicle charging stations in residential and commercial buildings and installing electric circuits to allow future conversions to highly efficient electrified equipment (Bresette, 2020). When implementing this new code, cities will have to decide how serious they are about electrification and the decarbonization of the building sector. They could add these (and other) provisions in, as was done by some California municipalities that banned gas connections to new buildings.

Two factors make the codes particularly powerful in achieving efficiency goals. First is the slow capital stock turnover or the time it takes to replace building technologies after they are adopted (Jaffe et al., 1999). As building systems can last decades, building codes may lock in fossil-fuel-dependent technologies for decades, delaying decarbonization efforts if they fail to adopt clean technologies. Second is the cost of retrofitting versus the cost of including technologies from the construction process. When it comes to new buildings, it is more cost-effective to include a new technology in the initial construction rather than by retrofitting (Billimoria et al., 2018).

11.3.1.2 Weatherization Programs for Low-Income Households

Many cities in the United States—Seattle, Baltimore, Austin (through its utility), and Cleveland—offer free or subsidized weatherization services to residents who meet income qualifications. Services may include energy audits, insulation, heating systems, air sealing, bathroom fans, and more. Seattle also offers low-interest loans to low- and moderate-income residents through its Home Repair Loan Program. Programs like these allow cities to make use of the most cost-effective and efficient lighting, insulation, and heating technologies, which can help reduce the energy burden on low-income communities.

11.3.1.3 Energy Consumption Building Benchmark and Building Energy Performance Standards

Many cities are leveraging the power of data by adopting energy benchmark programs that require buildings to disclose their energy consumption for comparison with other buildings in the city (U.S. Department of Energy, 2019). Below is a map

Fig. 11.1 Catagorizing policies related to performance requirements and efficiency benchmarking in US cities

from the Institute from Market Transformation (IMT) with the cities that have adopted benchmarking for different types of buildings. In some cases, the purpose of the benchmark is to allow buildings to track, monitor, and detect abnormal energy consumption while providing the incentives to improve energy efficiency. Since the data is public, no one wants to be seen as the laggard. In other cases, the data provided by the buildings establishes specific performance targets that the building must achieve by a specific date. Failing to achieve the target may result in fines. This kind of benchmarking and performance tracking, especially real-time tracking, is only now becoming possible, thanks to technological advancements in distributed and autonomous monitoring and sophisticated data aggregation and analysis software (Fig. 11.1).

11.3.1.4 Managing Traffic

Congestion generates costs for the city economy, increases gasoline use, and increases CO_2 emissions and air pollution. In New York City alone, congestion costs the metropolitan region's economy up to $15 billion annually, consuming nearly 300 million extra gallons of fuel (NYC Government, 2019). Tackling congestion is an effective way for cities to reduce emissions, improve air quality, and achieve climate targets while improving city livability. Many cities around the world curb traffic by restricting vehicle circulation on certain days by plate number (Mexico City, Bogota, Beijing, and Athens). The progress in image-based tolling and automatic license plate recognition technology, as well as the development of the smart transponders and GPS technologies (U.S. Department of Transportation, 2008) enabled the implementation of congestion charges to enter a city's congested zones (London, Stockholm, and Singapore), or even banning inefficient vehicles from certain areas by declaring Low Emissions Zones (LEZs), thus incentivizing drivers to upgrade their cars. These policies have been effective in changing vehicle use, and, in some cases, they have achieved a reduction in congestion and emissions. Careful steps need to be taken, however, to ensure that these policies do not create perverse incentives, such as in Mexico City where many drivers acquired a second car which was usually older (and thus more polluting) to circulate on those days when their first car had restrictions (Davis, 2008).

Even the most climate-committed cities in the United States have not implemented these vehicle restrictions yet, due in part to the popularity of car use and the relatively low concern with urban air pollution. That said, Chicago and Los Angeles have commissioned studies on congestion charges and New York City's congestion charge has been years in the making, with the proposal recently being blocked by the Trump administration. It is expected that under the Biden presidency, the NYC congestion charge will finally be approved (Snyder et al., 2021). In the past, the federal government expressed concerns over congestion prices on the basis that it would be regressive or disproportionately impact poor households (U.S. Department of Transportation, 2017). Evidence to the contrary shows that across the United States, poor, urban households generally do not use cars for transportation, but instead rely on public transit (King et al., 2019). Other initiatives that American cities have implemented for congestion are car-free zones, reduced parking spaces, and dedicated bus lanes in central areas.

As ride sharing, curbside deliveries, and autonomous vehicles all become prominent fixtures of society, city traffic and congestion policy have the potential to play an increasingly large role in urban energy policy. Local authorities in many cities have, for example, started to implement curbside restrictions and have made various attempts to control or limit the scope of ridesharing apps, such as Uber and Lyft (OECD, 2018).

11.3.2 Electricity Decarbonization

Even with dramatic improvements to energy efficiency, cities will always be major hubs of energy consumption. One of the most effective and technologically achievable strategies for reducing the carbon footprint of remaining energy demand is to shift electricity generation from coal and natural gas power plants to renewable sources like wind and solar. Two factors determine the control of a city over its electricity supply: ownership of the electric utility and the market structure in which the utility operates. In the United States, utilities can be owned by private investors, municipalities, cooperatives, or even by the federal government. They can operate in a restructured market where they cannot own generation assets, and therefore need to purchase power from the wholesale market. They may face competition when their customers have the option to choose an energy supplier. On the other side of the spectrum are the vertically integrated utilities. Many of these integrated utilities own generation assets or can directly sign power purchase agreements with merchant generators. They tend to have captive customers, as they operate in a retail sector that is not open to competition.

Municipally-owned, vertically-integrated electric utilities have control of their supply mix and can help local governments achieve climate targets. This is the case for the city of Los Angeles. In 2019, as part of the city's Green New Deal, the Los Angeles Water and Power Department was directed not to update or repower three coastal natural gas power plants—representing 38% of the city's natural gas portfolio—and replace them with renewable energy. The decision allowed LA to accelerate its transition to 100% renewable energy and put the city on track to meet its carbon-neutral target by 2050 (Stark, 2019).

According to the American Public Power Association, for the period 2005 to 2017, public-owned utilities decreased CO_2 emissions by 33% while the average emission reduction for all utilities (private and public) was only 24%. Moreover, a study about local clean energy policy decisions in the United States found that cities with investor-owned utilities adopt fewer renewable energy policies than cities served by public-owned utilities (Curley et al., 2021). Not surprisingly, some cities have attempted to reclaim public ownership of local electric and gas utilities as a method of implementing climate policies. In the past, municipalization efforts focused on reducing rates or improving reliability. Today, there are movements in cities like Boston, New York, San Francisco, and Chicago that seek municipal control of the utilities to reduce rates, improve utility worker conditions, expand low-income assistance, and accelerate decarbonization, among other objectives (Ambort, 2020) In many Californian cities, after the bankruptcy of Pacific Gas and Electric, municipalization seems to be a popular option.

But municipalization can be a lengthy and costly process that does not always achieve utility ownership change. Boulder, for example, spent more than 10 years fighting to buy their local electric utility from Xcel Energy. In November 2020, voters finally decided the city would enter into a 20-year franchise agreement with Xcel Energy, a move that paused the city's efforts. Despite this outcome, Xcel committed to reduce greenhouse gas emissions by 80% from 2005 levels and to work together with the city to implement energy-efficient initiatives that support Boulder's goal of 100% renewable energy by 2030 (Sakas, 2020).

Whether cities exist within a restructured market or they have a municipally-owned, vertically-integrated electric utility, the changing technology that cities can leverage to reduce grid emissions are often similar.

11.3.2.1 Renewable Power for Government Buildings

Even under conditions where cities have little control over the electricity supply, cities are large energy customers. Procuring clean energy for government buildings could significantly contribute to emission reduction targets. This is the case in Philadelphia, a city with a private electric utility operating in an unregulated market. In 2018, the city signed a power purchase agreement directly with a renewable energy supplier to ensure that at least 22% of the electricity consumed by city buildings would come from renewable energy (City of Philadelphia, 2020). Some other cities, like Phoenix, have used similar agreements to lease solar panels installed on government buildings. Although this can be a more expensive option, the onsite generation "behind the meter," helps to reduce peak consumption, decrease electricity bills, and meet climate goals (Miller, 2016). Enacting power purchase agreements with at-scale renewable energy providers is an example of how energy markets are adapting to the rapidly decreasing costs associated with solar and wind development. In the recent past, what would have been a high-risk proposition for both cities and renewable energy developers is now easily within reach.

11.3.2.2 Promoting Distributed Generation (DG) Through Community Solar Projects

The regulation of payments made or received by distributed generation customers is in the jurisdiction of state regulators and even the Federal Energy Regulatory Commission. In many places, net metering is limited to very small rooftop solar systems, preventing the participation of larger systems like those from schools, businesses, and local governments. In places where net metering covers community solar projects, cities are promoting the deployment of distributed generation by installing solar gardens to power up public housing buildings. In addition to the benefits of clean power supply, these projects create jobs and training.

As part of New York City's 100% renewable commitment, the city installed 25 megawatts of solar panels atop the city's public housing buildings, which were leased to community solar projects (Gilpin, 2017). Denver's housing authority also plans to have its own community solar garden. This project is expected to power up to 700 public housing units and low-income homes, while cutting energy bills by about 20 percent, offsetting over 54,000 tons of carbon emissions (U.S. Department of Transportation, 2017), and providing job training. St. Paul, Minnesota's public housing is building solar gardens for 10 of its public housing high-rises. The solar gardens are expected to save the housing agency $130,000 a year. Community solar is only possible with sophisticated tracking and metering of distributed generation and consumption by members of the participating community. The challenges faced by community solar, and the outdated regulations banning it in some states, will be increasingly addressed as data collection, smart metering, and real-time pricing structures are deployed more widely.

11.3.2.3 Building or Facilitating Solar and Storage Microgrids

In addition to generating clean energy, solar and storage microgrids can operate in isolation of the grid and bring resilience to the city in instances where there are major power disruptions. They can also facilitate the management of smart grids, as they can work as aggregators of small systems. Many cities have built microgrids to improve resilience after a major disaster. In Hartford, Connecticut, after Hurricane Irene in 2011 and a major snowstorm left 750,000 homes without power for 10 days, the city built a fuel cell-powered microgrid that began operating in March 2017 (Gies, 2017). More recently, cities in California are building solar and storage microgrids after experiencing long power cuts due to the deadly wildfires.

11.3.3 Electrification

Even after the decarbonization of all existing electricity demand, heating and transportation in cities would still contribute to global carbon emissions and jeopardize emissions targets. Leveraging emerging technologies to help shift heating and vehicle demand onto the electricity sector will be a critical strategy for system-wide decarbonization, especially in cities where density allows for a number of innovative strategies in transforming this existing use of distributed fuel.

When it comes to decarbonizing transportation, cities still lack a great deal of governance power. They do not, for example, have the authority to set gasoline taxes or establish fuel efficiency regulations for automakers. Those regulations fall under state and federal jurisdiction. However, when it comes to the electrification of transportation, cities can exercise a great deal of control by leveraging density and investing in emerging technologies.

11.3.3.1 Improving Transit

Efficient urban transit systems could be the best alternative to car use and could simultaneously achieve climate reduction goals. Yet transit ridership is significant in only a few U.S. cities. Before the pandemic, improving, modernizing, and electrifying transit systems and integrating micro-mobility were at the backbone of urban mobility strategies to increase ridership while tackling climate change and improving city livability. With the pandemic, this progress has been seriously delayed. Transit agencies are facing massive operating budget shortfalls as federal funding they received from initial relief packages is running out and funding from local taxes has plunged (Calvert, 2020). The financial hardship has forced drastic service cuts (Goldbaum & Wright, 2020) and the delay of capital projects. Transit system recovery will require a combined effort of federal, state, and local authorities.

11.3.3.2 Vehicle Electrification

Electrification, or the use of alternative fuels for the transportation sector, could be the fastest way to reduce emissions from vehicles, especially since scaling up centralized transit systems has been severely delayed. For cities, the decision to promote and support the adoption of electric vehicles is not always easy. The first consideration is the fuel mix: vehicle electrification may not result in an immediate reduction of emissions if the electricity or alternative fuel is produced from fossil fuels. In this case, vehicle electrification would only shift emissions from the city to the electricity producing region. Second, electric vehicles remain expensive and unaffordable for middle- and low-income populations (Breetz & Salon, 2018). EV promotion could be a regressive policy, primarily benefiting wealthier city dwellers. Third, even if vehicles are electric or powered by alternative and clean fuels, these cars can still pose additional congestion risk.

In the United States, apart from California, the adoption of electric vehicles remains minimal. Cities do not have the jurisdiction to license vehicles nor can they ban the registration of regular gasoline vehicles, which is the most effective policy to ensure EV adoption. Cities can build or support charging infrastructure in public premises; they can electrify their municipal fleet, including transit buses and school buses; and they can provide preferential parking for EVs. These measures certainly help boost EV adoption but are not enough to achieve complete electrification. A study in the United Kingdom, a country which also has a low penetration of EVs, analyzed the impact of local EV policies and found that many of the factors that influence EV adoption are beyond city control. They concluded that investing in charging infrastructure, however, may be the most effective policy, as the short range of EVs is perceived as the main limiting factor for EV purchases (Heidrich et al., 2017).

EVs technological progress and the increasing decarbonization of the grid will facilitate the adoption of EV policies at the local level.

11.3.4 Alternatives for Hard-to-Electrify Sectors

11.3.4.1 Hydrogen Cities

Hydrogen is starting to be considered a clean alternative fuel source, thanks to the technological progress in electrolyzers and fuel cells and the massive reduction in renewable energy generation costs. Although the cost of green hydrogen (produced with water and renewable energy) remains high compared to hydrogen produced with coal or natural gas, many countries have set ambitious green hydrogen adoption targets. Among these countries, South Korea's plan is focused on urban areas. Three initial pilot projects in the cities of Ansan, Ulsan, and Wanju will fuel cooling, heating, electricity, and transportation systems with green hydrogen, with an aim to transform 10% of urban areas to hydrogen power by 2030 and 30% by 2040. Each city will receive USD 24 million from the Korean government to support the development of infrastructure, while a fourth city will be a hydrogen R&D hub (FuelCellsWorks, 2020). While this is an initiative of the national level, it is expected that with further technological developments (costs reductions for hydrogen production, storage technology improvements, reduction in conversion costs of gas boilers), more cities will follow this transformation to achieve climate targets while creating green jobs.

11.3.4.2 Carbon Capture, Utilization, and Storage

Very few cities have implemented carbon capture, utilization, and storage (CCUS) solutions to fight climate change. Most operational CCUS projects capture carbon from power stations and use the carbon for enhanced oil recovery. Therefore, they are located in proximity to oil fields and not in urban environments. While the number of CCUS operational facilities has increased in the last several years, their cost remains a barrier for large-scale implementation in hard-to-decarbonize industries. But the urgency to reduce emissions has made the implementation of these technologies desirable and necessary, especially for cities with ambitious climate targets. In response to this perceived need for urban CCUS solutions, Amsterdam, Copenhagen, Helsinki, Oslo, and Stockholm have created the Carbon Neutral City Alliance to jointly review CCUS solutions. According to a report written by the Bellona Foundation in collaboration with these cities, some CCUS projects are in the process of being implemented (Bellona Foundation, 2020): Amsterdam, for example, plans to expand an existing CO_2 pipeline from Rotterdam to Amsterdam to procure CO_2 from industrial zones and dispatch it to greenhouses. The Netherlands also plans to build a connection pipeline from both the Rotterdam and Amsterdam industrial areas toward gas fields in the North Sea for storage. Oslo is considering a carbon capture and storage (CCS) facility for two of its waste incineration plants. Equinor would build a dedicated pipeline to transport the carbon to a storage site southeast in the North Sea. For cities without access to storage sites, the implementation of CCS is more challenging.

Copenhagen is currently collaborating with the local waste incineration plant on a carbon capture solution but still needs to prove the technical and legal option to do underground CO_2 storage. In Helsinki and Stockholm, the option of bio-energy with carbon capture and storage (BECCS) has been considered to reduce emissions from combined heat and power (CHP) plants. Since 2019, a test facility to capture carbon from bio-cogeneration became operational in Stockholm. The storage infrastructure for this facility will be located in Norway and won't be ready until 2024, as Sweden lacks adequate storage capacity.

11.4 Conclusion

In this chapter, we have offered a sampling of the many climate policy initiatives that cities are taking that either actively leverage or are indirectly supported by recent technological innovations in electricity generation, access, monitoring, alternative fuel synthesis, transportation, and building materials. These technology advancements are allowing cities to respond to global energy challenges with a level of autonomy and power that has not been possible since the regionalization of power grids and fuel supply. Cities are increasingly positioned to become leaders in climate change, not just because of local political realities but also because of universal technological realities.

References

Ambort, L. (2020). *Spreading like wildfire: An interest in making electric power public*. Retrieved from https://ilsr.org/municipalization-electric-utilities-update-2020/. Accessed 26 Mar Accessed 2021.

Bellona Foundation. (2020). *Cities aim at zero emissions: How carbon capture, storage and utilization can help cities go carbon neutral*. Retrieved 26 March, 2021, from http://carbonneutralcities.org/wp-content/uploads/2020/01/Report_How-Carbon-Capture-Storage-and-Utilisation-Can-Help-Cities-Go-Carbon-Neutral.pdf.

Billimoria, S., Guccione, L., Henchen, M., & Louis-Prescott, L. (2018). *The economics of electrifying buildings: How electric space and water heating supports decarbonization of residential buildings*. Retrieved 26 March 2021, from https://rmi.org/insight/the-economics-of-electrifying-buildings/#download-form.

Breetz, H., & Salon, D. (2018). Do electric vehicles need subsidies? Ownership costs for conventional, hybrid, and electric vehicles in 14 U.S. cities. *Energy Policy, 120*, 238–249.

Bresette, D. (2020). *Building energy codes leap forward on efficiency but stumble on electrification*. Retrieved 18 December, 2020, from https://www.eesi.org/articles/view/building-energy-codes-leap-forward-on-efficiency-but-stumble-on-electrification.

Calvert, S. (2020). *Public transit agencies slash services, staff as coronavirus keeps ridership low*. The Wall Street Journal. Retrieved from https://www.wsj.com/articles/public-transit-agencies-slash-services-staff-as-coronavirus-keeps-ridership-low-11606582853

City of Philadelphia. (2020). *City and ENGIE announce power purchase agreement staffing plans*. Retrieved 26 March, 2021, from https://www.phila.gov/2020-02-06-city-and-engie-announce-power-purchase-agreement-staffing-plans/.

Curley, C., Harrison, N., Kewei, C., & Zhou, S. (2021). Collaboration mitigates barriers of utility ownership on policy adoption: evidence from the United States. *Journal of Environmental Planning and Management, 64*, 124–144.

Davis, L. (2008). The effect of driving restrictions on air quality in Mexico City. *Journal of Political Economy, 116*(1), 38–81.

FuelCellsWorks. (2020). *Korean Government announces its selection of world's first hydrogen cities*. Retrieved 26 March 2021, from https://fuelcellsworks.com/news/korea/korean-govt-announces-its-selection-of-worlds-first-hydrogen-cities/.

Gies, E. (2017). Microgrids keep these cities running when the power goes out. *Inside Climate News*. https://insideclimatenews.org/news/04122017/microgrid-emergency-power-backup-renewable-energy-cities-electric-grid/

Gilpin, L. (2017). Community solar heads for rooftops of NYC's Public housing projects. *Inside Climate News*. https://insideclimatenews.org/news/22112017/community-solar-new-york-city-public-housing-projects-nycha-solstice/

Goldbaum, C., & Wright, W. (2020) "Existential Peril": Mass transit faces huge service cuts across U.S. *The New York Times*. https://www.nytimes.com/2020/12/06/nyregion/mass-transit-service-cuts-covid.html

Hedges & Company. (2020) *U.S. Vehicle Registration Statistics*. Retrieved 29 March 2021, from https://hedgescompany.com/automotive-market-research-statistics/auto-mailing-lists-and-marketing/.

Heidrich, O., Hill, G., Neaimeh, M., Huebner, Y., Blythe, P., & Dawson, R. (2017). How do cities support electric vehicles and what difference does it make? *Technological Forecasting and Social Change, 123*, 17–23.

Hughes, M., & Colijn, C. (2018). *Putting energy into implementation: Challenges to subnational participation in SDG 7*. Retrieved 29 March, 2021, from https://kleinmanenergy.upenn.edu/research/publications/putting-energy-into-implementation-challenges-to-subnational-participation-in-sdg-7/.

Jaffe, A., Newell, R., & Stavins, R. (1999). Energy-efficient technologies and climate change policies: Issues and evidence. *Resources for the Future*. Retrieved from https://media.rff.org/documents/RFF-CCIB-19.pdf.

King, D., Smart, M., & Manville, M. (2019). The poverty of the carless: Toward universal auto access. *Journal of Planning, Education and Research*. https://doi.org/10.1177/0739456X18823252

Miller, B. (2016). *Three ways local governments can use solar power for themselves*. Retrieved 12 December, 2020, from https://www.govtech.com/fs/Three-Ways-Local-Governments-Can-Use-Solar-Power-for-Themselves-Part-3-Phoenix.html.

NYC Government. (2019). One NY City of NY2050: Building a strong and Fair City. *Efficient Mobility*. Retrieved 12 December, 2020, from https://onenyc.cityofnewyork.us/wp-content/uploads/2019/05/OneNYC-2050-Efficient-Mobility.pdf.

OECD. (2018). *The shared-use city: Managing the curb*. Retrieved 26 March, 2021, from https://www.itf-oecd.org/sites/default/files/docs/shared-use-city-managing-curb_5.pdf.

Philadelphia Office of Sustainability. (2018). *City adopts new code for energy efficient and safe commercial buildings*. Retrieved 29 March, 2021, from https://www.phila.gov/2018-06-28-city-adopts-new-code-for-energy-efficient-and-safe-commercial-buildings/.

Routley, N. (2017). *Visualizing the trillion-fold increase in computing power*. Retrieved 29 March 2021, from https://www.visualcapitalist.com/visualizing-trillion-fold-increase-computing-power/.

Sakas, M. (2020). Boulder ends decade long pursuit of city-owned power utility. *CPR News*. Retrieved from https://www.cpr.org/2020/11/20/boulder-ends-decade-long-pursuit-of-city-owned-utility/.

Serpell, O. (2017). *Aligning local logic with global need The Kleinman Center for Energy Policy*. Retrieved 29 March, 2021, from https://kleinmanenergy.upenn.edu/research/publications/aligning-local-logic-with-global-need-a-path-towards-sturdier-mitigation-policy/.

Snyder, T., Muoio, D., & Kahn, D. (2021). Driving downtown? *Get Ready to Pay Extra*. Politico. Retrieved from https://www.politico.com/news/2021/03/21/congestion-pricing-biden-coronavirus-477288.

Stark, K. (2019). *Mayor Garcetti: LA Won't Invest $5 billion to rebuild coastal gas plants*. Retrieved 4 June, 2021, from https://www.greentechmedia.com/articles/read/garcetti-la-5-billion-rebuild-coastal-gas-plants.

U.S. Department of Energy. (2016a). *Building Energy Codes Program*. Retrieved 18 December, 2020, from https://www.energycodes.gov/about/why-building-energy-codes#:~:text=Building%20energy%20codes%2C%20which%20govern,money%20back%20into%20consumer's%20pockets.

U.S. Department of Energy. (2016b). *Saving energy and money with building energy codes in the United States*. Retrieved 18 December, 2020, from https://www.energy.gov/sites/prod/files/2016/12/f34/Codes%20Fact%20Sheet%2012-28-16.pdf.

U.S. Department of Energy. (2019). *Benchmarking and transparency: Resources for state and local leaders*. Retrieved 26 March, 2021, from https://www.energy.gov/eere/slsc/downloads/benchmarking-and-transparency-resources-state-and-local-leaders.

U.S. Department of Transportation. (2008). *Technologies that enable congestion pricing: A primer*. Retrieved 26 March, 2021, from https://ops.fhwa.dot.gov/publications/fhwahop08042/fhwahop08042.pdf.

U.S. Department of Transportation. (2017). *Income-based equity impacts of congestion pricing-A primer*. Retrieved 26 March, 2021, from https://ops.fhwa.dot.gov/publications/fhwahop08040/cp_prim5_03.htm.

Mark Alan Hughes is a professor of practice at the University of Pennsylvania Stuart Weitzman School of Design and founding faculty director of the Kleinman Center for Energy Policy. He is also a faculty fellow of the Penn Institute for Urban Research, a research fellow of the Wharton Risk Center, and a distinguished senior fellow at Penn's Fox Leadership Program. He has been a senior fellow at the Brookings Institution's Center for Urban and Metropolitan Policy, a senior adviser at the Ford Foundation, and a weekly opinion columnist for the Philadelphia Daily News. Hughes joined the Princeton faculty in 1986 at the age of 25, has taught at Penn since 1999, and is widely published in the leading academic journals of several disciplines, including Economic Geography, Urban Economics, Political Science Quarterly, Policy Analysis and Management, and the Journal of the American Planning Association, for which he won the National Planning Award in 1992.

Hughes holds a B.A. in art history and religion from Swarthmore and a Ph.D. in regional science from the University of Pennsylvania.

Angela Pachon is the research director at the Kleinman Center for Energy Policy. She is also the author and editor of various publications of the Center and has studied issues related to electricity markets in the U.S., gas policies in Pennsylvania, and climate policies in Latin America. Prior to the Kleinman Center, Pachon worked as a policy advisor at the Ontario Energy Board in Toronto, Canada, and as a consultant at NERA Economic Consulting in London, England, leading electricity sector reform projects in Europe, Africa, and the Middle East.

Pachon earned her master of science in local economic development from the London School of Economics and her B.S. degree from Pontificia Universidad Javeriana in Bogota, Colombia.

Oscar Serpell is the associate director of academic programing and student engagement at the Kleinman Center for Energy Policy. He oversees student engagement activities, new student programming, and alumni connections. He also participates in several key research projects at the Center, and writes blog posts and policy digests on timely energy policy topics.

Serpell has a master's degree in environmental studies and a B.A. in environmental management, both from the University of Pennsylvania.

Cornelia Colijn is the executive director of the University of Pennsylvania's Kleinman Center for Energy Policy. In this role, she is responsible for envisioning, planning, and managing the center's strategy, programming, and administration. She builds connections with students, faculty, and decision makers, ensuring that the center's research is used to inform sound policy outcomes. Prior to joining the Kleinman Center, Colijn was with Penn's Professional Programs in Earth and Environmental Science. Before that, she worked for several Philadelphia-based nonprofits, focusing on the ecological restoration of Philadelphia's extensive park system.

She holds a master's degree in applied geoscience and a bachelor's degree in earth and environmental science, both from the University of Pennsylvania.

Chapter 12
Sustainable Transportation

Ethan Elkind

Abstract Reducing emissions from transportation is critical to the success of tackling the broader global greenhouse gas emissions that cause climate change. Transportation is responsible for 14% of global greenhouse gas emissions, including 27% and increasing for the European Union and 28% for the United States, not including emissions from petroleum production. This chapter provides an overview of the law and policies that most directly affect greenhouse gas emissions from transportation, as well as the range of technologies available to do so. Specifically, it covers the various sources of transportation emissions, from per capita driving kilometers to burning fossil fuels used in a variety of transportation activities. It provides an overview of law and policy approaches that reduce driving kilometers and improve overall transportation efficiency, such as land use changes and transit investments. The chapter then summarizes the key clean fuel technologies that can reduce transportation emissions, including low-carbon biofuels, hydrogen, and battery electric vehicles, and offers examples of law and policies around the world that have proven to help expedite the deployment of these low- and zero-emission fuels. These mechanisms include mandates and incentives for clean transportation and subsidies and support for private sector action.

12.1 Introduction: The Importance of Transportation Sustainability for Addressing Climate Change

Few sectors of the economy touch our lives as directly and routinely as transportation. From the moment we leave our homes, transportation policies affect our daily lives. Can we walk or bicycle easily and safely to our destination? Is high-quality, affordable transit, like shuttles, buses, and rail, available nearby? Or do we need a private vehicle to access goods, services, and jobs? If we drive a vehicle, what fuel

E. Elkind (✉)
Center for Law, Energy & the Environment (CLEE),
University of California, Berkeley School of Law, Berkeley, CA, USA
e-mail: elkind@berkeley.edu

© The Author(s), under exclusive license to Springer Nature Switzerland AG 2022
S. Valaguzza, M. A. Hughes (eds.), *Interdisciplinary Approaches to Climate Change for Sustainable Growth*, Natural Resource Management and Policy 47,
https://doi.org/10.1007/978-3-030-87564-0_12

do we use to power it? What are the physical and traffic conditions of the roads and highways? How much of our income do we spend on this mobility? And how much of the air pollution in our community—as well as the greenhouse gases in the atmosphere—derive from burning fossil fuels to power this transportation?

All of these questions are largely determined by transportation law and policy at the national and subnational levels, and even by international and foreign policies that determine the range of transportation products available to us. But more significantly for the globe, reducing emissions from transportation is critical to the success of tackling the broader global greenhouse gas emissions that cause climate change. Transportation is responsible for 14% of global greenhouse gas emissions, including 27% and increasing for the European Union and 28% for the United States, not including emissions from petroleum production. In short, without emission reductions from this sector, meaningful action to stabilize Earth's climate will be infeasible.

This chapter provides an overview of the policies that most directly affect greenhouse gas emissions from transportation, as well as the range of technologies available to do so. The following sections

- Describe the various sources of transportation emissions, from per capita driving kilometers to burning fossil fuels used in a variety of transportation activities, from cars to airplanes
- Provide an overview of law and policy approaches that reduce driving kilometers and improve overall transportation efficiency

The chapter will then

- Summarize the key clean fuel technologies that can reduce transportation emissions, including low-carbon biofuels, hydrogen, and battery electric vehicles
- Offer examples of law and policies around the world that have proven to help expedite the deployment of these low- and zero-emission fuels

Many of the example law and policies are drawn from leading jurisdictions like California, which has one of the world's first and most long-standing program to reduce transportation emissions, as well as the European Union, China, and individual countries with successful policies on urban development, transit, and zero-emission vehicles. The legal regimes in these countries and jurisdictions around the world will of course vary in how to implement these changes, depending on the form and layers of government in any given nation or subnational entity. But the basic thrust of the policies—including land use and transit investment, mandates and incentives for clean transportation, and subsidies and support for private sector action—remains consistent.

12.2 Transportation Emissions Cause and Effects

Transportation emissions constitute a major portion of global greenhouse gas emissions, including 14% globally, 28% in the United States, and 27% in the European Union. In California, that percentage is even higher, at over 40%. When factoring industrial emissions from petroleum production, the percentage increases to 50%. In China, transportation produces 65% of emissions in major cities such as Shenzhen (UC Berkeley, 2020). These emissions increase as economies expand, unless governments and industry take action to develop alternative transportation options.

What exactly comprises transportation? As California tracks it for regulatory purposes, the sector includes passenger vehicles (the biggest portion of the overall sector), freight trucks, aviation, shipping, and other vehicles. The European Union also includes international aviation and maritime emissions.

Other than zero-emission vehicles, all of these transportation modes rely on technologies that burn fossil fuels like petroleum in internal combustion engines to power mobility. That petroleum has to be pumped from the ground, with associated methane and other emissions, and refined and delivered to fueling stations, using energy that may also come from burning fossil fuels. Beyond the greenhouse gas emissions, these technologies also emit particulate matter, nitrogen oxides (NOx), and volatile organic compounds that harm human health.

Emissions increase the more we burn fuel in these engines. More efficient engines can reduce the amount of burning and therefore the associated emissions required to move people and goods the same distance. But as more people and items travel greater distances—an inevitable result of an increasingly global economy—emissions continue to rise globally. The metric used to track these emissions is often expressed as vehicle miles traveled (VMT) or vehicle kilometers traveled (VKT).

Globally, VKT generally increases when the economy performs well. If fuel is relatively inexpensive, people tend to travel and ship goods more. In these times, the economy provides more jobs that require commuting and the extra income allows people to travel for leisure and otherwise spend money traveling, purchasing, and shipping items. When the economy contracts, VKT can decrease—sometimes dramatically, as in the case of the COVID-19 pandemic—as people lack money to travel or jobs to access.

Notably, not all people travel equally. Lower-income people often cannot afford private vehicles and, depending on the country and context, take public transit or walk or bike for mobility in greater numbers. Higher-income individuals tend to travel much more than people in other income groups, be it by airplane, rail, or more driving kilometers. They also purchase more goods, with increases in freight emissions. As countries like India and China develop and experience improved living standards, these transportation emissions increase. Ultimately, much like the broader issue of more industrialized nations contributing disproportionately to climate change, within nations it is often the wealthiest who are disproportionately responsible for transportation emissions.

But at the same time, transportation often represents a greater share of the budget for low-income earners, so policies that reduce the need to travel can disproportionately benefit low-income earners. In order to achieve environmental and equity goals, policy responses to address transportation emissions should therefore disproportionately influence high-income residents' travel behavior, while also relieving the economic burdens of transportation on low-income residents. Ideally, successful policies will improve mobility access and opportunity for advancement and a better quality of life.

Furthermore, the harmful air pollution from burning transportation fuels, beyond greenhouse gas emissions, often disproportionately affects low-income neighborhoods. These neighborhoods are often located in polluted areas, such as next to highways or airports. The associated pollution can include particulate matter and other pollutants that cause respiratory ailments, smog and other public health harms, as well as visual blight. Decreasing transportation emissions can therefore greatly benefit low-income people, not just by addressing climate change but by decreasing toxic air pollution.

As policy makers develop climate policies to address transportation, they will need to be mindful of these equity impacts. Doing so is not just moral and just, it can also represent smart political strategy. As more residents experience the benefits of climate policies, such as clean transportation options or improved mobility, they will be more likely to support the continuance of these policies, leading to a virtuous cycle that maintains their durability and expansion over time.

12.3 Law and Policy to Reduce Driving Kilometers and Improve Transportation Efficiency

For policy makers, the first step to reducing transportation emissions is to improve transportation efficiency. In other words, the goal is to achieve more mobility for the same amount of fossil fuel or other energy source used. But the primary means to accomplish this goal does not involve direct regulation of transportation: it involves how we design, build, and retrofit our neighborhoods.

12.3.1 Orienting Land Use Around Walking, Biking, and Transit

Developing more compact cities and towns, featuring clusters of home in walkable, bikeable neighborhoods, with easy access to high-quality transit, is a proven strategy for reducing per capita carbon emissions. While these communities reduce emissions in multiple ways, including from smaller homes that use less energy and water, the main advantage is the reduction in need for transportation. Residents in these densely populated areas can more easily walk, bike, or take transit for mobility. As a result, our most densely populated cities in places like the United States, Europe, China, and South America typically feature far fewer per capita emissions than residents of areas that are auto-oriented and low population density.

Leaders in jurisdictions that want to reduce these emissions through land use changes have many law and policy tools to do so. First and foremost, they can design new communities around walkability and transit by ensuring through local land use policies that new development features a mix of uses (such as residential and commercial combined, as opposed to mandating low-density, single-family-only neighborhoods), is oriented toward transit and pedestrian infrastructure, and does not provide excessive on-site parking spaces to encourage automobile ownership and use. Land use policies should allow for apartments near transit stops and concentrate office development in walkable, transit-friendly neighborhoods.

Land use leaders can also retrofit existing areas to feature more compact, convenient "infill" housing, accompanied by centralized office and retail uses. Often the law and policy approaches involve investing in transit and pedestrian infrastructure, streamlining permitting for redevelopment of existing buildings, and relaxing restrictions on redevelopment in key areas. To ensure housing is affordable, government leaders can subside units and require them to be rented or sold to low-income residents, such as with revenue from tax assessments or bond sales, as well as tax credits.

To avoid creating inequitable outcomes, policy makers need to ensure that redevelopment of existing parcels does not displace low-income renters. They can protect these residents by discouraging or banning redevelopment in certain situations with high risk of significant displacement. They can also develop policies to ensure that current low-income renters will have access to comparably priced and situated units nearby. In the long term, increased production of housing will in general lower prices for all through expanded supply, while also creating a pipeline of affordable market-rate units as new units age and decrease in price over time.

12.3.2 Improving Transit Infrastructure

Policy makers can also reduce emissions by investing in transit infrastructure to serve existing or newly planned communities. If they locate transit near concentrations of jobs and homes, residents are more likely to use these forms of mobility, particularly rail transit. Complementary policies can facilitate easy and safe pedestrian and bike access to these transit stops, with less land devoted to parking for private vehicles. In many countries, the most cost-effective investments are bus-only lanes on existing roadways, where high-capacity buses have access to dedicated lanes to ensure fast, efficient, and reliable service outside of congested lanes. Rail transit can also be useful for high-density cities but is otherwise more expensive and time-consuming to build.

To make these investments, national and subnational leaders need to direct transportation revenues primarily to transit, pedestrian, and bicycle infrastructure. Funding for highway and automobile infrastructure should focus on repairing existing capacity first, with expansion only for lanes dedicated to high-occupancy vehicles.

High-speed rail can serve as an important additional strategy for diverting air travel to more sustainable land-based means, particularly if deployed between major cities that are located too far for convenient driving but too close to make air travel efficient. High-speed rail is generally defined as electric railways that travel in excess of 250 km/h (160 mph). One global success story is the Madrid to Barcelona high-speed rail line, which helped absorb and obviate much of the air traffic between those cities. Some high-speed rail lines are profitable and can be operated by private entities, with the public sector paying for construction. Federal, state, and local leaders can finance these investments through sale of bonds or by directing transportation revenue to support construction. However, high-speed rail is only cost-effective if serving densely populated cities at speeds and prices that are competitive with air travel.

12.3.3 Congestion Pricing and Mileage/Kilometer Fees

For jurisdictions with heavy traffic and automobile dependence, policy makers may want to consider pricing strategies to discourage excessive automobile usage and encourage carpooling and other more sustainable modes of travel. The most common method is congestion pricing, which levies electronic tolls on vehicles entering cordoned, congested areas during times of peak traffic, priced variably to optimize the number of people entering these designated city centers. London and Singapore already implement such a policy. These pricing strategies can exempt high-occupant vehicles and also be adjusted based on income. Low-income drivers could potentially receive rebates or reduced tolls based on how the agencies collect and redistribute the toll revenue. Furthermore, the revenue from the tolls can fund transit alternatives, which can also benefit low-income travelers who depend on it for mobility.

Another policy option to discourage excessive vehicle kilometers traveled is to replace or supplement petroleum taxes that fund road infrastructure with a mileage/kilometer-based fee. With these fees, drivers are charged based on the number of miles or kilometers they drive, as opposed to the amount of fuel they purchase. These kilometers can be tracked electronically, such as via a global positioning device on the vehicles, smartphone tracking, or odometer reporting by drivers at the end of each billing or tax cycle. National and subnational leaders may need to experiment with different technology options and pilot programs to ensure that the programs can be effective and can adequately replace the revenue lost from taxing the sale of petroleum.

Ultimately, as vehicles switch to electricity and other zero- or low-carbon fuels over time (discussed below), this change to road pricing and kilometer/mileage fees may be necessary anyway. If petroleum taxes are the main source of infrastructure funding, as in the United States, these revenue streams will decrease to almost zero in the coming decades if transportation policies to reduce emissions by switching to clean fuels prove effective. As a result, mileage/kilometer fees or congestion pricing could be critical to providing a replacement revenue stream commensurate with actual road usage. And if drivers pay per kilometer, they may be more efficient in how they travel, leading them to avoid extraneous trips or carpool if possible.

A similar principle could apply to aviation travel, which represents a relatively minor but still significant share of global greenhouse gas emissions. Governments could encourage or require airlines not to offer reward programs and inexpensive tickets for those who fly frequently. Business traveler incentives in particular could end, given how much revenue and flying results from this type of travel. Alternatively, governments could levy higher fees on business travel to discourage excess flying, which could help reduce aviation emissions.

Ultimately, the land use and transportation policies discussed in this section often represent the "low-hanging fruit" of climate policy because they do not involve deployment of new technology. Instead, they represent tried-and-true methods of designing cities and neighborhoods. Before the rise of the automobile in the early twentieth century, all cities and towns were designed around walkability and mass transit, whether horse-drawn or electric train-powered. Today, many cities either take advantage of that legacy design or lack the incomes to build around the automobile.

However, as incomes rise and more countries develop economically, pressure will build to replicate the auto-oriented model of growth. But to be sustainable from a climate, land use, and pollution perspective, new cities will need to be compact and walkable, and existing urbanized areas will need to be retrofitted to achieve these aims. The benefits reach beyond climate, from improved public health from greater exercise, higher quality of life from reduced travel time and more interactions with neighbors, and saved money on transportation costs. Political leaders will need to take these impacts and benefit into account in making what can sometimes be challenging—but ultimately necessary—political decisions to grow their cities and towns more sustainably.

12.4 Law and Policy to Improve Fuel Economy and Reduce Its Carbon Content

Even with more compact, transit-friendly cities and towns, a functioning economy and society will still need to power transportation with fuels, whether for leisure, work, or commerce. From a climate and air quality perspective, these fuels will need to be clean. But switching the entire global vehicle fleet entirely to zero- or low-emission fuels will take decades, even in the best-case scenario. As a result, policy makers will need to act in the meantime to increase the fuel economy of existing fossil fuel-powered vehicles. They can accomplish these changes by setting fuel economy standards on automakers that increase in stringency over time.

One prominent example of a national fuel economy program and the innovation it can induce comes from the United States. Through the Corporate Average Fuel Economy (CAFE) standards, the federal government sets a fleet-wide target for all automakers to meet in terms of how many gallons of gasoline are required for each kilometer driven. The program is subject to much political wrangling, depending on which political party is in power. For example, the Obama Administration imposed standards that covered model years 2020–2025, which required fuel economy improvements in passenger automobiles between 4 to 5% per year in order to achieve average fuel economy levels of about 55 miles per gallon by 2025. The Trump Administration then proposed to reduce those fuel economy improvements to roughly 2% per year. President Biden is now reversing those rollbacks and potentially ratcheting up the standards from the preceding Obama-era rule (Elkind, 2020).

Despite the political challenges in the United States, the program represents an important model for other countries to consider if they do not already have such standards. If they choose to set their standards at the US levels, they will have the benefit of accessing the fuel-efficient products that automakers are already developing to serve the US market. Countries that harmonize their standards with the United States can therefore save their residents money on fuel purchases and reduce pollution and greenhouse gas emissions at the same time.

Notably, this federal program, authorized in part by the national Clean Air Act, included a special carve-out for California to exceed these standards, given the state's history of severe air pollution and action on regulatory measures. The state must first obtain a waiver from the federal government to go beyond these national standards, and other states then have the opportunity to join California's standards. As a result, the program allows for innovation in national action that can first begin within one state willing to experiment. Other countries can similarly provide flexibility to their subnational entities (provinces or states) to exceed national standards, balancing the need for the industry to have consistent standards across nations against the need for local innovation.

12.5 Key Technologies to Boost Sustainable Fuels

There are three primary types of low- and zero-carbon fuels. If all transportation sources switched to using these fuels over fossil fuels, the world could largely eliminate the greenhouse gas emissions from this sector that are contributing to climate change.

12.5.1 Low-Carbon Liquid Biofuels

Liquid biofuels, derived from plant matter, can substitute for or blend with petroleum-based fuels to reduce the carbon content. The most common liquid biofuels are ethanol and biomass-based diesel, with ethanol most typically blended into gasoline up to 10% and biodiesel blended up to 20%. Ethanol producers derive it from soluble carbohydrates, including corn, sugarcane, sorghum, and wheat, as well as insoluble carbohydrates from the fibrous parts of plants (this biomass can also be used to generate electricity).

But not all biofuels are created equal. Their carbon content almost entirely depends on the source of the feedstock to make the biofuel. For example, feedstocks that come from distant shores can have much higher emissions due to the emissions from shipping. In addition, feedstocks can be quite carbon intensive if they come from plants that are grown specifically to make biofuel and that in the process disturb natural lands that otherwise function as carbon sinks or convert agricultural land from food production.

On the other hand, biofuels made from feedstocks that would otherwise decompose or possibly burn, which would result in potent gases like methane, can be low or even negative emissions. By harvesting these fuels, like certain forest or agricultural waste products, biofuel providers can avoid these decomposition emissions and instead potentially offset fossil fuel emissions from petroleum (or to generate electricity that offsets power from coal-, diesel-, or natural gas-fired facilities). Examples of low-carbon biofuels can include biodiesel derived from used cooking oil or distiller corn oil, which is a co-product of ethanol production. These fuels can provide greenhouse gas benefits from 86 to 96% compared to petroleum diesel. The feedstocks could also be from sources local to the biofuel producer, which avoids or reduces emissions from shipping. As a result, policy makers should prioritize low-carbon biofuels in any incentive or mandate programs (Elkind, 2015).

12.5.2 Hydrogen

Hydrogen represents a promising low- or zero-carbon fuel for a number of mobility options. It can be used to generate electricity for transport when it powers fuel cells. Fuel cells work like batteries to produce electricity and heat, but they do not need to be recharged as long as the hydrogen fuel is available. The hydrogen fuel cell separates hydrogen molecules to create a flow of electricity, creating a by-product of water and heat. While these fuel cells produce no carbon emissions, the overall carbon footprint of the technology depends on the source of the hydrogen. If the hydrogen gas is produced by fossil fuel-powered energy, then it will potentially have a significant carbon impact. However, if the hydrogen is produced by excess solar and wind power, or other zero-carbon sources, it is truly a low- or zero-emission fuel.

Hydrogen is likely best suited for long-distance zero-emission transportation, such as long-haul trucking or transoceanic air travel, due to infrastructure constraints on the fueling and relative lack of energy efficiency for shorter trips, for which batteries are at this point better suited. The cost of producing hydrogen will need to decrease in the next decade for it to become commercially competitive with other sources of energy, though notably hydrogen can also be used for producing electricity for the grid and as a form of long-duration energy storage, if generated by surplus renewables for later use during times of low-renewable energy production.

12.5.3 Battery Electric Vehicles

By far and away the most dominant, functional, and cost-competitive form of zero-emission transportation comes from battery electric vehicles. While the precise carbon benefits of battery electricity over internal combustion engines depend on the source of the electricity, these vehicles represent a significant emission saving over petroleum even when powered by dirtier electricity grids. Overall, some estimates place electric vehicle life-cycle emissions at approximately 50% fewer greenhouse gases per kilometer traveled than internal combustion engines. The benefits range from 25 to 28% fewer emissions in jurisdictions with fossil fuel-based electricity supplies, and up to 72–85% lower emissions in areas with high renewable energy penetration. These benefits should become ever more widespread as electricity grids globally become cleaner with the success of renewable energy technologies. This fueling infrastructure for battery electric vehicles is also less of a challenge compared to hydrogen, given the ubiquity of electricity in our buildings and our wired landscapes.

The dominance of this form of zero-emission fueling is due to the massive price decreases over the past decade and energy density of lithium-ion batteries. First used widely in consumer electronics, this battery chemistry has now revolutionized zero-emission transportation and enabled it to become cost-competitive with internal combustion engines. Electric mobility brings additional consumer benefits as well, including faster vehicle acceleration, reduced maintenance cost, and generally superior driving performance compared to gas engines.

These vehicles have become especially popular in places like California, Norway, and China in part due to consumer incentives and mandates to boost supply. In fact, China is the number 1 market in the world for plug-in electric vehicles, accounting for 54% of the world's 2.26 million vehicles sold in 2019 (UC Berkeley, 2020). Similarly, California accounted for nearly half of the plug-in vehicle sales in 2019 in the United States, with 156,101 vehicles sold and approximately 700,000 vehicles on the state's roads. To put these numbers in context, plug-in vehicles represented just 8.26% of California's annual new vehicle sales in 2019.

Battery electric vehicles are not limited just to passenger vehicles. E-scooters, e-bicycles, e-buses, and now e-trucks are all becoming commercially available due to this revolution in battery technology and cost. And while not powered by batteries, electromobility in the form of electric trains and high-speed rail also forms an important component of zero-emission transportation.

Notably, and as a source of some controversy, these batteries depend on key minerals, which typically include lithium, cobalt, nickel, graphite, and manganese (though battery manufacturers plan to reduce, if not eliminate, cobalt in the coming years). Mining for these minerals can create local environmental impacts, as well as raise human rights concerns in certain places like the Democratic Republic of Congo, which dominate the cobalt supply. Many companies are committed to sustainability practices that seek to remedy these abuses, just as has been attempted with mining and oil and gas production in other contexts. But concerns remain among human rights and other advocates over the transparency of industry adherence to these standards.

In addition, some national security experts and industry leaders worry about the reliability of the battery supply chain, given the small number of companies involved that are largely concentrated in Japan, South Korea, and especially China. The market analysis firm Benchmark Mineral Intelligence estimated that in 2019 Chinese actors controlled 23% of upstream mining (of lithium, cobalt, graphite, and manganese), 80% of chemical refining, 66% of cathode and anode production, and 73% of lithium-ion battery cell manufacturing. This geographic concentration of production can undermine the reliability of global supplies if domestic policies in these countries change to limit production or foreign access. To address supply chain sustainability and reliability, electric vehicle stakeholders will need to promote more deployment worldwide of mining, processing, and manufacturing. They will also need to ensure greater transparency, coordination, and enforcement to strengthen existing due diligence and disclosure standards (Elkind et al., 2020).

12.6 Law and Policies to Boost Deployment of Low- and Zero-Emission Fuels

The transition to zero-emission fuels and transportation options did not begin with an unfettered free market. If left to their own devices, fossil fuel companies and transportation companies would otherwise be financially encouraged to maintain the status quo as long as they are not held liable for the damages from the pollution they help induce or cause. By not paying for the externalities of their products, they have made significant profits from fossil fuels.

Instead, the launch of the zero-emission transportation revolution underway has been the result of careful, deliberate, and sustained policy action, often at the subnational level, to encourage private sector innovation. The market has responded accordingly. This section will discuss the general policy approaches to boosting these clean energy markets, which generally center around industry mandates, consumer incentives, subsidies for infrastructure and product deployment, and regulatory streamlining, including electricity rate reform.

12.6.1 Mandates

Mandates have been key globally to launching the market for zero-emission vehicles. In the United States, the process began in earnest with California's zero-emission vehicle (ZEV) regulation. The rule requires automakers to ensure that a minimum percentage of their sales are zero-emission vehicles. If the companies have no such vehicles to sell, they can purchase credits from automakers with surplus sales. The state aims to have 1.5 million zero-emission vehicles on the road by 2025 and 5 million by 2030. Following this goal, manufacturers are subject to a 22% credit requirement in 2025, which is expected to result in a zero-emission vehicle market share (as opposed to a plug-in vehicle share, which could include hybrid vehicles) of about 8% by 2025.

As a result of this policy, traditional automakers began producing the initial battery electric and hydrogen fuel cell vehicles. Eventually, all-electric companies like Tesla have been able to earn profits on the sale of credits. Other states, representing a total of approximately 40% of the US market, have joined California's ZEV regulation under the aforementioned authority of the US Clean Air Act. The ZEV mandate has ultimately been the state's foundational policy to boost these vehicle technologies, leading to the state's dominant position in the national industry and relatively high sales.

As another example of mandates, China currently aims to have 25% of the country's total vehicle sales come from zero-emission vehicles by 2025. This proposed target is set to increase to 40–50% of total sales by 2030. To achieve these targets, China developed a "new energy vehicle credits" (NEV) rule in 2017 that equates to a requirement that approximately 3–4% of new vehicles be zero emission by 2019, with a higher percentage in 2020. To complement this national policy, Hainan Province similarly set a target for 100% of new vehicle sales by 2030 to be zero emission (UC Berkeley, 2020).

The European Union has also legislated mandates for zero- and low-emission cars, dating to 2009, and since 2011 for new vans. As part of its goal of achieving net-zero carbon emissions by 2050, with an interim 2030 goal of reducing emissions 55%, the EU has set zero- and low-emission vehicle targets for 2020, 2025, and 2030. The targets all represent percentage reductions from a 2021 baseline. For new cars, they must achieve a 15% reduction by 2025 and 37.5% reduction by 2030 and beyond, while vans must have a 15% reduction by 2025 and 31% by 2030 and beyond. Manufacturers meet these targets based on their EU fleet-wide emissions. The EU is further creating an industry crediting system to be introduced in 2025 (European Commission, 2020).

Beyond the ZEV regulation, California pioneered another mandate that has boosted low-carbon biofuels and electric vehicles alike. The "low carbon fuel standard" regulation, pursuant to the state's climate change law, requires providers of fuel to reduce the carbon content by increasingly stringent amounts over time, with a 2030 goal of reducing the carbon content of fuel by 20% from the 2010 baseline year. As with the ZEV mandate, fuel providers who cannot meet this standard can buy credits from providers who have an excess. The policy deems electricity to be a low-carbon fuel when it powers an electric vehicle. As a result, the state allows electric utilities to be considered a fuel provider and to sell credits to oil companies, resulting in revenue that has been returned by law to electric vehicle purchasers as a cash rebate.

The California program has also encouraged fossil fuel providers to purchase and blend more low-carbon biofuels into their mix, particularly biodiesel. The result has been a modest but steady increase in the low-carbon biofuel industry in California. National and subnational leaders alike should consider similar mandates for fuel providers to reduce the carbon content of their fuels. These policies have resulted in a strong market for low-carbon biofuels, which can boost opportunities for the agricultural and restaurant sectors to convert or sell waste products for fuel production. The program has also generated an important source of revenue for the state and credit providers that can support additional low-carbon investments. And since low-carbon biofuels can be blended with fossil fuels, residents can usually continue using their current internal combustion engine vehicles while benefitting from a reduction in carbon pollution from the fuel they burn.

Policy makers must pay careful attention, however, to the source of the biofuel. California uses a complex "carbon intensity" score to grade each source of biofuel in order to determine its carbon savings, based primarily on feedstock and shipping distances, as well as any land conversion to grow the biomass. At the national level, the United States has a renewable fuels mandate, which requires petroleum fuel providers to include a certain amount of biofuels in their product. However, the national law does not use a carbon screen like California, and as a result, the program can encourage biofuels that are otherwise not necessarily beneficial from a climate perspective.

12.6.2 Incentives

Mandates to produce zero-emission vehicles can help bring products to market, but consumers may still require incentives to be able to choose them over internal combustion engine models. These incentives can include cash rebates, tax reductions, or other privileges, such as free or reduced-cost parking or express lane access. Governments around the world have utilized all of these approaches. In Norway, the government slashed import taxes on electric vehicles, leading to a surge in sales. California has also dedicated proceeds from its cap-and-trade carbon pollution program on industrial emitters to fund cash rebates for electric vehicle and hydrogen fuel cell vehicle purchases. In addition, as mentioned, electric utilities have dedicated low-carbon fuel standard revenues to consumers as cash rebates. The United States also offered tax credits for zero-emission vehicle purchases based on manufacturer sales, though the program has expired for some leading manufactures like Tesla and General Motors, as their sales have exceeded the tax credit cap. And California offers carpool and express lane access for plug-in electric vehicles, while individual cities like Santa Monica and Honolulu have provided parking incentives for drivers.

Policy makers seeking to increase incentives for zero-emission vehicle purchases should look to these examples around the world. Simple cash rebates available at the point of purchase are often the most effective incentives. Financing programs, including leases, can be helpful to encourage consumers wary of high upfront costs or long-term investments. Ultimately, battery electric vehicles are often cheaper to own over the life of the vehicle due to reduced fuel costs (electricity is usually less expensive than petroleum) and fewer maintenance needs (fewer moving parts in a battery electric vehicle compared to an internal combustion engine). So if consumers receive financial assistance to afford the upfront cost, they will usually benefit in the long term. Many industry analysts anticipate that electric vehicles will soon reach cost parity with legacy gas-powered vehicles. But until that point, and even beyond, these consumer incentives will be crucial for helping the world meet its ambitious long-term climate goals through zero-emission vehicles.

12.6.3 Subsidies

Subsidies for private industry can take multiple forms, including direct cash payments, tax credits or reductions, low- or no-cost loans, and paying for partial costs of a project or product. In the case of hydrogen and battery electric vehicles, industry has received multiple types of support, both for vehicle development and for supporting infrastructure. Tesla, for example, received a low-interest loan from the US federal government to purchase its California manufacturing facility. The US government also has an active research and development program at the Department of Energy, while the California Energy Commission similarly funds early-stage clean technology research. Funded projects have spurred advancements in both battery and hydrogen fuel cell technologies.

In addition, zero-emission vehicles require a new infrastructure deployment of charging and fueling stations, and governments at multiple levels have subsidized this development. The federal and state governments in the United States, along with municipal utilities, have directly funded new electric vehicle charging stations through grants and other programs. Tax credits are also available for some purchasers/installers. In some cases, litigation settlements have provided funding. For example, the Volkswagen diesel emission cheating scandal, in which the company installed "defeat devices" to avoid having regulators detect their vehicles' noncompliance with emission standards around the world, resulted in a settlement that directed almost billions of dollars toward EV charging infrastructure and other EV-related investments in the United States and Europe, with almost 1 billion dollars alone going to California.

Governments have also authorized their electric utilities, which they regulate as monopolies, to invest in electric vehicle charging infrastructure. Utilities in California and other states can pay for the necessary wiring and infrastructure upgrades to enable private charging companies to install the actual stations at a lower cost, thereby allowing them to avoid paying for the attendant electric infrastructure upgrades. Utilities can then recover the cost of these investments, along with a reasonable profit, from ratepayers. Some studies indicate that ratepayers may be compensated for these investments in the form of lower overall rates, as more electric vehicle charging means increased sales for utilities with otherwise flat fixed costs, putting downward pressure overall on rates for all ratepayers (Baumhefner, 2020).

Successful law and policy solutions to subsidize the electric vehicle and zero-emission transportation industry have taken many forms, most prominently public sector research and development support for technology and manufacturing innovation and subsidies for infrastructure deployment. The need is pressing: many consumers cite perceptions of a lack of access to available charging stations as a reason not to purchase an electric vehicle. This problem is particularly acute for residents in urban areas, such as in multifamily buildings, if they lack a dedicated parking space for their vehicle. Policy makers can address these fears over charging needs by subsidizing infrastructure deployment in key places, such as fast-charging at retail destinations or chargers at work or on city streets via lampposts and other existing infrastructure. Ultimately, policy makers need to ensure that any subsidies are carefully directed at maximizing sales to ensure the most rapid fleet turnover to clean vehicles as possible.

12.6.4 Regulatory Streamlining and Electricity Rate Reform

Regulatory reform to promote infrastructure deployment, including revised electricity rates, represents another important policy tool. In many jurisdictions, permitting barriers can stifle the deployment of zero-emission vehicle infrastructure and sometimes the vehicles themselves, as well as inhibit access to low- or zero-carbon feedstocks. Various regulations that never contemplated such new technologies can sometimes inadvertently create barriers, adding time, uncertainty, and costs that can discourage investments. These challenges can be particularly difficult for developers who need to obtain permits from multiple agencies or secure updated building or other codes to incorporate newer clean technologies. Similarly, low-carbon biofuel producers often need access to feedstocks and permission to site facilities to convert this biomass into liquid fuels.

Some jurisdictions, including subnational states and cities, have developed model codes or exemptions for permitting these technologies, such as electric vehicle charging and biofuel production facilities. Policy makers can conduct outreach and advance planning to catalog existing barriers and applicable regulations and determine optimal sites. Model building and other codes can also be adopted at the local level to streamline installation and permitting of these facilities as well.

In some cities, disputes have arisen over micro-mobility technologies such as e-scooters, primarily based on complaints about their ubiquity on city sidewalks and potential danger to pedestrians from speeding users. Cities can develop programs to clarify rules for e-scooter companies in terms of where they can be stored and ridden. Ideally, e-scooters could make common cause with bicyclists to advocate for expanded infrastructure such as dedicated e-scooter and bicycle lanes.

Electricity rate reform can also be a top priority for encouraging more infrastructure deployment of clean technologies like electric vehicle charging stations. In many countries and jurisdictions, these rates were not designed with electric vehicle charging in mind, particularly fast-charging (high-powered charging that can add 100 miles or more of electric vehicle range in 20 minutes or so). Charging site owners can incur significant expenses from commercial electricity rates in particular. Commercial charging generally involves payment by time of use. These rates therefore reward electric vehicle owners or site hosts who charge during hours when the cost of energy is lowest but punish those who are unable to moderate demand successfully. In addition, large commercial and industrial rates often entail "demand charges" that place additional costs on the maximum load drawn by a customer during the billing period. Many electric vehicle charging sites that have high but infrequent demand, coupled with inconsistent low-energy utilization (particularly for fast charging), face high exposure to these demand charges. As a result, the charges can sometimes severely undercut the economics of infrastructure deployment and operation.

Agencies that regulate electric utilities and approve rates could encourage utilities to design vehicle-grid integration (VGI) rates in ways that motivate drivers to charge during hours when the electricity grid has excess capacity or to alleviate local distribution-level constraints, rather than exacerbating the system-wide peak demand (particularly in jurisdictions with supply constraints on their electricity). These options can include employing real-time or demand response rates that send out price signals about when the grid has available capacity to charge vehicles inexpensively. Utility regulators can also adopt incentives for utilities to improve the overall utility "load factor" (determined by the ratio of average electricity demand to peak electricity demand, as a way to measure asset utilization) to encourage utilities to utilize their assets better through performance-based ratemaking. This type of rate could motivate utilities to adopt incentive programs that encourage optimal charging at all sites in their service territory, potentially without the need for charging-specific rates. In addition, regulators and policy makers could allow demonstration or pilot rates for utilities to gather data on what rates and degrees of utility investment might best encourage optimal infrastructure deployment, as well as data on utilization, maintenance, and reliability of the charging station (Elkind, 2017).

As the clean transportation revolution unfolds, new infrastructure will be required to service it. This infrastructure can bring multiple benefits in terms of construction and maintenance jobs and increased municipal and private sector revenues, as well as reduced pollution. But national and subnational leaders will need to address any permitting, electricity rate design, and other regulatory challenges that stand in the way to the extent that legacy legal regimes that never contemplated these clean technologies might otherwise be creating unintended barriers to achieving environmental aims.

12.7 Conclusion: Smart Policy and Private Sector Innovation Are Key to a Zero-Emission Transportation Future

The zero-emission transportation future is crucial for the larger effort to reduce greenhouse gas emissions. Much of the attendant law and policy needs involve tried-and-true approaches, particularly on orienting land use around transit and pedestrian and bicycle infrastructure. Some of these steps may be politically challenging but have otherwise proven themselves in low-carbon cities and towns all over the world, most of which were built before the automobile and therefore oriented to pedestrian scale.

The technology solutions are more complex. The challenge is the dispersed nature of the world's vehicles, requiring hundreds of millions of consumers around the globe to opt in for more efficient travel and land use patterns and zero-emission technology. It also requires mobilizing a diverse industry, including battery manufacturers, automakers, charging companies, and electric utilities, to support the deployment. Furthermore, many of the law and policy actions require alignment across multiple levels of government, from tax and funding decisions at the national level to permitting and rebate policy at the subnational levels.

The good news for climate advocates is that the industry is scaling and innovating rapidly, the result of careful and deliberate policy around the globe. Central to this progress has been declining prices for lithium-ion batteries. Policy makers can continue this progress by focusing on transportation policies such as mandates, incentives, subsidies, and permit streamlining and electricity rate reform.

Ultimately, a low- or zero-carbon transportation sector across the globe will not just bring climate benefits. These changes, such as smart land use and zero-emission technologies, also provide critical co-benefits. They include reduced costs for consumers, cleaner skies, improved public health, saved commuting time that results in more leisure and family hours, and new industries and jobs. A zero-emission transportation future, if done well through smart law and policy, can therefore improve our quality of life as much as the climate on which we depend for survival and prosperity.

References

Baumhefner, M. (2020). *Electric vehicles are driving rates down.* Expert blog. Natural Resources Defense Council. Retrieved March 25, 2021, from https://www.nrdc.org/experts/max-baumhefner/electric-vehicles-are-driving-rates-down

Elkind, E. N. (2015). *Planting fuels: How California can boost local, low-carbon biofuel production.* UC Berkeley & UCLA Schools of Law. Retrieved March 25, 2021, from https://www.law.berkeley.edu/wp-content/uploads/2015/11/Planting-Fuels.pdf

Elkind, E. N. (2017). *Plugging away: How to boost electric vehicle charging infrastructure.* UC Berkeley & UCLA Schools of Law. Retrieved March 25, 2021, from https://www.law.berkeley.edu/wp-content/uploads/2017/06/Plugging-Away-June-2017.pdf

Elkind, E. N. (2020). Trump's flawed rollback of fuel economy rules. In *The regulatory review*. University of Pennsylvania Law School. Retrieved March 25, 2021, from https://www.theregreview.org/2020/05/18/elkind-trumps-flawed-rollback-of-fuel-economy-rules

Elkind, E. N., Heller, P., & Lamm, T. (2020). *Sustainable drive, sustainable supply: Priorities to improve the electric vehicle battery supply chain.* UC Berkeley School of Law, Natural Resources Governance Institute. Retrieved March 25, 2021, from https://www.law.berkeley.edu/wp-content/uploads/2020/07/Sustainable-Drive-Sustainable-Supply-July-2020.pdf

California China Climate Institute, UC Berkeley. (2020). *Driving to zero: California and China's critical partnership on zero emission vehicles.* Retrieved March 25, 2021, from https://ccci.berkeley.edu/sites/default/files/ZEV%20Paper%20-%20September2020.pdf

European Commission. (2020). CO_2 *emission performance standards for cars and vans.* Retrieved March 25, 2021, from https://ec.europa.eu/clima/policies/transport/vehicles/regulation_en

Ethan Elkind is the director of the Climate Program at the UC Berkeley School of Law's Center for Law, Energy and the Environment (CLEE), with a joint appointment at the UCLA School of Law. He researches and writes on law and policies that address climate change. He previously taught at the UCLA law school's Frank Wells Environmental Law Clinic and served as an environmental law research fellow. His book Railtown on the history of the modern Los Angeles Metro Rail system was published by University of California Press in January 2014. He is also a co-host of the weekly call-in radio show "State of the Bay" on the San Francisco NPR affiliate KALW 91.7 FM, airing Monday nights at 6 pm. He received his B.A. with honors from Brown University and graduated Order of the Coif from the UCLA School of Law.

Chapter 13
From Green to Social Procurement

Laura Carpineti

Abstract Public procurement has recently been considered one of the key actors for achieving sustainable goals and contributing to mitigation and adaptation policies against climate change. In fact, public procurement effectively could play this key role, due to his huge impact on global GDP.

This chapter shows that public procurement is at a mature level in achieving green products, but there is still a long road ahead when adopting public procurement to enhance effective equal rights and to protect minorities. In fact, if green procurement is at an advanced stage of the process, since there are solid international recognized standards and a huge set of tools for monitoring results, it is still difficult for public authorities awarding contracts to verify that contracts respects human rights and work conditions belong the whole supply chain via public procurement, due to lack of specific competences, the high costs of monitoring, and few international standards.

The chapter, after a deep description of tools in greening products works and services, emphasizes how to abate barriers via social procurement, giving best practices at international level for evaluating competitors in a sustainable way. Finally, some recommendations are given to policy makers in order to set up a sort of revolution of "sustainable procurement."

13.1 Introduction

All public institutions globally, from international institutions to local municipalities, are stressing in recent years the role of public procurement policies in achieving tangible sustainability results against climate change (Arrowsmith et al., 2000).

So, since the meaning of sustainability in public procurement, in the past, was addressed by governments and by public procurers only to green aspects and to

---

L. Carpineti (✉)
Martino & Partners—University of Milan, Milan, Italy
e-mail: laura.carpineti@martinopartners.com

legal "battles" and rules to face corruption, today public procurement seems to be recognized as a key element for implementing mitigation and adaptation policies.

Effectively, public procurement plays a crucial role in contributing to the international Gross Domestic Product (GDP): the OECD estimates that 12% is the amount spent in OECD countries on public procurement. Moreover, looking at the European borders, the estimated value of tenders published in TED, in 2017, amounts to 545.4 billion euros.

It seems no coincidence that the World Bank, one of the largest worldwide institutions providing grants for developing programs via public procurement, launched its "Climate Change Adaptation Program."

Furthermore, according to the World Bank, public procurement is a necessary strategic instrument of development to promote good governance and to implant effective and efficient use of public resources. Thus, public procurement is a crucial component of democratic governance, poverty reduction, and, especially for the purpose of this book, sustainable development (Valaguzza, 2016; Schooner & Speide, 2020; Fontana et al., 2020). Additionally, the literature talked about "Global Revolution," referring to regulation of public procurement as a global phenomenon, in which states adopt formal rules to award contracts in order to promote domestic objectives. As an example, the emerging economies of China and Indonesia, also with the support of international aid, implemented a new regulatory system on public procurement.

Besides the recognized importance of public procurement at an economical level, in the past two decades public procurement also started to be considered worldwide as a key factor for the promotion of sustainability, in parallel with the new change of perspective of sustainability that moved in the private sector. If in the past sustainability was considered simply as a set of philanthropic activities, nowadays sustainably enters into the business strategies, with possible negative consequences in case of overlooking.

Simultaneously, concepts of protection of the environment and safeguarding the community advanced also in public procurement strategies.

As we will see, public procurement can effectively enhance the acquisition of products, services, and infrastructures with a lower environmental impact (then supporting mitigation and adaptation strategies) and, at the same time, internalize social distortions connected to the supply chain. Moreover, the most far-sighted public authorities can effectively enhance competition, quality, and sustainability by setting goals and adopting flexible contracts and efficient supply relationship management (Trepte, 2004; Valaguzza, 2018; Klinger, 2020).

The United Nations provides the following definition of sustainable procurement: (UN, 2021). "Procurement is called sustainable when it integrates requirements, specifications and criteria that are compatible and in favor of the protection of the environment, of social progress and in support of economic development, namely by seeking resource efficiency, improving the quality of products and services and ultimately optimizing costs." And more: "Sustainable procurement means making sure that the products and services we buy are as sustainable as possible, with the lowest environmental impact and most positive social results."

For instance, if we consider that modern slavery affects over 45.8 million people in 167 countries, public procurement can request to suppliers to preserve human rights at work, for example, child labor, health and safety, working hours, remuneration, etc.

For this reason, it is important to consider public procurement as an active and strategic tool to face climate change. In this regard, the European Green Deal states that "the EU's trade policy facilitates trade and investment in green goods and services and promotes climate-friendly public procurement (…) Public authorities, including the EU institutions, should lead by example and ensure that their procurement is green. The Commission will propose further legislation and guidance on green public purchasing."

In fact, the strategic sectors individuated by the EU Green New Deal Communication (European Commission, 2019) for transforming the economy will affect also public procurement activities, such as

- Supply clean, affordable, and secure energy: public procurement for innovation
- Mobilize industry for a clean and circular economy: evaluating offers in terms of LCC
- Build and renovate in an energy and resource-efficient way: public procurement of infrastructures and public buildings
- A zero-pollution ambition for a toxic-free environment: procurement of green busses
- From farm to fork: procurement of catering services for schools and hospitals

In the next paragraphs, we will see how public buyers may enhance the transition and contribute to a successful mitigation and adaptation policy to climate change.

13.2 Toolkit for Understanding Sustainability in Public Procurement

Before starting the discussion about sustainable procurement as a key element for implementing mitigation and adaptation policies to climate change, it is crucial to introduce three key concepts that are strictly connected to the procurement process and that must be recognized in order to better appreciate the ensuing paragraphs. These concepts are supply chain, LCC, and circular economy (Lee, 2021).

13.2.1 Supply Chain

According to the definition given by Chartered Institute of Purchasing Management (CIPS 2020 web page: "What is a supply chain?"): "A Supply Chain is a focus on the core activities within our organization required to convert raw materials or component parts through to finished products or services."

Commonly, the supply chain can be defined as a "to-do list" of activities within an organization, required to convert raw materials into finished products, works, or services. This transformation process from raw materials to delivery to final consumers involves the procurement sector as well. More precisely, the procurement sector has the role of selecting, in the supply chain model, subcontractors and raw materials in order to assure the best quality at the lowest prices by fixing adequate Service-Level Agreements (SLAs).

Indeed, supply chain models involve several stakeholders: the consumers, which in the public procurement sector consist in the public authority that purchases for itself or for the community, the retailers, the carriers, the suppliers, the subcontractors, and the transformers of raw materials.

It is important to stress the fact that any single product category has a different supply chain. Actually, it is crucial to keep in mind that to enhance a sustainable procurement system means to analyze in depth any single segment of the supply chain in order to achieve energy transition, transparency, and equitable working conditions. In the following paragraphs, we will see how difficult it is to monitor the supply chain, including actors contributing to the realization and distribution of a commodity in terms of sustainability, with particular regards to the respect of disadvantaged categories.

13.2.2 Life-Cycle Costing (LCC)

Life-cycle costing (LCC) was adopted for the first time in history in 1966 in the United States by the Department of Defense. LCC was applied in the procurement of military equipment as the acquisition costs only accounted for a small part of the total cost for the weapons systems, while operation and support costs comprised as much as 75% of the total.

LCC means to consider all the costs that will be incurred during the lifetime of the product, work, or service, such as purchasing price, delivery price, installation price, insurance, operating costs, including energy, fuel, and water use, spares, end-of-life costs, such as decommissioning or disposal, and residual value, namely the revenues from sale of product (Perera et al., 2009; Estevan & Schaefer, 2017).

Therefore, when procuring a commodity, it is incorrect in terms of LCC to take into account only the purchasing price. Differently, a strategic buyer should carefully analyze the costs generated by the life of each product, starting from the purchasing phase until the decommissioning one.

Thus, also from the perspective of the European Commission, "by applying LCC, public purchasers take into account the costs of resource use, maintenance and disposal which are not reflected in the purchase price "(European Commission, 2021c)."

Furthermore, LCC may also include the cost of externalities such as greenhouse gas emissions and water pollution. Only by including these externalities into the LCC analysis is it possible to adopt this tool as an "environmental measure" of impacts of commodities acquired.

Economists Gluck and Bauman (Gluch & Baumann, 2004) stated that traditional LCC does not become an environmental tool just because it contains the words "life cycle." It cannot be adopted only to analyze past, present, and future costs in order to choose the most cost-effective option, but it must also include the environmental impacts by "internalizing" the cost associated to pollution and to sustainability.

13.2.3 Circular Economy

The current economic growth was powered by a linear "take, make, waste" model of commodities. However, this model has become unsustainable in a system of limited raw materials that is experiencing a continuous environmental degradation. The response to this problem is the adoption of a new model, named "circular economy," as the sustainable answer to decoupling continued economic growth.

This sustainable model aims at eliminating wastes and increasing productivity by focusing on the "reuse" and "recycling of materials" instead of their waste, the design of products to emphasize longevity and repair, and the creation of new business models including the sharing economy and the development of local closed-loop systems. In other words, a product that is considered as waste in the linear economy becomes a raw material in the circular model.

As we will analyze in depth in Sect. 13.5.2, the concept of circular economy may be a key point element for comparing suppliers when making a tendering procedure.

13.3 Green and Social Procurement: The State of the Art

13.3.1 Sustainable Procurement: Respect of the Environment and of the Community

Sustainable procurement encompasses both green procurement and social procurement. According to the importance of the procurement expenditure over GDP, it can be a strategic tool for public administrations in putting in action mitigation and adaptation policies against climate change. In several international organizations, we saw a progressive increase in the interest of adopting public procurement as a policy tool. In fact, public procurement plays not only the role of addressing the critical question of climate change, but it can also increase in developing countries the issues of fair working conditions and fair-trade public contracts. In fact, multilateral donors and international organizations lead state-aid procurement initiatives to support countries implementing effective reforms of their systems (La Chimia & Trepte, 2019). When discussing sustainable procurement, green procurement and social procurement are often associated. Nevertheless, the two concepts have completely different meanings (Arrowsmith & Kunzlik, 2009).

In fact, it is possible to launch a green procurement procedure without considering any social aspect, and vice versa.

As an example, a purchasing authority can procure the greenest smartphone in terms of consumption and duration of the battery, thus reaching the green goal by protecting the environment without considering the labor rights-related aspect of the extraction of the raw materials (such as coltan) that is needed to produce the electronic components of the devices.

Differently, the same purchasing authority requiring a supply of tomatoes may impose the respect of labor rights of the workers from the economic operators, but neglect to set environmental specifications in the tender documents (e.g., the ban on products grown with the use of pesticides).

For this reason, it is important to distinguish between the two different ambits. Green public procurement means paying attention to the environmental impact of a product, service, or work throughout the whole supply chain. Conversely, social procurement means paying attention to an equal and inclusive process with reference to disadvantaged categories of firms or people. Consequently, a public buyer must consider social and green procurement equally central in the designing of a procurement strategy oriented to climate change mitigation and adaptation.

Sustainable procurement may be thus intended as the intersection of (a) economics (products, services, and works being economically affordable), (b) environment (products, services, and works respecting the environment), and (c) social requirements (products, services, and works safeguarding health and security of the community and of workers) (Ethics & Sustainability in Procurement, 2019).

However, it is crucial to admit that, nowadays, green procurement is rapidly advancing; at least in developed countries, a lot of work remains to be done in terms of social and inclusive procurement.

In fact, purchasing authorities have several tools and methodologies to evaluate the carbon footprint, the water footprint, or other ecological impacts of a commodity, linked to green standards, likely labels, and certification systems. Differently, in terms of respect to social aspects, the public buyer does not have enough knowledge and tools at this moment to evaluate economic operators in terms of social impact of their products, although some interesting examples can be observed internationally (see Sect. 13.7).

With this conceptual premise, it is possible to look at the procedural sequence of social and green procurement.

13.3.2 Standards in Procuring Green and Social Solutions

When defining sustainable procurement, a useful tool that can be easily adopted by procurers is the request of labels, assuring, by the audit of third independent authorities, that the procuring goods is made by respecting some standards in terms of environment or in terms of social aspects.

As an example, the purchasing authority can evaluate suppliers offering a specific good or service in terms of the possession of the certification "ISO 45001," the international standard for occupational health and safety management. The title by the supplier of this process certification assures that he improves occupational health and safety management, eliminates hazards, and minimizes risks for his workers (ISO 45001:2018, 2021).

Another example of standard assessing social procurement is the SA8000, based on internationally recognized standards of decent work, including the Universal Declaration of Human Rights. (SA8000, 2021)

Among the most important standards in evaluating the environmental impacts of a production process is the certification ISO 14001:2015. This international standard assesses (ISO 14000, 2021) "environmental responsibilities in a systematic manner that contributes to the environmental pillar of sustainability," , consistently with SDG nos. 1 (no poverty), 2 (zero hunger), 3 (good health and well-being), 4 (quality education), 6 (clean water and sanitation), 7 (affordable and clean energy), 8 (decent work and economic growth), 9 (industry, innovation, and infrastructure), 12 (responsible consumption and production), 13 (climate action), 14 (life below water), and 15 (life on land).

Another interesting experience of procuring standards is the one promoted by the Italian Antitrust Authority (Italian Antitrust Authority, 2021) that goes under the name of "Transparency Rating." The rating consists of a "label" given by the Authority to those Italian companies respecting some minimal parameters in terms of anticorruption measures and transparency. The label opens from the rating of 1 star till a maximum of 3 stars. According to the Italian legal framework on public procurement, the transparency rating can be adopted by procuring authorities to compare suppliers in technical terms (see Sect. 13.5.2).

As this example demonstrates, evaluating suppliers and ranking them in terms of the possession of governmental and/or private international sustainable standards can be adopted by purchasing authorities in order to implement their sustainability goals, and, on the other hand, to stimulate the market to be more sustainable (De Grauwe, 2017; Kelman, 1990).

13.4 Enhancing the Transition: Procuring Innovation

Innovative public procurement is famously defined as follows: "When a public agency acts to purchase, or place an order for, a product—service, good or system—that does not yet exist, but which could probably be developed within a reasonable period of time, based on additional or new innovative work by the organization(s) undertaking to produce, supply, and sell the product being purchased" (Edquist et al., 2000).

In order to provide a very simple and intuitive example, the purchasing authority asks for innovative procurement to develop a product, a service, or an innovative infrastructural system or mechanism that has not been invented, designed, or made operational yet.

The basic principle of innovative procurement is to let the suppliers compete on the idea behind the achievement of the results by

- Making a competitive procedure: fixing the needs to be satisfied without describing the specification behind the needs. The market will offer the better innovative solution fitting the need.
- Giving a monetary prize to competitors in order to stimulate them to submit their best innovative idea.

It is quite evident the impact that innovative procurement may have in mitigation and adaptation policies to climate change: innovative procurement is a tool given to the public authorities, at any level, for "pushing" the market in exploring new solutions of solving environmental and/or social constraints (Edler & Georghiou, 2007; Trybus, 2006; Thai & Piga, 2007).

Considering the starting point for achieving the result of innovative solutions, there are different legal instruments that can be adopted in innovative public procurement.

In particular, the most used and common tool among purchasing authorities and European level are public procurement of innovative solutions (PPI) and pre-commercial procurement (PCP) (European Commission, 2021a, b). The first one is aimed at strengthening an already existing innovative solution—for example, at prototype level—which is not yet available on a large-scale commercial basis. The second one is usually adopted when the solution does not exist yet: PCP enables public procurers to compare alternative potential solution approaches in Research and Innovation (R&I) and filter out the best possible solutions that the market can deliver to satisfy the public need.

Considering that innovative procurement can push the market to offer new solutions to face climate change, the European Union has been increasingly promoting this practice in the last decade, also with substantial grants pertaining to the Horizon 2020 package.

An interesting example is an EU-funded project named "4Cities," which brings together six leading European cities looking for artificial intelligence (AI) solutions to accelerate carbon neutrality. In December 2020, AI4Cities launched its international request for tenders to go through a pre-commercial procurement to acquire innovative solutions in the fields of energy and mobility. These solutions should use artificial intelligence or other enabling technologies—such as big data applications, 5G, edge computing, and IoT—and contribute to the reduction of CO_2 emissions in the six cities and regions participating in the project.

As previously anticipated, this tendering procedure gives a very clear example of designing "functions" that are requested to the market in terms of submitting innovative solutions to "fill the gap." The tendering documentation describes the following functions:

- Function 1: CO_2 emissions reduction—The solution has potential to reduce CO_2 emissions in the city where it's deployed.

    - Requirements (must have): 1.1 The solution reduces CO_2 emissions in the field of mobility and/or energy. 1.2 Methodology for measuring the emission reductions.
    - Requirements (nice to have): 1.3 Estimation of how much the solution reduces CO_2 emissions including a detailed description of what that estimation is based on.

A second example that is interesting to present for its social outcomes is the "e-Care" project, which aims at delivering disruptive digital solutions for the prevention and comprehensive management of frailty, through the implementation of a pre-commercial procurement scheme.

From an analysis of the examples given, it emerges that innovative public procurement requires the acquisition of advanced technical expertise before the formal procurement phase in order to assess market difficulties to join the innovative solutions.

Undoubtedly, when the contracting authority lacks such experience in the subject matter of the innovative contract, the preliminary market consultation phase plays the pivotal role of effectively and efficiently qualifying innovative procedures.

In this context, it is interesting to mention the experience on procurement for innovation that was recently auctioned by an Italian municipality located in Sardinia (Municipality of Fonni): the goal of the innovation procurement was the development of new technologies for economic and energy traceability and optimization of the solid waste life cycle. The procurement procedure was anticipated by a preliminary open market consultation phase, which gave to the municipality the opportunity of reaching the following results: innovation and modernization in the management system of public services, increase of the percentage of differentiated waste collection, optimization of the management process and delivery of wastes, increase in the level of satisfaction of the citizen, enhanced transparency and full involvement of the Citizens, and amelioration of the urban decorum.

13.5 Evaluate Suppliers' Performance for a Sustainable Procurement

Public demand has the power of steering the market toward the satisfaction of needs. For this reason, if public procurement addresses the market to offer progressively more sustainable products and services, it can effectively enhance climate mitigation through public procurement policy.

Purchasing authorities can generally award a public contract based on three alternative criteria: the "lowest price" criterion, the "most economically advantageous tender," and the quality criterion at a fixed cost.

13.5.1 The Lowest Price Criterion

Professors Schooner and Markus Speide, in a recent article published by the George Washington University law school, sentenced that "successfully establishing a sustainable procurement regime will require dramatic change, including, among other things, overcoming the persistent tyranny of low price" since procurement professionals are "obsessed" by savings on their annual budgets (Schooner & Speide, 2020).

Starting from the analysis of implications in choosing the lowest price criterion, this mechanism, basically, scores different offers only in terms of savings that the public authority may achieve. This awarding criterion is generally used when the product categories to procure are standardized or do not forecast innovation processes (e.g., printing paper; see Carpineti et al., 2006).

For instance, the purchasing authority choosing to buy a certain commodity (e.g., stationery) can impose to "green" the delivery service of orders by obliging the supplier to use only electric vehicles or bicycles. In addition, it can be requested that products delivered (pens, notebooks, paper for printers, etc.) must possess standards, likely eco-labels and eco-certifications.

The adoption of the lowest price criterion can also guarantee to achieve a minimum level of social procurement. Continuing with the example of the stationery, the buyer can decide, in the tendering procedure designing, to invite only those suppliers certified with SA8000 (see Sect. 13.3.2). It means that only certified suppliers could make an offer.

Additionally, social criteria can be added in the framework contract: the supplier awarding the contract is obliged to monitor if their subcontractors respect the minimal salary conditions of their employees.

Nevertheless, it is necessary to stress that the choice of defining a scoring rule based on price must be accompanied by a sound market analysis (see Sect. 13.6). Without a deep knowledge of the market, the risk for the purchasing authority is to fix compulsory requirements in terms of service-level agreement or in terms of technical specifications, which significantly reduces the number of participants and, thus, the competition level.

If, for instance, the procurement authority designed a participation requirement based on SA8000, and only one firm possessed this specification in the relevant geographical area, during the procurement process, the results would be to receive only one offer replying to the tender, probably with a poor result in terms of discount.

Additionally, it is also possible to consider the case of absence of economic operators able to respond to the tendering procedure, given that the admission criteria are achievable by no one. Under this circumstance, there is an evident loss of efficiency since the purchasing authority is obliged to redesigning the technical specifications and to reopen the competitive phase.

In conclusion, we must conclude that, on the one hand, a competition among suppliers based on price only can certainly guarantee the achievement of some results in terms of green and social procurement since there are compulsory and fixed SLA and technical specifications that the winning supplier is obliged to guarantee. Nevertheless, on the other hand, competitors submitting their offer during the tendering phase are not incentivized to offer higher quality since they are only evaluated in terms of price reduction from the reserve price.

13.5.2 Criteria Based on Evaluation of Quality

The second criteria that can be chosen to compare offers during a tendering procedure are those based on a comparison of price for the quality.

Then, the procurement authority can choose to score offers by mixing technical points and economical points, assigned to competitors on the basis of the quality and the price offered. Continuing with the example of procurement of paper for printers, the buyer could design the tendering procedure by assigning a maximum of 30 points to the best price and a maximum of 70 points to the quality.

Therefore, starting with the analysis of price evaluation, the lowest price obtains 30 points, and worst prices, offered by the other competitors, achieve proportionally less economical points because of a scoring rule defined in the tendering documentation.

For example, the purchasing authority can define the following scoring rule:

PE = 30 points * (lowest price/price offered).

Supplier A offers 30$; supplier B offers 90$; supplier C offers 50$. The lowest price is equal to 30$. So:

- Supplier A achieves the following economical points: 30 points *(30$/30$) = 30 * 1 = 30 points.
- Supplier B achieves the following economical points: 30 points *(30$/90$) = 30 * 0.75 = 10 points.
- Supplier C achieves the following economical points: 30 points *(30$/50$) = 30 * 0.60 = 18 points.

In addition, suppliers are scored based on the quality if the offered product. For example, the purchasing authority designs the following subcriteria to assign the total amount of 70 points:

- 20 of them are achieved by the competitors involving women in their organizational process for delivering the contract at managerial level
- 10 points are assigned to products/services reducing the CO_2 emissions during the life-cycle costing
- 10 points are assigned to suppliers that better implement an effective health and safety system to reduce the risks of workers
- 10 points are assigned to the carbon footprint of the paper
- 10 points are assigned to firms offering products carrying eco-labels and other recognized sustainable certifications
- 10 points are assigned to firms having the SA8000 certification

According to the given example, it is immediately clear that the risk to reduce the competition because of restrictive admission criteria and technical specifications, given by the lowest price criterion, is mitigated by the most economically advantageous offer criterion.

In fact, according to the previous example, the purchasing authority does not oblige suppliers to possess the SA8000 certification in order to participate but gives a prize in terms of technical points to the bidders possessing him. This means that bidders without certification can make an offer, but they will not obtain the 10 points promised to the SA8000.

So, on the one hand, the purchasing authority loses the certainty of a winning certified supplier because it may happen that the winner does not have the certification of interest, but, on the other hand, it can afford the goal of augmenting participation and competition in offering additional quality.

As a consequence of adopting the scoring rule of MEAT, bidders are effectively "pushed" to offer quality instead of simply reducing the price since higher quality means higher number of points.

Coming back to the example of suppliers A, B, and C, we can imagine that, after the evaluation of the purchasing authority,

- A achieves 20 technical points
- B achieves 75 technical points
- C achieves 70 technical points

Adding the technical points to the economical ones, the result is A = 30 economical points + 20 technical points = total score: 50 points; B = 10 economical points + 75 technical points = total score: 85 points. C = 18 economical points + 70 technical points = 88 points.

So, C is the winning supplier since it achieves the highest total score. In contrast, in the scenario of the lowest price the results would have been the opposite, with supplier A winning the contract at the lowest price.

For this reason, the recent legal framework in Europe emphasized to give priority to the adoption of the most economically advantageous tender.

Then, leaving aside the criterion of lower price, it is important to stress that, during the tendering procedure, the comparison of offers largely based on quality aspects is, for purchasing authorities, a strategic tool for promoting environmental and social sustainability, and then for implementing mitigation and adaptation policies to climate change.

In addition, scoring the suppliers not only in terms of price offered, but also in terms of additional quality can incentivize them to invest in innovation instead than in lowering the price.

13.5.3 Quality Criterion at a Fixed Cost

The third criteria are the one based on quality only. This criterion fixes the price in the tendering documentation and reports it into the final contract without any reduction or discount. Therefore, the suppliers are evaluated only in terms of higher quality offered.

Coming back to the example of Sect. 13.5.2, adopting this scoring rule, the supplier C would have been the winner, achieving 75 technical points, against A with 20 points and C with 70 points.

13.5.4 How to Design Evaluation Criteria for Assessing Sustainability

Once defined the choice of adopting criteria 2 and 3, the MEAT and the fixed price, the subsequent strategic choice that should be made by the procuring entity is the definition of items to be evaluated.

The awarding criteria shall be set on the basis of two important features:

- Items' objective and indexes of an effective competition among suppliers
- Items closely related to the object of the contract in order to ensure conditions of transparency, nondiscrimination, and equal treatment

For this reason, it is crucial to point out that awarding criteria based on evaluation of quality are necessarily linked to the evaluation of LCC (see Sect. 13.2.2).

As an example, the purchasing authority wishes to move from fossil fuel electricity consumption to electricity 100% based on renewable primary energy. Public procurement strategy offers two solutions to solve the problem. The first one is based on the lowest price. So, the tendering procedure defines as a minimum quality requested 100% of renewable electricity and suppliers are forbidden to offer electricity based on fossil fuel energy. But we already said that this solution can be chosen only if the market is ready to reply to the request. If we assume, hypothetically, that the market is not ready to offer this product, the solution is the MEAT or the fixed price criterion. This new strategic choice elicits the suppliers to win the contract to offer renewable energy.

13.6 Preliminary Market Consultation

Before launching any tendering procedure intended for asking the market to concur under certain specifications and tendering rules, it is crucial that public procurements have a clear and sound knowledge of the needs that must be satisfied and of the potentially interested suppliers. Indeed, before launching the official tendering procedure, it is strongly recommended to open a preliminary market consultation.

Undeniably, throughout the preliminary market consultation, public buyers get a better knowledge of the market structure, the available technologies and services, and, especially, the innovation level of the market in satisfying specific needs linked to sustainable and social procurement.

As an example, moving back to the designing strategy of designing technical specifications to be fixed, it is imperative to know if a determined parameter is achievable only by a restricted number of suppliers, or, gravely, by a unique competitor. In fact, the public purchaser, in fixing a specification possessed by a single supplier within the relevant market as entry access, automatically generates a monopolistic state.

On the other side, the only supplier that is able to meet the criterion can rationally choose not to offer a discount to the purchasing authority.

In order to avoid the risk of designing a procurement strategy that reduces competition and/or does not answer to the needs of the contracting authority, the preliminary market consultation assesses whether the preliminary strategy of procurers elaborated on the basis of needs analysis only is reasonable for the market under a technical and economic point of view. More precisely, a sound and well-designed preliminary market consultation can guarantee the assessment of an effective competition in the market, with an adequate number of firms responding to the desired requisites fixed by the public buyer.

The adoption of a preliminary market consultation phase is particularly necessary and recommended when the purchasing authority must launch complex public procurement requests for quotations.

In fact, the importance of a preliminary market consultation is strategic in order to assess if the market already elaborated solutions that are sufficient to satisfy the needs of the purchasing authority. In other words, to launch a complex tendering procedure without a previous interaction with the market may generate negative effects in terms of loss of direct and indirect costs.

Under the EU Directive on public procurement, it is necessary to keep in mind that the preliminary market consultation must guarantee the respect of transparency and fair competition principles. In fact, Art. 40 of the EU Directive n. 24/2014 states that "before launching a procurement procedure, contracting authorities may conduct market consultations with a view to preparing the procurement and informing economic operators of their procurement plans and requirements. For this purpose, contracting authorities may for example seek or accept advice from independent experts or authorities or from market participants. That advice may be used in the planning and conduct of the procurement procedure, provided that such advice does not have the effect of distorting competition and does not result in a violation of the principles of non-discrimination and transparency."

This result can be operationally achieved by putting in action the following methods:

- To set up online platforms for communicating the initiative to the market and guarantee transparency principles
- To adopt desk-based contacts, usually in the form of a questionnaire for economic operators
- To do telephone interviews, based on standard questions, possibly conducted by a third party to ensure equal treatment
- To organize live events at which economic operators are invited to participate
- To report activities on the results of market consultation to the website in order to guarantee transparency

13.7 Abating Barriers with Social Procurement

Aiming at maximizing the best value for money, public procurement has an implicit conflict with social and labors goals: internalizing costs of decent work conditions and inclusion of disadvantaged categories may "damage" the economic efficiency of the procurement process, and, then, it may increase costs and reduce competitiveness in the market. The next sections explore how disadvantaged categories may be beneficiary of a sustainable procurement practice (Kashiwagi & Savicky, 2003; Storteboom et al., 2017).

13.7.1 Small- and Medium-Size Enterprises

Small- and medium-size enterprises are universally recognized as a "disadvantaged group" with regard to access to public contracts. Only in Europe SMEs represent 99% of all businesses in the European Union.

The trend is similar in the United States, where small businesses consist of two-thirds of all new jobs in recent decades. Small businesses account for 98% of all identified US exporters and support nearly 4 million jobs in communities across America through both direct and indirect exports.

Consequently, the European Union and the United States are strengthening the cooperation to enhance the participation of SMEs in trade between the United States and the European Union.

Once again, public procurement, in terms of sustainability, can do its part in supporting SMEs to access public contracts. According to the present scenario, we can argue that the United States set up a legal framework that surely incentivizes SMEs to negotiate with the public sector. The most common program is known as the Small Business Administration Act, (Office of the United States Trade Representative-Executive Office of the President, 2021). Enhanced in 1953, and that, as we will see later on, over the years, extended the concept of supporting small business as prime contractors and subcontractors, also to women-owned small businesses, small disadvantaged businesses, service-disabled veteran-owned small businesses, and small businesses located in historically underutilized business zones.

In practical terms, the federal government encourages small businesses to take on contracting opportunities, and each federal agency has a statutory annual goal for awarding at least 23% of prime contracts for small businesses.

Unfortunately, at the present date, the legal European framework on public procurement in favor of the SMEs is still at an embryonal level if compared to the SBA. Even if the European Directives consider SMEs crucial for the economic sector, there are no effective solutions to reserve contracts to SMEs. At the moment, purchasing authorities are recommended to split their contracts into lots in order to enhance the access to those contracts to SMEs.

The EU Directive n. 2/2014 states that "Member States should remain free to go further in their efforts to facilitate the involvement of SMEs in the public procurement market, by extending the scope of the obligation to consider the appropriateness of dividing contracts into lots to smaller contracts, by requiring contracting authorities to provide a justification for a decision not to divide contracts into lots or by rendering a division into lots obligatory under certain conditions. With the same purpose, Member States should also be free to provide mechanisms for direct payments to subcontractors" (see Whereas n. 78).

However, this legal framework does not force the public sector to reserve a portion of contract exclusively to SMEs, as has been done in the United States for decades.

13.7.2 Gender Equality

As already discussed, public procurement policies can enhance sustainability and equity toward disadvantaged groups, such as the access of women to business. Taking into account the 17 Sustainable Development Goals (SDGs), fixed in 2015 by the United Nations, Goal no. 5 is addressed to gender equality (United Nations Department of Economic and Social Affairs, 2021): "Achieve gender equality and empower all women and girls": "Gender inequalities are still deep-rooted in every society. Women suffer from lack of access to decent work and face occupational segregation and gender wage gaps. In many situations, they are denied access to basic education and health care and are victims of violence and discrimination. They are under-represented in political and economic decision-making processes."

Indeed, the international community unanimously recognizes that women, with respect to male counterparts, regularly encounter supplementary difficulties because of social, economic, cultural, and/or legal discrimination against women.

Data retrieved from the United Nation website show that

- 35% of all SMEs are owned by women
- Women-owned SMEs produce around 20% of GDP
- 8–10 million of formal SMEs in emerging markets are owned by women, which represents 31–38% of all formal SMEs in those markets
- If women played an identical role in labor markets to that of men, as much as USD 28 trillion, or 26%, could be added to the global annual GDP by 2025

Public procurement can, once again, play an effective and active role in promoting gender equality since, through appropriate evaluation criteria, it is possible to evaluate suppliers guaranteeing gender equality.

Nevertheless, this is not an ordinary strategy, and unfortunately, only recommendations from politics and few case studies are available.

As a primary problem, public procurers must face the so-called "tokenism practice": firms apparently give power to women within a business organization only as a façade and give the appearance that women are being treated fairly, but without accompanying this apparent power with effective tools, such as adequate salary, real responsibilities, and authority within the organization.

As an effect, the purchasing authority must evaluate suppliers not only in terms of a higher number of women involved in the business with managerial responsibilities, but also verifying and assessing if the salary is adequate. As a result, public buyers should carefully request to competitors adequate proofs, such as the payroll of employees, the organizational model, and the official business procedures, confirming the effectiveness of gender equality, with the risk of overlapping problems of privacy and of incurring in litigation with the suppliers (Corvaglia, 2017; Chin, 2017).

Another example of interest, in order to enhance gender equality via public procurement, is the evaluation of firms in terms of "woman-owned business." (U.S. Small Business Administration, 2021b). This recent approach was adopted by the United Nation, which defines a woman-owned business as a vendor, which includes

- At least 51% independent ownership by one or more women (or woman sole proprietorship)
- Unconditional control by one or more women over both long-term decision-making and the day-to-day management and administration of the business operations
- Independence from non-women-owned businesses

Indeed, the public procurement sector can overcome the barriers of gender equality by evaluating vendors in terms of higher involvement of woman-owned business in their supply chain (based on the previous definition).

In the United States, under the Small Business Act (see Sect. 13.7.1), there are also projects aimed at encouraging women-owned small businesses (WOSBs) to compete for federal contract: under the WOSB Federal Contracting Program, the federal government wants to achieve the goal of awarding each year at least 5% of all federal contracting dollars to women-owned small businesses. Under this program, eligible firms achieve support and training as well as access to a loan program.

Moreover, the program establishes an "independent certification" to achieve the status of WOSBs, which guarantees the access to specific procurement categories (U.S. Small Business Administration, 2021a). As a result, the federal government as well as the Small Business Act limits competition to certain contracts to businesses that participate in the WOSB Federal Contracting Program. The contracts are identified among those specific industries where women-owned small businesses (WOSBs) are underrepresented, like building construction, support activities for forestry, structural steel and precast concrete, plumbing, heating, and air-conditioning.

At European level, as well as for the Small Business Act case parallel, there is not an effective legal framework guaranteeing the access of women to procurement. At the moment, the European Institute for Gender Equality (European Institute for Gender Equality, 2019) was set up and gives some recommendations to contracting authorities in order to enhance gender equality in terms of

- A gender-balanced composition of the project team and beneficiaries
- The balanced presence of women and men in decision-making positions

13.7.3 Young Professionals

Young professionals can also be considered a disadvantageous category, in the context of public contracts, since usually purchasing authorities select suppliers also because of past performances and of similar previous experiences.

An example can be taken from the Italian legal framework, which as far as the engineering and architecture services are concerned imposes to the providers to include at least a young professional in the team, defined as a professional achieving the authorization to signing technical projects less than 5 years from the publication of the tendering procedure.

13.7.4 Other Minorities

Besides SMEs, women, and young professionals, public procurement can also enhance sustainability by protecting and favoring other groups, allowing them additional opportunities to be awarded public contracts.

Relevant examples and good practices are coming from the private sector, where there is an existing network with the purpose of supporting minorities at the business level. For instance, since in the United Kingdom over 300,000 ethnic minority-owned businesses (EMBs) represent over 7% of all SMEs, the UK MSDUK (MSDUK, 2021) has the goal of bringing together innovative and high growth ethnic minority-owned businesses with global corporations committed toward creating an inclusive and diverse supply chain.

Another interesting example comes from the South Africa Supplier Diversity Council (SASDC, 2021), a corporate-led initiative, bringing together private sector companies to promote supplier diversity as a business strategy, to achieve competitiveness and long-term sustainability, especially by supporting "black suppliers," suffering for the cultural heritage of the apartheid history.

SASDC has definitely the goal of developing and growing black SMEs with the ultimate goal of including them into their supply chain in a sustainable way by improving their performances and their business capability.

Similarly, looking at the public sector, the South African government recently implemented grants to black business suppliers in order to improve their competitiveness and sustainability.

Then, the so-called Broad-Based Black Economic Empowerment (BBBEE) set some rules for enhancing the access of black businesses to South African public contracts. The BBBEE is an integration program launched in 2003 by the South African government to reconcile South Africans and addresses the inequalities of apartheid by attempting to compensate for land that was repossessed from Africans. The program encourages businesses to integrate black people in the workspace, support black businesses, and give back to poor black communities affected by land repossession. Businesses are awarded points, which they can claim on a BBBEE certificate that entitles them to a greater chance of obtaining government contracts. Under the BBE, certain industries require an entity to have a particular percentage of black ownership or BEE level in order to receive and maintain a license to operate.

13.7.5 *Command and Control: The Case of Switzerland*

The government of Switzerland is applying a tangible command and control policy for forcing sustainable public procurement: the Confederation is obliged to award contracts for services only to companies guaranteeing compliance with the Federal Act on Public Procurement (Federal Office for Gender Equality, 2021). This covers working conditions, industrial safety regulations, and equal pay for men and women.

At the same time, the federal government has implemented a monitoring system aimed at verifying if the rules are respected by vendors, with audit mechanisms, penalties, and sanctions, having the effect of excluding bidders from the tendering procedure or solving the contracts.

13.8 Centralization Versus Decentralization: Opportunities for Sustainability

Public procurement can be implemented at a local, central or international level, depending on the type of purchasing authority and of the end users of the product/service or work being procured.

Nowadays, it is very rare to observe completely centralized procurement experiences or completely decentralized ones. In contrast, there are, predominantly, hybrid models of centralization/decentralization, in which single purchasing authorities group together in order to satisfy their needs in one shot. Alternatively, they could delegate a third legal entity to solve the role or award the tendering procedure at their place, and they sign the final contracts with the winning supplier. In this second case, usually there are central public procurement authorities awarding framework contracts addressed to satisfy needs of a group of local authorities.

The most extreme form of centralized procurement is the so-called "cross-border procurement": purchasing authorities from different countries bundle their needs for pursuing a unique supranational tendering procedure and achieve a unique supplier. An example of this cross-border procurement is provided in Sect. 13.4, where the experience of a group of cities launching together a pre-commercial procurement is described.

Fixing the appropriate level of centralization versus decentralization is a challenging aspect of procurement strategy designing, which has positive and negative aspects.

Primary, central public procurement procedures may achieve higher performances in terms of economies of scale. Central public procurement tendering procedures are bundling single needs and the request for quotation is launched at higher monetary value, which is the sum of the single needs behind the centralized tendering procedure. As a consequence, competing suppliers can have the chance to increase profits by awarding a single tendering procedure instead of competing for several single procurement processes. So, they are stimulated to be more aggressive in terms of discount and quality for awarding the contract.

For this reason, considering the context of public procurement for innovation, centralization of public procurement can play a primary role, since it may easily attract the interest of the market for pursuing innovative goals in terms of sustainability.

Secondly, centralization can assure a higher level of transparency since centralized contracts are usually auctioned via open tendering procedure in which any interested supplier may submit an offer.

A third and most important advantage that is recognized to central public procurement—considering the context of this book—is the warranty to establish standards for the specific commodity that is procured.

Assuming to be a central purchasing body launching a tendering procedure for the acquisition of laptops, which will be sold to several local authorities within a territory. The consequence is to guarantee laptops with the same level of performance from north to south of the territory.

Contextualizing this positive effect of centralization into sustainability, the adoption of centralized procurement can accelerate and multiply the adoption of green labeling and environmental standards for the inclusion of green specification and sustainable SLA into framework contracts.

Conversely, central public procurement is "accused" of disorienting SMEs in participating and awarding centralized contracts since they do not have the technical and fiscal capacity for submitting offers.

Moreover, considering the negative effects of centralized public procurement, in the case in which a specific commodity must be tailored according to the specific needs, a unique procurement procedure for the awarding of the same standard may risk missing the goal of satisfying needs of authorities' beneficiary of the tendering procedure as end user of the commodity. In this case, the goods or service risk to be unsuitable and to generate, instead of saving for economies of scale, an effect of overspending.

In conclusion, the centralization of public procurement can be considered a strategic tool for promoting sustainability, but it should be carefully evaluated with respect to the commodity to be acquired, also in terms of innovation, and with respect to end users of the commodity.

13.9 The Role of e-Procurement for Promoting Sustainability

e-Procurement is defined as the procurement that is performed online through a web interface and other networked systems. e-Procurement moves the paper-based and traditional procedure to online systems, and it may include spending analysis, sourcing strategies, e-auctions, procure-to-pay, and contract management. So, interactions and transactions between buyers and suppliers are completely automated.

At the same level, public e-procurement combines the general principles of transparency, fair competition, and efficiency into e-public procurement platforms. According to the OECD, e-procurement does not only increase efficiency by facilitating to the market the supply to public tenders, but it also improves transparency by holding public authorities more accountable. In the context of sustainability, e-procurement is a strategic toolbox given to purchasing authorities to promote sustainability.

Preliminarily, public authorities that set an electronic vendor rating and an e-catalogue, both based on the sustainability of products and suppliers (e.g., certifications, labels, etc.), can achieve their sustainability goals straightforwardly and in a more transparent way.

Moreover, suppliers interacting with the public sector with the public e-procurement tools know faster and more easily need to be satisfied and can have a preliminary idea of sustainability strategies adopted.

Additionally, as an indirect effect, public e-procurement achieves faster results in terms of process management of the awarding of contracts and of the evaluation of suppliers since it completely dematerializes the traditional paper-based procurement procedures.

Consequently, the purchasing authorities implementing e-procurement can also achieve savings in the CO_2 emissions linked to the reduction of energy consumption and paper and of storage spaces.

Last but not least, an effective vendor rating system in which suppliers are classified on the basis of their sustainability performances can afford an objectively monitoring and evaluation method on the basis of objective sustainable criteria.

Sustainable indexes applied to e-procurement platforms in ranking suppliers also create a virtuous circle, enhancing suppliers to be more conscious and to better performance in terms of resource consumptions, eco-designing of products, and innovation in green technology in order to award additional contracts with the purchasing authority. An efficient vendor rating can also promote sustainable suppliers like faster interactions with SMEs, women-owned firms, and disadvantaged categories.

13.10 Monitoring and Performing Contracts: The Real Challenge for an Effective Sustainable Procurement

Any project, program, and monitoring system aimed at progressively evaluating the achieved results is essential to measure effective progress toward any organization and to make adjustments where improvement is required.

With a particular focus on public procurement, evaluation and monitoring of performances is a crucial point in order to achieve several results: first, to check if promises made by suppliers during the negotiation phase to get the contract are really respected; second, with a focus on sustainable procurement, to measure if the likelihood of sustainability is respected.

More importantly, when analyzing the performances in terms of sustainable procurement, the most difficult challenge that must be faced by the procurer (both public and private) is to monitor the performances of the whole supply chain. Who can guarantee that the same working conditions are equally granted both by the final producer and by the raw materials extractor? The already-mentioned case of the coltan extraction in Congo without any respect of salaries and children is internationally renowned and was often denounced by international NGOs (Tsongo, 2020).

Nevertheless, for purchasing authorities, often without appropriate financial and human resources, it is simply unbearable to monitor the whole supply chain of their vendors. In this context, certification systems can play an important role in requiring certifications also for the members of the supply chain.

13.11 Concluding Remarks: Which Suggestions to Policy Makers?

This chapter demonstrated how public procurement could have a tangible effect on mitigation policies to climate change. Surely, public procurement plays a substantial role in this historical moment: under the Next Generation EU funding program, more than 50% of the amount will support modernization, through fighting climate change, promoting research and innovation, preparedness, recovery and resilience, biodiversity protection, and gender equality. It is evident that these results will be achieved only if the public sectors will implement efficient strategies in sustainability.

In fact, the adoption of procurement procedures for selecting suppliers focused on qualitative criteria based on the reduction of pollution of commodities acquired and on the social promotion belonging to the supply chain can effectively enhance sustainability. We also analyzed that standards, likely certification systems, and labels managed by the market (e.g., ISO 14001) or promoted by public authorities (e.g., ecolabel) associated with a sound monitoring system can effectively contribute to enhancing the green economy of public contracts.

Another tool that could be used for launching green transition is the procurement of innovative solutions. So, policy makers probably should invest more in training systems for innovation procurement, and, probably, the centralization mechanisms of innovation procurement could achieve this result.

Moreover, the lack of competences and of financial resources is also challenging when considering the importance of monitoring the performances of suppliers during the execution phase in terms of sustainability. For this reason, input to policy makers could be to set up an independent office or society, under the public control, having the mission of monitoring randomly suppliers working with the public sector in order to assess if the respect of social procurement is guaranteed on the whole supply chain. It certainly requires an innovative political view and huge investments in terms of technical and human resources, but an effective green new deal cannot be achieved only with promises of suppliers but also with the evaluation of facts.

In conclusion, we learned that, while green procurement is at an advanced stage, since contracting authorities have several standardized tools to evaluate and monitor green quality and environmental impact of commodities, sustainable procurement is still at an embryonal phase. In fact, investigating good practices in favoring disadvantaged categories through public procurement, apart from the international certification SA8000 and the ISO 45001, uniquely some experiences in the United States and South Africa, seems to be effectively giving a clear legal framework at governmental level to their purchasing authorities.

Indeed, the WOSB Federal Contracting Program, promoted in the United States, guarantees a certified system to make it eligible to award public contracts to women-owned small businesses. Looking at the South African government, the BBE initiative aims to concretely create a governmental mechanism to sustain black-owned firms accessing public procurement.

In contrast, in Europe it seems that, while green procurement is at a mature phase, apart from recommendations and good practices launched independently by "virtuous purchasing authorities," there is not an effective compulsory obligation given to the public sector for applying social procurement.

Nevertheless, the most important challenge that public governments and policy makers must face is to apply command and control policies to impose sustainable prices to the market and to impose a regulatory approach to integrate sustainability. It is meant to internalize into the final cost, by a correct LCC analysis, the negative externalities given by polluting products, unfair competition due to slavery, and decent work condition, while protecting and safeguarding minorities, likely SMEs, women, young professionals, etc. Indeed, policy makers would take in place a strict legal framework fixing new entry barriers to suppliers willing to sign public contracts in order to enhance a "360° sustainable procurement." In fact, the experience suggests that the market alone is not able to abate barriers and a more ambitious legal framework, probably, could reach faster and better, actions of filling the gap.

We strongly hope that the good practices illustrated in this chapter for enhancing an effective social procurement at a large scale would amplify the debate on this topic and that the next EU Directive on public procurement will be sharpened on that. We also hope that a real sustainable policy will be taken into account by national governments in the Next Generation EU funds in order to definitively enforce the transition from a simple "green public procurement" to an effective sustainable procurement, considering also those social aspects that could eliminate barriers of disadvantaged categories and minorities.

Last but not least, climate change is showing us that single initiatives are not sufficient. For this reason, the last suggestion that could be given to policy makers is to build up a community for sustainable procurement working at the international level that is aimed at, on the one hand, setting legal frameworks and monitoring systems addressed to sustainability and, on the other hand, to train and educate public buyers to a revolutionary of "sustainable procurement."

References

Arrowsmith, S., & Kunzlik, P. (2009). *Social and environmental policies in EC procurement law: New directives and new directions.* Cambridge University Press.

Arrowsmith, S., et al. (2000). *Regulating public procurement. National and international perspective.* Kluwer Law International.

Carpineti, L., Piga, G., & Zanza, M. (2006). *Benchmarking european public procurement practices: Purchasing of 'Fix-Line Telephone Services' and 'Paper for Printers'.* Available via SSRN. Retrieved February 15, 2021, from, https://papers.ssrn.com/sol3/papers.cfm?abstract_id=934504

Chartered Institute of Procurement and Supply. (2020). What is a supply chain? *CIPS.* Retrieved February 15, 2021, from https://www.cips.org/knowledge/procurement-topics-and-skills/supply-chain-management/what-is-a-supply-chain/

Chin, K. (2017). *The power of procurement: how to source from women owned businesses.* Available via UN Women. Retrieved January 12, 2021, from https://www.unwomen.org/en/digital-library/publications/2017/3/the-power-of-procurement

Corvaglia, M. A. (2017). *Public procurement and labour rights. Toward coherence in international instruments of procurement regulation* (pp. 1–26). Bloomsbury Publishing.

De Grauwe, P. (2017). *The limits of the market. The pendulum between the government and the market* (p. 1). Oxford Scholarship Online.

Edler, J., & Georghiou, L. (2007). Public Procurement and innovation—Resurrecting the demand side. *Research Policy, 36*(7), 949–963.

Edquist, C., Hommen, L., & Tsipouri, L. (2000). *Public technology procurement and innovation* (pp. 2–16). Kluwer Academic.

Estevan, H., & Schaefer, B. (2017). Life cycle costing state of the art report. Available via ICLEI. Retrieved December 17, 2021, from https://sppregions.eu/fileadmin/user_upload/Life_Cycle_Costing_SoA_Report.pdf

Ethics & Sustainability in Procurement. (2019). *Available via procurement business skills.* Retrieved January 15, 2021, from https://www.procurementbusinessskills.com/blog/services/ethics-sustainability-in-procurement/

European Commission. (2019). Communication from the Commission to the European Parliament, the European Council, the Council, the European Economic and Social Committee and the Committee of the Regions, the European Green Deal, COM(2019) 640 final, Brussels, European Commission. Available via European Commission. Retrieved January 15, 2021, from https://eur-lex.europa.eu/legal-content/EN/LSU/?uri=COM:2019:640:FIN

European Commission. (2021a). Public Procurement of innovative solutions. Available via European Commission. Retrieved January 15, 2021, from https://ec.europa.eu/digital-single-market/en/public-procurement-innovative-solution

European Commission. (2021b). Life cycle costing. Available via European Commission. Retrieved January 14, 2021, from https://ec.europa.eu/environment/gpp/lcc.htm

European Commission. (2021c) Recovery plan for Europe. Available via European Commission. Retrieved January 20, 2021, from https://ec.europa.eu/info/strategy/recovery-plan-europe_en

European Institution for Gender Equality. (2019). Gender procurement. Available via EIGE. Retrieved January 20, 2021, from https://eige.europa.eu/gender-mainstreaming/methods-tools/gender-procurement

Federal Office for Gender Equality. (2021). *Government controls in public procurement.* Available via EBG. Retrieved March 10, 2021, from https://www.ebg.admin.ch/ebg/en/home/topics/work/equal-pay/government-controls-in-public-procurement.html

Fontana S, Menato A, Barni A (2020) A methodology to integrate sustainability evaluations into vendor rating. *Paper presented at ifip international conference on advances in production management systems*, Novi Sad, Serbia, 30 August–3 September 2020.

Gluch, P., & Baumann, H. (2004). The life cycle costing approach: a conceptual discussion of its usefulness for environmental decision-making. *Building and Environment, 39,* 571–580.

ISO 14000. (2021). *Family environmental management.* Retrieved February 17, 2021, from https://www.iso.org/iso-14001-environmental-management.html

ISO 45001:2018. (2021). *Occupational health and safety management systems—Requirements with guidance for use.* Retrieved February 17, 2021, from https://www.iso.org/standard/63787.html

Italian Antitrust Authority. (2021). *Rating di Legalità.* Retrieved March 12, 2021, from https://www.agcm.it/competenze/rating-di-legalita/

Kashiwagi, D., & Savicky, J. (2003). The cost of 'best value' construction. *Journal of Facilities Management, 2,* 285–297.

Kelman, S. (1990). *Procurement and public management: The fear of discretion and the quality of government performance* (Vol. 10, pp. 490–493). AEI Press.

Klinger, D. U. (2020). *Measuring what matters in public procurement law: Efficiency, quality and more.* Yale Law School.

La Chimia, A., & Trepte, P. (2019). *Public procurement and aid effectiveness: A roadmap under construction* (pp. 22–27). Bloomsbury Publishing PLC.

Lee, G. (2021). *The state of circular economy in America: Trends, opportunities, and challenges.* Via Circular COLab. Retrieved March 15, 2021, from https://static1.squarespace.com/static/5a6ca9a2f14aa140556104c0/t/5c7e8c8de5e5f08439d71784/1551797396989/US+CE+Study_FINAL_2018.pdf

MSDUK. (2021). *About minority supplier development UK.* Retrieved January 12, 2021, from https://www.msduk.org.uk/about-us

Office of the United States Trade Representative-Executive Office of the President. (2021). *Small businesses.* Retrieved February 10, 2021, from https://ustr.gov/issue-areas/small-business

Perera, O., Morton, B., & Perfrement, T. (2009) *Life cycle costing in sustainable public procurement: A question of value.* Available via IISD. Retrieved March 18, 2021, from https://www.iisd.org/system/files/publications/life_cycle_costing.pdf

SA8000. (2021). *About SA8000.* Available via SA800. Retrieved March 1, 2021, from https://sa-intl.org/programs/sa8000/

SASDC. (2021). *The South African supplier diversity council.* Available via SASDC. Retrieved January 9, 2021, from https://www.sasdc.org.za/index.php/about-us

Schooner, S. L., & Speide, M. (2020). Warming up to sustainable procurement. GW law school public law and legal theory paper no. 2020-70. GW Legal Studies Research Paper No. 2020-70.

Storteboom, A., et al. (2017). Best value procurement—The practical approach in The Netherlands. *Procedia Computer Science, 121*, 398–406.

Thai, K. V., & Piga, G. (2007). Advancing public procurement: Practices, innovation and knowledge sharing. *Journal of Public Procurement, 7*, 280–286.

Trepte, P. (2004). *Regulating procurement: Understanding the ends and means of public procurement regulation.* Oxford (pp. 7–9).

Trybus, M. (2006). Improving the efficiency of the public procurement systems in the context of the European Union enlargement process. *Public Contract Law Journal, 35*(3), 409–425.

Tsongo, E. (2020). *Children in the democratic Republic of Congo mine for coltan and face abuse to supply smartphone industry.* Available via ABC. Retrieved January 9, 2021, from https://www.abc.net.au/news/2020-03-01/tech-companies-rely-child-labour-abuse-to-mine-coltan-in-congo/11855258

U.S. Small Business Administration. (2021a). *Qualifying NAICS for the women-owned small business federal contracting program.* Available via SBA. Retrieved April 9, 2021, from https://www.sba.gov/document/support%2D%2Dqualifying-naics-women-owned-small-business-federal-contracting-program

U.S. Small Business Administration. (2021b). *Women-owned small business federal contracting program.* Available via SBA. Retrieved April 9, 2021, from https://www.sba.gov/federal-contracting/contracting-assistance-programs/women-owned-small-business-federal-contracting-program

UN. (2021). *What is sustainable procurement?* Sustainable Procurement Tools Via UN. Retrieved February 8, 2021, from https://www.ungm.org/Shared/KnowledgeCenter/Pages/PT_SUST

United Nations Department of Economic and Social Affairs. (2021). *Achieve gender equality and empower all women and girls.* Available via UN. Retrieved January 8, 2021, from https://sdgs.un.org/goals/goal5

Valaguzza, S. (2016). *Sustainable Development in Public Contracts. An example of Strategic Regulation.* Editoriale Scientifica.

Valaguzza, S. (2018). *Procuring for Value. Governare per contratto creare valore attraverso i contratti pubblici.* Editoriale Scientifica.

Laura Carpineti is BD in Economics and CIPS Graduated in Public Procurement. Her fields of expertise are strategic and sustainable procurement, including demand, stakeholders, and market analysis within a risk management framework for optimization of procurement effort. She is currently senior manager at Martino & Partners, consulting company on public procurement. After 12 years of experience in centralized procurement at national and regional level (Consip ltd and Lombardy Region), she was head of procurement and contracts division at CAP Holding ltd, one of the biggest Italian public companies in treating the water life cycle in the metropolitan area of Milan. She collaborates occasionally with the universities of Rome Tor Vergata and Macerata (Italy) in teaching procurement strategies. Co-author of the chapter with 2 other colleagues: G. Piga and M. Zanza of a chapter in in N. Dimitri, G. Piga, and G. Spagnolo (Eds.), Handbook of Procurement (pp. 14–44), Cambridge, UK: Cambridge University Press. She firstly introduced in Italy gender procurement mechanisms for comparing suppliers.

Chapter 14
Climate, Sustainability, and Waste: EU and US Regulatory Approaches Compared

Madeline June Kass

Abstract This chapter examines the interrelated topics of climate change, sustainability, and waste by comparing the regulatory approaches of the European Union (EU) and the United States (US). Americans and Europeans produce vast quantities of nonhazardous, unwanted, and unusable material. In the EU, individual waste contributions are almost double the global average. In the US, the numbers are even higher, constituting more than four times the global average. The interrelationship between US and EU regulation of materials and waste and the problem of climate change centers on greenhouse gas-emitting activities at various points along the product to waste continuum (including natural resource extraction, goods production and use, recycling and reuse, waste collection and disposal, and materials transport). The US mostly follows a traditional linear approach to waste management. In contrast, the EU embraces and is moving towards a circular model of materials management. This chapter describes and compares both approaches along with their implications for climate change and sustainability.

14.1 Introduction

This chapter considers the interrelated topics of climate change, sustainability, and waste by comparing approaches for solid waste regulation in the European Union (EU) and the United States (US). Narrowly conceived, *waste regulation* refers to legal requirements for management of unwanted or unusable material, typically from its collection to its disposal. The nature of the EU and US approaches, however, encompasses a wider spectrum of laws and policies. Broadly conceived,

---

M. J. Kass (✉)
Seattle University School of Law, Seattle, WA, USA
e-mail: kassm@seattleu.edu

© The Author(s), under exclusive license to Springer Nature Switzerland AG 2022
S. Valaguzza, M. A. Hughes (eds.), *Interdisciplinary Approaches to Climate Change for Sustainable Growth*, Natural Resource Management and Policy 47,
https://doi.org/10.1007/978-3-030-87564-0_14

> **Important Note on Terminology and Data**
>
> The word waste can refer to a multitude of unwanted or discarded substances, including solid materials (such as abandoned packaging, spent batteries, or discarded clothing) and liquids (such as used household cleaning fluids, discarded chemicals, or stormwater). Additionally, waste may be categorized as hazardous or nonhazardous, or classified according to origin (originating from consumers, construction, or industry). Legal definitions of waste are similarly myriad, differentiated, and confounding.
>
> Due to the wide variation in the types and definitions of waste, unless specifically noted, this chapter uses the term waste in a nonlegal, nontechnical manner to refer to commonly understood forms of nonhazardous solid waste generated by consumers—what you ordinarily think of as trash, garbage, or rubbish (often referred to as municipal solid waste or MSW).
>
> Also, due differences in data collection methods, data availability, and measurement conventions, solid waste data, including the data contained this chapter, may not be fully comparable and should be considered with a degree of prudence.
>
> Lastly, throughout this chapter, the phrase waste regulation will be used when the discussion focuses narrowly on regulatory approaches for disposal. The phrase materials regulation will be used when the discussion examines regulatory approaches to waste management more expansively.

waste regulation refers not only to management of product recycling, recovery, or disposal, but also materials extraction, design, production, transport, use, and reuse. Accordingly, another term for nonhazardous waste management is *materials regulation*.

14.2 Linking Climate, Sustainability, and Waste

14.2.1 Waste by the Numbers

Americans and Europeans produce vast quantities of nonhazardous, unwanted and unusable material. According to data published in 2018, the world produces approximately 2 billion metric tons of waste every year, which comes out to roughly 0.74 kg (close to the weight of two soccer balls) per person every day. In the EU, individual waste contributions increase to approximately 1.4 kg (almost four soccer balls) per person per day, almost double the global average. The US numbers rise even higher, averaging at about 3.2 kg (about eight soccer balls) for every American per day, or more than four times the global average.

US and EU consumers use much of this material only briefly before discarding it. For example, consumers dispose of packaging materials (such as fast-food containers, snack wrappers, plastic water bottles, and the mailing envelopes for stuff bought online) almost immediately. In the EU and the US, such packaging comprises between 30 and 40% of total waste. Food and clothing items tend to remain in use somewhat longer than packaging materials, but have shorter lifespans compared to durable goods (such as cars, appliances, and furniture).

Researchers estimate that by 2050 global generation of waste could reach 3.4 billion metric tons annually. Expressed more direly, by 2050 our oceans may contain more plastic than fish. All of this waste can and does pose environmental challenges, including contributing to climate change.

14.2.2 *The Interconnection Between Climate Change, Waste, and Sustainability*

The connections between waste, climate change, and sustainability exist within a societal web of materials use and environmental, energy, and economic impacts. In the environmental realm, our use of materials contributes to many problems, including land contamination and habitat degradation, air pollution, freshwater and ocean contamination, harm to wildlife and marine life, and climate change. As explained in a report issued by the US Environmental Protection Agency (US EPA), "Our use of materials now challenges the capacity of the Earth—air, water and land—to withstand the many resulting environmental problems. This situation fundamentally affects many other aspects of our future, such as the economy, energy and climate."

Scientists attribute climate change to human activities that emit greenhouse gases (GHG) into the atmosphere. Based on the work of more than 1000 independent scientific experts from around the world, the Intergovernmental Panel on Climate Change (IPCC) concluded that there is a greater than 95% probability that human activities have warmed our planet. The IPCC also found that "[c]ontinued emission of greenhouse gases will cause further warming and long-lasting changes in all components of the climate system, increasing the likelihood of severe, pervasive and irreversible impacts for people and ecosystems."

The interrelationship between climate change and waste centers on GHG-emitting activities at various points along the product to waste continuum. These activities include natural resource extraction and processing, goods production and use, waste collection and disposal, and materials transport. Resource extraction is carbon-intensive because mining of materials requires energy. Much of the energy needed comes from burning of fossil fuels, which release GHG into the atmosphere contributing to climate change. Extraction and transportation also require land clearing for site access. Land clearing may require removal of trees, plants, and soil (so-called carbon sinks) that would otherwise capture and store carbon from the atmosphere. Next, processing of raw materials, manufacturing of goods, and transportation of new products require fossil fuels for energy and release GHG to the atmosphere.

Climate change impacts continue once goods are no longer wanted or useful. Waste incineration, the burning of rubbish, emits carbon dioxide to the atmosphere. Organic waste decomposing in landfills generates carbon dioxide along with methane, an even more potent GHG. Land clearing for and transportation to and from landfills also contribute to climate change. Waste composting and recycling generate GHG, but at lower levels than extraction and manufacturing. For example, recycling aluminum cans uses 95% less energy than making new cans from raw materials. Thus, all stages in the product to waste cycle contribute to climate change.

There are links as well between sustainability, waste regulation, and climate change. As a general principle, *sustainability* aims at averting depletion of natural resources to achieve long-term ecological, economic, and social balance for the survival and benefit of current and future generations. There are many definitions of sustainability. A definition adopted by the Academic Advisory Committee for the Office of Sustainability at the University of Alberta neatly captures the key concepts: "Sustainability is the process of living within the limits of available physical, natural and social resources in ways that allow the living systems in which humans are embedded to thrive in perpetuity."

The EU and US regulatory approaches for materials and waste management reflect each authority's commitment to, and aspirations for, greater sustainability. Regulations to minimize extraction of raw materials, extend product usefulness, reduce waste, and reuse and recycle materials seek a more sustainable—less wasteful—world. A more sustainable world is one that limits GHG emissions that cause climate change.

14.2.3 *Economic and Waste Regulation Models*

In the US, waste regulation and economic growth have generally followed a *traditional linear model*. In this model, materials typically flow in one direction—from extraction to production to use, and finally, to waste. US waste regulations focus predominantly on managing unwanted and unusable waste material at the end of the line.

THE TRADITIONAL LINEAR MODEL

RAW MATERIALS → PROCESSING & PRODUCTION → USE → WASTE

In the EU, waste regulation and the economy are shifting toward a *more circular model*. The EU's regulatory system encourages a continuous flow of materials.

Used materials are fed back into processing and production and the goal is zero waste.

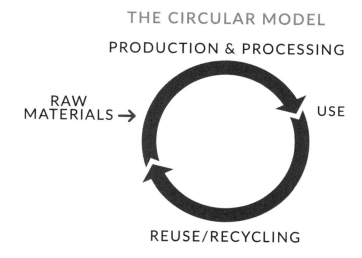

In practice, US waste management is not strictly linear because recycling puts used materials back into use. Conversely, the EU approach is not entirely circular because not all waste is reused.

14.3 The Traditional Linear Economy and the United States

14.3.1 *The Linear Model*

In the traditional linear model, material flows start with extraction of natural resources and end with the disposal of waste. Extracted materials move to processing and manufacture, then to use, and finally to disposal. For example, a smartphone in the linear economy is one designed to be replaced often, made with raw materials, and discarded for a replacement. Some experts refer to this one-directional flow as "Take, Make, Waste." Catherine Weetman (2021), author of *A Circular Economy Handbook*, explains: "We take materials, make something, use it and then dispose of it."

Early on, this linear model seemed sensible because natural resources were abundant, land for waste storage was plentiful, and waste disposal appeared mostly benign. The US economy grew and benefited from extracting more and more materials to make more and more stuff for more and more economic gain.

This mindset began shifting in the 1970s as it became apparent that many raw materials relied on for survival and quality of life are finite (such as metals, minerals, and fossil fuels) while others are constrained by limits on regeneration (forests need time to regrow and food production requires land). It has also become apparent that the ever-increasing extraction of raw materials and corresponding escalation in waste disposal create pollution, degrade habitats and ecosystems, pose human health risks, and contribute to climate change. With shifting attitudes, the US moved to regulate waste in more comprehensive ways.

14.3.2 Waste Laws in the United States

US law regulates a wide array of material flow activities from the point of extraction of raw materials to the disposal of waste. At the extraction stage, US laws regulate mining practices, air emissions, and water pollution discharges. During processing and manufacture, again there are laws regulating air emissions, water discharges, as well as the transportation, treatment, storage, and disposal of hazardous wastes (and the remediation of hazardous waste contaminated areas). However, none of these laws specifically target materials that end up as ordinary garbage or address GHG emissions associated with such waste. Instead, to manage ordinary solid waste, the US relies on a complex web of state and local government laws in conjunction with a federal law known as the Resource Conservation and Recovery Act.

14.3.3 The US Resource Conservation and Recovery Act

The Resource Conservation and Recovery Act of 1976 (RCRA n.d.) is the foremost national law regulating solid waste—hazardous and non-hazardous—in the United States. RCRA Subtitle C addresses *hazardous waste* with a linear, "cradle to grave" approach to management. RCRA regulates hazardous materials from extraction (the cradle) through production and use (the life) and finally at the point of treatment or disposal (the grave).

For *nonhazardous solid waste*, RCRA predominantly focuses on disposal, the grave. RCRA Subtitle D prohibits dumping, sets criteria for landfills, and encourages states to adopt environmentally sound waste management practices. Much of Subtitle D pertains to municipal landfills. Subtitle D provisions detail where landfills may be located (e.g., , not in earthquake prone areas); mandate installation of liners to contain leachate (to keep liquid contaminated by waste from entering groundwater); contain procedures to monitor groundwater (to confirm the liners work); specify controls for disease vectors, fires, odors, and blowing litter (such as daily soil covering of deposited waste); mandate controls and monitoring of methane gas (to avoid explosions, not for minimization of GHG); and include procedures for recordkeeping, landfill closure, and post-closure care (for when the landfill no longer accepts new waste).

RCRA's provisions reflect the era in which it was enacted. The US EPA, the federal agency tasked with RCRA's implementation, described the state of pre-RCRA waste management in the US as follows: "So widespread was pollution from waste that favorite 'swimming holes' were no longer safe for swimming and town well water was no longer safe for drinking. Unsightly dumps marred the countryside and waterways. Dumps not only spoiled the land and the water, but they also were vectors for disease, providing safe habitats for rats, flies, mosquitoes, and other vermin. They frequently burned or caused extensive damage to surrounding areas."

Taking stock of all this environmental damage, Congress passed the Solid Waste Disposal Act (SWDA) in 1965. [SWDA subsequently amended and renamed RCRA] formed the framework for states to better control the disposal of trash from all sources. SWDA set minimum safety requirements for local landfills. Even with SWDA in place, trash still overflowed from landfills and dumps. In the decade between 1950 and 1960, the amount of trash individuals created increased 60 percent."

As amended, RCRA sets out three overarching goals to address the identified waste problems. First, RCRA aims to protect human health and the environment by proper waste management. Second, RCRA seeks to reduce or eliminate waste expeditiously. Third, RCRA sets as a goal the conservation of energy and natural resources through waste recycling and recovery. Notably, these goals reflect RCRA's focus on the end stage of the linear model. RCRA's provisions apply to the waste stream and do not address environmental problems posed by materials extraction, processing, or use. Moreover, despite RCRA's stated goals, the nonhazardous solid waste provisions (Subtitle D) focus predominantly on human health risks and environmental hazards of landfilling, and only indirectly impact materials conservation, waste reduction, and climate change. RCRA leaves for the states and local governments the work of conserving materials and reducing waste.

14.3.4 State and Local Waste Regulation in the United States

Under RCRA's cooperative framework, states and municipalities take on the lion's share of solid waste management and federal authorities assist by issuing guidelines and providing technical support. State, city, and town governments managed garbage collection, disposal, recycling, and composting. As a result of these efforts and according to 2018 US EPA data, approximately 23% of garbage in the US is recycled and about 9% composted (for a total recycling and composting rate of 32.1%).

Due to the numerous state and local authorities managing waste, there exists a hodgepodge of approaches across the country. By way of illustration, the City of Seattle, one of the greener cities in the US, separately collects residential garbage, recycling (paper, cardboard, glass, metal, and certain plastics), and composting (food scraps and yard trimmings). To encourage waste reduction, Seattle charges higher fees for larger amounts of garbage (based on can size). To encourage composting, Seattle charges less for composting service than ordinary garbage service. To encourage recycling, Seattle imposes no additional cost for recycling services (recycling is free) and a Seattle ordinance bans residents from putting recyclables in with garbage—repeat offenders are subject to a $50 fine. With this set of strategies, Seattle recycles about 60% of its waste (about twice the national average). Seattle is also one of several hundred US cities making efforts to remove single-use plastics from its waste stream by banning single-use plastic bags, straws, and utensils.

In contrast, the City of Indianapolis sets a relatively low, flat fee for garbage, does not provide recycling to all households, and discourages recycling by charging residents for collection. Near the bottom for US cities, Indianapolis recycles only 7% of its residential garbage. The City also diverts waste for incineration creating racial justice and GHG emission concerns. At the state level, Indiana (in which Indianapolis exists) is one of 15 US states that has passed laws disallowing the taxing or restriction of single-use plastic bags, making it harder for cities to reduce and manage plastic wastes.

14.3.5 Waste Management Trends in the United States

Despite the existing, mostly linear regulatory approach, US policy envisions a more sustainable, material-efficient future. The US EPA has long advocated a shift away from narrowly conceived "waste" management and toward the more expansive conception of "materials" management. US EPA policy recommendations include managing materials on a "life-cycle" basis that involves examining environmental impacts from initial extraction of raw materials to final disposal, not merely at the point products become waste. A life-cycle approach, according to US EPA, would create opportunities to conserve resources, reduce costs, and address environmental impacts, including climate change. As an illustration, a life-cycle analysis of drinking water shows use of tap water preferable to single-use plastic bottles—regardless of whether the bottles are recycled—due to the emissions, resources, and energy associated with making water bottles. Adoption of life cycle approaches can help move the US in the direction of more circular and sustainable materials management, but alone they do not compel circularity.

Taking the lead on waste reduction, a growing number of US communities have embraced the concept of a *zero-waste* future. Los Angeles, for example, aspires to achieve zero waste by 2025, New York City by 2030, and Austin by 2040. Although the definition of zero waste varies, common themes include pledges to reduce and divert at least 90% of waste by recycling and composting materials otherwise destined to landfills; placement of greater emphasis on repairability, reuse, and recycling; application of whole life or life-cycle analysis; and reconceptualizing waste as valuable rather than valueless material.

14.4 The Circular Economy and the European Union

14.4.1 The Circular Model

In a circular model, materials flow in a continuous loop. Resources move from processing and production to use, and then recirculate back into the loop. For example, a smartphone in a circular economy is one designed with replaceable, repairable, and upgradable parts and with dismantlable, reusable, and recyclable components. The goal is to keep products (like smartphones) in use longer and then reuse, recycle, and recover them to achieve zero waste. EU law defines a circular economy as one that "minimises resource input, waste, emissions and energy leakage. It can be achieved through long-lasting design, maintenance, repair, reuse and recycling. It contrasts to a linear economy which extracts resources, uses them, then throws them away."

The concept resembles earth's natural cycles of renewal and regeneration. According to Weetman (2021), in a circular economy all waste becomes food for new products or nature.

Given the specter of climate change and other looming ecological crises, global interest in the circular economy model continues to grow. In particular, the EU has embraced this model as central to addressing climate change and for securing the region's economic future. In 2015, the EU Commission adopted its first Circular Economy Action Plan (CEAP), "Closing the Loop—An EU Action Plan for the Circular Economy." According to the 2015 CEAP, the "transition to a more circular economy, where the value of products, materials and resources is maintained in the economy for as long as possible, and the generation of waste minimized, is an essential contribution to the EU's efforts to develop a sustainable, low carbon, resource efficient and competitive economy. Such transition is the opportunity to transform our economy and generate new and sustainable competitive advantages for Europe."

The 2015 CEAP proposed over 50 actions for "closing the loop" of product life cycles.

In 2020, the EU adopted a new CEAP, the "Circular Economy Plan for a Cleaner and More Competitive Europe." The 2020 CEAP contains initiatives targeting product design, promoting circular economy processes, fostering sustainable consumption, and ensuring resources are kept in the EU economy for as long as possible. The 2020 CEAP aims to "accelerate the transformational change required by the European Green Deal," build on circular economy actions already in place, and "transform consumption patterns so that no waste is produced in the first place."

Note, in addition to the CEAPs, the EU has issued policy communications touching on EU waste management, climate change, and the circular economy that are beyond the scope of this chapter (such as the European Green Deal, the EU Methane Strategy, the Critical Raw Materials Strategy, and policies for improving energy performance in building renovations).

14.4.2 Circular Economy Regulatory Measures in the European Union

The EU adopted a package of legislative measures to implement the proposals of the first Circular Economy Plan in 2018. The package included legal mandates on EU countries (Member States) called directives. *EU Directives* set out binding targets (regulatory ends) but leave to each individual Member State the choice of how to achieve the targets (the regulatory means). Member States implement their approaches through national legislation. The 2018 directives for solid waste management, landfills, and packaging are summarized below.

The *Waste Framework Directive* (WF Directive), amended by EU Directive 2018/851, establishes a regulatory structure for EU nonhazardous solid waste management. The framework's primary objectives are avoiding creation of waste (prevention) and renewal of waste as a resource (reuse and recycling). The preamble of this EU Directive declares: "Waste management in the Union should be improved and transformed into sustainable material management, with a view to protecting, preserving and improving the quality of the environment, protecting human health, ensuring prudent, efficient and rational utilisation of natural resources, promoting the principles of the circular economy, enhancing the use of renewable energy, increasing energy efficiency, reducing the dependence of the Union on imported resources, providing new economic opportunities and contributing to long-term competitiveness. (…) The more efficient use of resources would also bring substantial net savings for Union businesses, public authorities and consumers, while reducing total annual greenhouse gas emissions."

As amended, the WF Directive includes increasingly stringent targets for reuse and recycling of waste as proposed in the first CEAP. EU Member States must increase reuse and recycling to a minimum of 55% by 2025, to a minimum of 60% by 2030, and to 65% by 2035. The WF Directive also establishes a specific biowaste target. By December 2023, biowaste must be separated and recycled at source or collected separately. The WF Directive further provides that Member States must—at a minimum—take a long list of measures to prevent waste. However, these generally worded prevention measures appear more aspirational in nature than the specific target values for recycle, reuse, and biowaste. For example, Member States must "promote and support" sustainable production and consumption; "encourage" design and manufacturing of more durable products; "aim" to halt marine litter; "promote" redistribution of unsold food products; and "support" information campaigns to raise awareness about waste prevention.

The WF Directive incorporates and clarifies key regulatory principles such as the Waste Hierarchy, Polluter Pays, and Extended Producer Responsibility principles. The *Waste Hierarchy* sets the waste management priorities to be applied by Member State in the EU. The WF Directive establishes that the most preferred options are waste prevention and reuse, followed by preparing for reuse, recycling, and composting, and then energy recovery, with landfills serving only as a last resort. The *Polluter Pays Principle* seeks to impose the costs of environmental harms on the entities responsible for the harm rather than on the public at large (in economic jargon, to internalize the externalities of pollution). In the context of the WF Directive, the principle seeks to impose certain costs of waste management on waste producers. A related concept, *Extended Producer Responsibility*, makes product producers (such as manufacturers and importers) financially or organizationally responsible for collection, recycling, and disposal of their products. Although Extended Producer Responsibility obligations already attach to certain products in the EU (such as batteries, electronics, and packaging), the WF Directive establishes general minimum standards for such producer responsibilities and encourages Member States to expand application of such producer responsibilities to achieve framework objectives. The Directive does not mandate but does encourage Member States to take measures that promote reuse and repair networks, consumer sustainability awareness, and reuse of products with critical raw materials.

The *Landfill Directive*, as amended by EU Directive 2018/850, phases in landfilling restrictions and reductions to implement elements of the CEAP and to address adverse impacts of landfill waste on the environment, including climate change. As amended, the Landfill Directive declares: "With a view to supporting the Union's transition to a circular economy (…) the aim of this Directive is to ensure a progressive reduction of landfilling of waste, in particular of waste that is suitable for recycling or other recovery, and, by way of stringent operational and technical requirements on the waste and landfills, to provide for measures, procedures and guidance to prevent or reduce as far as possible negative effects on the environment, in particular the pollution of surface water, groundwater, soil and air, and on the global environment, including the greenhouse effect, as well as any resulting risk to human health, from landfilling of waste."

The Landfill Directive requires that all waste suitable for recycling or other recovery be excluded from landfills by 2030 and that landfill waste be reduced 10% or less of generated solid waste by 2035. This directive additionally requires Member States to minimize biodegradable waste to landfills.

The *Packaging and Packaging Waste Directive* (PPW Directive), as amended by EU Directive 2018/852, addresses both packaging design and packaging waste management. The directive defines packaging as "all products made of any materials of any nature to be used for the containment, protection, handling, delivery and presentation of goods, from raw materials to processed goods, from the producer to the user or the consumer."

Examples include aluminum soda cans, glass milk bottles, plastic candy wrappers, and paper food containers.

As amended, the PPW Directive sets as its "first priority" preventing packaging waste. Toward this end, the PPW Directive requires that Member States implement measures to prevent generation of packaging waste and increase reusable packaging. The directive leaves to Member States the choice of what measures to adopt, offering recommendations rather than explicit targets or programs. For example, the PPW Directive recommends, but does not demand, use of economic incentives, extended producer responsibility schemes, deposit–return programs, and the setting of targets as possible measures to achieve reduction and reuse goals. Member States are expected to establish extended producer responsibility schemes for all packaging by the end of 2024.

The PPW Directive also contains measures aimed at recovery of packaging wastes. Here, the directive establishes explicit recycling targets and measures. The directive sets a minimum target for all packaging at 65% by 2025 and to 70% by 2030. The directive also sets targets for the various types of packing wastes. For example, Member States must increase recycling of plastic packaging waste to 50% by 2025 and to 55% by 2030 and improve recycling of glass packaging waste to 70% by 2025 and to 75% by 2030.

Looking at plastics in particular, the EU adopted circular economy directives both before and after the 2018 legislative package enactments. In 2015, the EU put in place the *Plastic Bags Directive* (PB Directive), EU 2015/720, requiring Member States to take measures to address adverse impacts on wildlife, marine life, and humans posed by lightweight plastic bags. In 2019, the European Union passed the *Single Use Plastic Directive* (SUP Directive), EU 2019/904, to address single-use plastic items beyond just plastic bags. *Single-use plastics* are items used just once or for only a short time, such as plastic plates, straws, stirrers, beverage containers, and wrappers, as well as less obvious items such as balloons, cigarette butts, and wet wipes. The SUP Directive bans certain plastics items, sets targets for recycled content and separate collection of plastic bottles, and requires labeling to inform consumers of appropriate disposal options and negative environmental impacts of improper disposal. There are also extended producer responsibility provisions to impose on producers' cleanup costs for certain plastic pollution (such as plastic fishing gear) and costs for consumer education programs.

Note that a number of other EU Directives—beyond the scope of this chapter—regulate waste materials and address climate change. Examples include directives relating to batteries, end-of-life vehicles, electronic equipment, biodegradable animal by-products, mining wastes, industrial emissions, energy-related products, and waste shipping.

14.4.3 Regulatory Trends in the European Union

The 2020 CEAP outlines future strategies designed to move the European Union closer to a circular model. According to the drafters, "This Circular Economy Action Plan provides a future-oriented agenda for achieving a clean and more competitive Europe in co-creation with economic actors, consumers, citizens and civil society organisations. It aims at accelerating the transformational change required by the European Green Deal, while building on circular economy actions implemented since 2015."

As a component of the EU's overarching commitments to dealing with climate change and attaining sustainable growth, the 2020 CEAP contains 35 new actions targeting products throughout their life cycles, including proposals for new EU law.

14 Climate, Sustainability, and Waste: EU and US Regulatory Approaches Compared

One such proposal is a *sustainable product policy legislative initiative*. This proposed legislation would integrate eco-design into a broader range of products to enhance product sustainability. New regulatory directives would be adopted to extend product life (rather than merely regulating wastes). The proposed directives will aim to improve product durability, reusability, upgradability, and reparability; reduce product carbon and environmental footprints; restrict premature obsolescence and single-use products; ban destruction of unsold durable goods; and encourage producer life-cycle responsibilities.

Another set of ideas for achieving climate-neutrality and enhancing circularity focus on production processes. The 2020 CEAP envisions integrating circular economy objectives into the existing *Industrial Emission Directive*, 2010/75/EU. Some proposed strategies include facilitating industrial symbiosis, adopting a bioeconomy action plan, and promoting green and digital technologies. Again, such actions look higher up the waste hierarchy (rather than focusing on regulating wastes).

14.5 Comparisons

The US and EU regulatory approaches to waste management have significant similarities and important differences.

With respect to *similarities*, both the US and EU approaches delegate and decentralize significant authority for waste management. In the US, states and local jurisdictions retain responsibility for garbage collection, disposal, recycling, and composting. In the EU, the use of directives (rather than regulations) positions primary responsibility for design, implementation, and enforcement of circular economy measures on Member States. On the plus side, decentralizing regulation allows state entities to formulate measures that best suit their populations and businesses and take into account local conditions and local knowledge. The flexibility to adopt innovative approaches allows for experimentation, adaptation, and cooperative learning as to best practices (assuming information is shared). A negative consequence of decentralizing responsibility is lack of harmonization. Lack of a standard approach can result in inconsistent implementation, enforcement, and outcomes. For example, while some regions move boldly forward with recycling efforts, others lag far behind. In the US, the City of San Francisco has achieved an 80% recycling (diversion) rate, one of the best in the nation, while the City of Chicago recycles only about 9% of its garbage, well below under the national average. In the EU, Germany recycles almost 70% of its garbage, but Malta's recycling rate is less than 10%.

A second commonality is reliance on waste hierarchies. The waste hierarchy developed by the US Environmental Protection Agency (US EPA) prioritizes waste prevention (reduction and reuse) as the most environmentally preferred strategy, followed by recycling and composting, energy recovery. At the bottom of the hierarchy is treatment and disposal. The EU Waste Framework Directive's waste hierarchy similarly prioritizes prevention, followed by preparing for reuse, recycling, recovery, and, lastly, disposal. Both rank waste prevention as the most environmentally preferred strategy and disposal as the least environmentally preferred strategy.

Third, US and EU regulatory approaches both rely to a great extent on recycling and composting. In the US, overall recycling efforts have increased from practicably no recycling in the 1960s to around 32% in 2018, with most Americans having access to some type of recycling program. In the EU, mandated recycling targets have helped to achieve an overall recycling rate of about 46%. Relatedly, both the US and EU plastics recycling efforts have drastically slowed in recent years. China's 2018 ban on importing foreign wastes, including plastics ("National Sword" policy), has hindered plastics recycling in both US and the EU and led to more landfilling and incineration of plastics globally. Prior to China's ban, 70% of plastics collected for recycling in the US and 95% in the EU were sold and shipped to Chinese processors.

Finally, US and EU aspirations for the future align, at least in part. Both US and EU policy pronouncements support and encourage reconceptualizing waste management as materials management. Regulators in both the US and the EU envision future regulatory approaches that integrate materials management across entire product life cycles. Progressive cities' governments in both the US and the EU are increasingly committing to working toward zero-waste futures. In at least these ways, US and EU policies appear to converge toward greater sustainability and circularity despite the EU's faster pace toward and more obvious embrace of these goals.

There are also important *differences*. First, the EU and US regulatory approaches diverge on their declared purposes. The EU approach seeks to unite and align environmental protection and economic growth (win-win strategy). Circular approaches seek to reduce GHG emissions, be sustainable, and keep businesses profitable. In contrast, the US regulatory approach serves a narrower purpose – protection of human health and the environment, disconnecting environmental and economic goals. Deregulation advocates have used this divide to characterize environmental and climate regulation as job destroying and antigrowth (win-lose dichotomy).

Second, the EU and US regulatory approaches differ with respect to their primary level of decision-making. Comparable to a top-down management style, EU institutions formulate the overarching policy direction for materials regulation; establish uniform definitions, guidelines, and principles; and mandate the legal end points to be implemented and accomplished by Member States. In contrast, RCRA authorizes the US EPA (2009) to set minimum landfill criteria, offer guidance, and provide technical assistance, but largely leaves states and localities to regulate nonhazardous waste as they see fit.

Third, although the EU and US regulatory approaches rely heavily on recycling, the nature and extent of that reliance differ. According to several reports, the US failed to rank in the top 20 recycling nations globally and was "out-recycled" by most of Western Europe, as well as several countries in Asia, and Australia. At the same time, scholars note that the EU approach is more than recycling (and more even than recycling on steroids). EU boasts a regulatory strategy that moves beyond waste recycling to all stages of the material life cycle.

Fourth, the EU and US waste management approaches diverge in their international involvement. Engagement in the Basel Convention on the Control of Transboundary Movements of Hazardous Wastes and their Disposal (1989) (Basel Convention) presents one such example. Among other things, parties to the Basel Convention agree to implement environmentally sound management measures related to cross-border movements of covered wastes (e.g., plastic waste), abide by notice and informed consent requirements, and accept several trade bans related to such waste. The EU is a party to the Basal Convention, the US is not. As another example, the EU signed a memorandum of understanding on the circular economy with China in 2018 to promote cooperation in areas such as eco-design, eco-labeling, extended producer responsibility, and green supply chains.

Finally, the US remains mostly a linear, "take, make, take, waste" economy. The EU, in contrast, is at least poised toward a transitioning toward a circular, "closed loop" economy. Despite US efforts to promote the waste hierarchy and its touting of preventive and sustainable waste management, there are few federal mandates implementing such policies. In contrast, the EU has moved forward with concrete commitments (financial and legal) and implemented circular economy targets and programs, even if some EU policies and programs may be more aspirational than entirely achievable.

14.6 Conclusion

Traditional resource management approaches contribute to climate change and are unsustainable. As a consequence, they threaten our future and the future of generations to come. The time for course correction is now. The regulatory choices we make about materials management can meaningfully reduce GHG emissions and move us toward a more sustainable world (or not). In making these critical choices, the US and the EU could and should learn from each other. We can discover the regulatory approaches that work best, which do not work, and how to best move forward.

References

Basel Convention on the Control of Transboundary Movement of Hazardous Wastes and their Disposal. (22 March 1989). (1673 U.N.T.S. 126)

European Parliament and the Council. (2015). *Directive 2015/720 of the European Parliament and of the Council amending Directive 94/62/EC on reducing consumption of lightweight plastic carrier bags*. Retrieved June 8, 2021, from https://eur-lex.europa.eu/legal-content/EN/TXT/?uri=celex%3A32015L0720

European Commission (2020). *Circular economy action plan: For a cleaner and more competitive Europe*. Retrieved June 8, 2021, from https://ec.europa.eu/environment/strategy/circular-economy-action-plan_it#:~:text=The%20EU's%20new%20circular%20action,(CEAP)%20in%20March%202020.&text=It%20is%20also%20a%20prerequisite,and%20to%20halt%20biodiversity%20loss

Resource Conservation and Recovery Act of 1976. (n.d.). 42 U.S.C. §§ 6901-6992k.

U.S. EPA. (2009) *Sustainable materials management: The road ahead*. Retrieved June 8, 2021, from https://www.epa.gov/smm/sustainable-materials-management-road-ahead

Weetman, C. (2021). *A circular economy handbook: How to build a more resilient, competitive and sustainable business*. Kogan Page.

Madeline June Kass is a distinguished scholar in residence and visiting professor at Seattle University School of Law and professor emeritus at Thomas Jefferson School of Law. She teaches comparative US/EU environmental law, US environmental law, US natural resources law, US wildlife and marine life law, torts, and professional responsibility. Professor Kass serves on the Council of the American Bar Association Section of Environmental, Energy, and Resources (ABA SEER) and the editorial board of the ABA's quarterly publication *Natural Resources & the Environment* (NR&E). She is also an IUCN World Commission on Environmental Law member and vice-chair and former chair of ABA SEER's International Environment and Resources Law Committee. In past years, she has served as chair of the Association of American Law Schools (AALS) Natural Resources & Energy Section, chair of ABA SEER's Endangered Species Committee, and as a fulbright scholar at the College of Europe. Prior to entering academia, Professor Kass practiced law for close to a decade in the Seattle offices of Preston Gates & Ellis (now K&L Gates) and Heller Ehrman White & McAuliffe. She guest lectures and has published numerous articles on environmental, natural resources, and endangered species law and policy topics. Professor Kass holds a BS degree from Tufts University, a MES from Yale University School of the Environment, and a JD from University of California Berkeley Law. Most importantly, she loves dogs and enjoys playing "old lady" soccer in Seattle.

Chapter 15
Construction Industry and Sustainability

Adriana Spassova

Abstract Aside from the potential for building over wild habitats, the construction industry is one of the most resource-consuming sectors. Considering a building's full life cycle, it accounts for approximately half of all extracted materials, half of total energy consumption, one-third of water consumption and one-third of waste generation.

Sustainable construction balances environmental, social, and economic issues to ensure a viable and valuable industry for future generations.

Sustainable construction methods include:

- Using renewable and recyclable resources
- Reducing energy consumption and waste
- Creating a healthy, environmentally friendly habitat
- Protecting the natural environment

The sustainable construction has a lot more to do with:

- The design
- Management of construction
- The choice of materials

Design and construction bring improvement in four areas: energy, waste, water, and pollution. The construction industry is beginning to recognize its role and respond with initiatives and innovations to make whole cities carbon neutral.

We have to act with a sense of urgency, since obviously we cannot reach by 2030 the UN goals for sustainable urbanization, affordable housing, and basic services for everyone. Sustainability in construction can be enabled through regulations, voluntary commitment and certification, financial instruments, lean practices, digitalization, collaborative procurement, education, and innovations.

A. Spassova (✉)
EQE Control OOD, Sofia, Bulgaria
e-mail: aks@eqe.bg

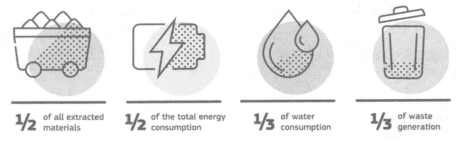

Fig. 15.1 Impact of the construction industry
Source: Level(s) The European framework for sustainable buildings, https://ec.europa.eu/environment/topics/circular-economy/levels_en

15.1 Construction Industry: The Elephant in the Glass Shop?

The *Oxford Dictionary* defines sustainability as the capacity to be maintained at a certain rate or level conserving an ecological balance by avoiding depletion of natural resources.

Contributing 13% of global GDP, construction is the largest industry in the world [24].

It is responsible for a higher level of carbon emissions than both international air and sea transport combined.

Aside from the potential for building over wild habitats, this is one of the most resource-consuming sectors. Considering a building's full life cycle, the construction industry accounts for approximately half of all extracted materials, half of total energy consumption, one-third of water consumption, and one-third of waste generation (Fig. 15.1).

15.2 How Can the Construction Industry Improve Sustainability?

Sustainable construction balances environmental, social, and economic issues to ensure a viable and valuable industry for future generations.

Sustainable construction methods include using renewable and recyclable resources, reducing energy consumption and waste, creating a healthy, environmentally friendly habitat, and protecting the natural environment.

Committing to zero net emissions of greenhouse gases in 2050, the United Nations Sustainable Development Goals (SDGs), the Paris Agreement on climate change, and the European Green Deal reinforce the status of sustainable development as a public and private sector imperative.

Several SDGs are of particular relevance for the construction sector, with their emphasis on access to housing and infrastructure and improvements in energy efficiency. It is possible to refer, for instance, to

- Goal 7 "Ensure access to affordable, reliable, sustainable and modern energy for all," which includes "by 2030, increase substantially the share of renewable energy in the global energy mix" (7.2) and "double the global rate of improvement in energy efficiency" (7.3)
- Goal 9 "Build resilient infrastructure, promote inclusive and sustainable industrialization and foster innovation," which includes "by 2030, upgrade infrastructure and retrofit industries to make them sustainable, with increased resource-use efficiency and greater adoption of clean and environmentally sound technologies and industrial processes" (9.4)
- Goal 11 "Make cities and human settlements inclusive, safe, resilient and sustainable," which includes "by 2030, ensure access for all to adequate, safe and affordable housing and basic services and upgrade slums" (11.1) and "by 2030, enhance inclusive and sustainable urbanization and capacity for participatory, integrated and sustainable human settlement planning and management" (11.3)

More in detail, design and construction bring improvement in four main areas: energy, waste, water, and pollution. This may be achieved, for instance, by an efficient use of energy and water with sustainable drainage systems, rainwater harvesting, solar panels, gas-power tri-generation plants, extra heat from data centers used by houses, central energy management for all systems in a building/ city, etc. Alternative methods are smart waste management and recycling (e.g., timber, concrete bricks, tiles, steel), reuse of concrete, steel, hardwood timber, and off-site fabrication, including modular and movable buildings.

Sustainable planning and sustainable land use may also be highly effective in selecting long-lasting and renewable materials and targeting minimum CO_2 emissions.

The construction industry is beginning to recognize its role and respond with initiatives and innovations to make whole cities carbon neutral.

The EU Roadmap for moving to a competitive low-carbon economy in 2050 estimated that emissions from the built environment could be reduced by around 90% by 2050 through the introduction of passive housing technology in new buildings, the refurbishment of old buildings to improve their energy performance, and the substitution of fossil fuels by renewable energy sources.

Sustainability in construction can be enabled through regulations, voluntary commitment and certification, financial instruments, lean practices, digitalization, collaborative procurement, education, and innovations.

Example: Atkins designed the Dubai Opera (completed in 2016) as a flexible building that can accommodate several different types of modes, avoiding the need to build separate venues for different events. It can convert from a theater into a concert hall and into a flat floor offering 2000 sqm for exhibitions and gala events. The concept is based around a dhow boat with a restricted base and a broad crown, allowing the building to cast a shadow in itself at peak periods in the day. The façade is made of glass panels with antireflective coating to mitigate solar radiations. 5 km of aluminum louvers create shade.

15.3 Strategic Planning for Whole-Life Sustainability

Sustainability must be considered from the very earliest stages as the potential environmental impacts are very significant. Once it has been decided to build a new building, as opposed to rehabilitation/refurbishing an existing building, a very significant commitment to consuming resources has already been made. Designers and contractors may be able to help limit that consumption, but they cannot change the overall commitment.

Key decisions may be picked up by an environmental impact assessment on larger projects, but even then, this can be used to justify the impact of the project to the public and authorities, rather than to make strategic decisions related to the most essential sustainability outcomes.

Strategic sustainability decisions come at the project initiation stage, where there is a minimum constraint on the project. Large-scale projects such as entire city developments, major water or wastewater schemes on a regional scale, new transport systems, nuclear, wind, or hydroelectric power proposals have the greatest sustainability impact. In the conceptual phase, there is potential for

- Sustainable thinking to be incorporated in all aspects of the planned development
- Selecting development site locations
- Identifying benefits and adverse impacts for communities and environments
- Identifying current and future vulnerabilities and building in resilience
- Deciding whether it would be more sustainable in the long term not to proceed with the development

Much value engineering can be done upfront to influence the cost and impact. The life-cycle approach is essential to reduce long-term costs for the projects. Sustainable choices can be made along the way: selection of materials, prefabrication, etc.

From a holistic point of view, to drive the sustainability agenda, consultants shall advise the clients upfront, from the idea, to develop and compare alternatives, using the available sustainability tools [18].

The International Federation of Consulting Engineers (FIDIC) advocates that life-cycle thinking is very important from the earliest stage of a project. FIDIC's 2021 State of the World report "Time to $Tn-vest" [20] recommends a renewed and massive global effort to improve infrastructure spending. Global infrastructure could require at least $7 trillion spending on it every year if the world is to address the growing climate emergency and recover from the effects of the Covid-19 pandemic. If the investment is not made and/or if the need is not lowered through innovation, meeting the UN's SDGs will be placed in jeopardy.

To address the challenge, FIDIC recommends (i) a renewed global effort to improve infrastructure spending to meet the investment challenge facing the world; (ii) to create global financial mechanisms to support sustainable investment and end all carbon-intensive investments and the burning of fossil fuels; and (iii) improved yearly global monitoring to ensure that infrastructure investment and maintenance spending is actually being undertaken in a sustainable way to meet the investment gap caused by years of underinvestment.

Circular economy thinking is needed to find the best solution and identify what services are necessary, more than what projects. Maintenance is underestimated, while construction is overestimated. Not building something may be a solution in many cases.

15.4 Construction Projects for Sustainable Energy Production and Energy Efficiency

15.4.1 Construction of Renewable Energy Projects

Electricity is a key enabler in the transition from fossil fuels and decarbonizing. Renewables like photovoltaic plants, wind farms, and biomass plants are being constructed, but there are a lot of hurdles:

- The capacity constraints within the transmission network have to be overcome
- The higher costs have to be reduced via innovations and green energy tariffs
- Renewables suffer from intermittency, and so baseload capacity is essential. This requires significant capital investment for the adoption of Carbon Capture, Utilization and Storage (CCUS) technologies [17]

The abovementioned problems do not exist for small-size projects, enabling passive houses: solar panels on facades and roofs of buildings, producing energy for own consumption and possibly for the neighborhood. Passive houses often rely on their own geothermal energy sources for heating and cooling.

15.4.2 Construction of Smart Energy Infrastructure

Grid optimization and aggregation initiatives are together worth a projected $440 billion to industry and $1.2 trillion to society through enabling smarter saving choices for customers, new job creation and lower peak demand, reducing carbon emissions. The digital transformation of the electric grid will play a key role in the transition.

Increased cross-border and regional cooperation will help achieve the benefits of the clean energy transition at affordable prices.

The European Green Deal envisages that the regulatory framework should foster the deployment of innovative technologies and infrastructure, such as smart grids, hydrogen networks or carbon capture, storage and utilization, and energy storage, also enabling sector integration. Some existing infrastructure and assets will require upgrading to remain fit for purpose and climate resilient.

15.4.3 Energy Efficiency of Buildings and Infrastructure

Energy savings through design and construction may be achieved by selecting energy-efficient construction products, plant and equipment, implementing retrofits of building shell and systems; and adopting building management systems (BMS), using building information modeling (BIM), setting forth higher energy efficiency requirements for new built, including certification and modernizing district heating systems.

If the most stringent current minimum energy performance requirements (MEPS) for product energy efficiency had been harmonized globally at this point in time, global final energy consumption would be 9% lower and energy consumption due specifically to products would be 21% lower [9].

The EC 2015 Final Report "Savings and benefits of global regulations for energy efficient products" [9] suggests that the use of the latest sustainable technologies in construction could deliver a remarkable €410 billion a year in savings on global energy spending.

Example: Residential building, about 15 km away from the Norwegian capital Oslo, built in 1980, today produces 220,000 kWh of electricity per year. As this amount of energy is more than the energy consumed by the building, what is left is used to charge electric cars in the adjacent parking lot. The roof is entirely covered with solar panels, the windows have been changed, and the glass from the old ones is reused inside the building for making interior doors. Heat pumps have been installed to provide the heating of the building. This was implemented during the renovation completed in 2014. Thanks to it, the building today fully meets the definition of a "Powerhouse"—or "active" building that produces more energy from renewable sources than it consumes.

15.5 Construction Projects for Sustainable Water Use

15.5.1 Demand for Sustainable Water Infrastructure

Providing clean water and basic sanitation to those who do not have it now is the most important challenge. Fresh water in all forms (e.g., water, wastewater, stormwater) is a valuable resource that needs to be effectively used and reused [21].

Unless more substantial fresh water demand reduction strategies are implemented, especially those related to agriculture, population growth, and increased wealth in developing countries, this will likely overwhelm efforts to reduce water consumption, resulting in a continued increase in the global demand for fresh water.

Depending on location, water quality problems can include untreated or inadequately treated wastewater; combined sewer overflows; deterioration of watersheds due to development, deforestation, and natural disasters like fires and floods; the emergence of contaminants, some of which difficult to treat; and naturally occurring constituents such as salinity and metals including arsenic.

15.5.2 Integrated Resource Management

Integrated resource management must ensure that the right projects are identified and sustainability considered to take into account future and present needs.

Integrated water resource management (IWRM) can address multiple issues simultaneously, enabling cost-effective programs. Ideally, such integrated systems would be river basin-wide and include multiple countries. Without IWRM at a broad enough scale, the risk of poorly conceived projects that do not address the big picture increases. Even when executed locally, IWRM can reduce cost.

Smart water networks may reduce demand and improve water quality and reduce energy usage, which can be as much as 30% of a utility's cost. They use remote sensing to detect meter inaccuracy and leaks, allowing early rectification; also, automated pressure regulation systems lower energy costs and reduce leakage volume.

15.5.3 Sustainable Construction of Water Infrastructure

Capital programs are needed for sustainable design and construction of

- Clean water supply infrastructure
- Basic water treatment and advanced treatment facilities, including desalinization, where water quality has deteriorated or oceans need to be used to supply arid coastal areas
- Reservoirs, wells, dams, pipelines, and ancillary facilities
- Wastewater conveyance and treatment systems
- Rainwater/stormwater collection systems
- Reuse of water/wastewater

Besides the new build, existing leaking pipelines and deteriorated wastewater treatment plants need renewal/rehabilitation, including new energy-efficient equipment and smart management devices.

More direct and indirect reuse options and conservation programs aimed at reducing waste and potable water consumption must be considered during the strategic planning phase, targeting less costly local or decentralized options.

15.5.4 Reuse

Valuable constituents within wastewater, including potential energy, biosolids and nutrients, can be harnessed, recovered, and used.

Local systems with various available and emerging membrane treatment processes, including microfiltration, ultrafiltration, and reverse osmosis systems, are advancing reuse because of their superior reliability and improving cost-effectiveness. The automated nature of such systems makes them amenable to remote monitoring and operation, which can facilitate the operation of decentralized systems.

Wastewater is reused directly after advanced treatment or indirectly after discharge to local surface water bodies or groundwater aquifers, and then recycled back into the water supply. Currently, indirect reuse is more prevalent than direct reuse. Beyond demand reduction, reuse helps utilities reduce the expense of removing nutrients at wastewater treatment plants to meet increasingly stringent discharge limits for nitrogen and phosphorous compounds. An increasing number of local waste treatment systems are separating gray water from black water and reusing it at the same location in a variety of ways, such as toilet flushing, nutrient capture, and creation of building materials from solids. Other conventional reuse systems are recreational impoundments, golf course irrigation, crop irrigation, snow making, street cleaning, boiler cooling, groundwater enhancement, and stream flow augmentation. Innovative approaches to reuse are occurring throughout the world.

Stormwater should be used to recharge groundwater aquifers. Decentralized or hybrid distribution and collection systems should be considered.

15.5.5 Water and Energy

Looking at energy and water, large amounts of one are generally required to produce and distribute large amounts of the other, be it water for cooling in power stations to produce electricity, or the electricity produced to power pumps and systems to move large volumes of water to users.

Wastewater is used to generate energy and fertilizer.

Flowing water is captured and turned into electricity in hydropower projects. They are renewable energy projects, but their environmental impact assessment must consider how the construction interferes with the climate, land, urban territories, and biodiversity.

Example: The $24-billion project Three Gorges Dam in China created environmental and social problems. The Chinese government ordered some 1.2 million people in two cities and 116 towns clustered on the banks of the Yangtze to be evacuated to other areas before construction. The biodiversity is threatened as the dam floods some habitats, reduces water flow to others, and alters weather patterns. The dam further imperils delicate fish populations in the Yangtze. The massive hydroelectric dam may be triggering landslides and earthquakes [22].

15.5.6 Flood Protection

Flood protection, including floating structures, early warning systems, and dams of all sizes, needs to be designed and build. Dams are critical in the flood control and mitigation element of resilience. To address sea-level rise and flooding due to more severe storms, coastal communities have been constructing levees, dikes, seawalls, massive tide gates, and other similar low-elevation dams. This trend will likely accelerate as the threat from sea-level rise and climate change intensifies.

Example: St. Petersburg, Russia, has constructed a multipurpose flood control system to protect from flooding from the Baltic Sea up to 5.4 m above sea level. The multibillion-dollar project, which comprises flood gates, sluices, embankments, and appurtenances, includes a portion of the city's ring road and some pollution control facilities [19].

15.6 Waste Management Projects

Instead of being an environmental and health hazard, solid waste can be separated, recycled, reused, and produce energy.

The domestic solid waste projects, designed and constructed for sustainable outcomes, include the closure of existing landfills and the construction of new sustainable landfills, sorting and recycling facilities or composting plants, installations for the demolition of large waste, facilities for the temporary storage of hazardous waste, and incinerators.

The EU's approach to waste policy is driven by the "waste hierarchy" that sets out the following fundamental priorities: prevention, preparing for reuse; recycling; and other recovery (e.g., energy recovery) and disposal.

Cleaning existing illegal dump sites, closing of landfills, and construction of regional sanitary landfills is the first step to reducing the environmental burden of domestic waste (preventing contamination of the soil and groundwater and reducing methane emissions). The policy on organic waste is to reduce landfilling, especially biodegradable organic waste. The new EU waste legislation adopted in 2018 as part of the Circular Economy Action Plan includes 2035 targets for municipal waste of a minimum 65% recycling and a maximum 10% landfilling.

Separation facilities, landfills, and waste to energy projects are implemented for the industrial waste.

Hot cells, processing plants, and storage facilities are designed and constructed for the radioactive waste.

Example: Amager-Bakke is a state-of-the-art incinerator burning nonrecyclable waste from homes and businesses and functioning as a recreational facility: its sloped roof doubles as a year-round artificial ski slope, hiking trail, and climbing wall [33].

15.7 The Construction Industry Fights Pollution

15.7.1 Cleaning Up the Construction Industry

The construction industry is a major source of pollution, responsible for around 4% of particulate emissions, more water pollution incidents than any other industry, and thousands of noise complaints every year. Although construction activities also pollute the soil, the main areas of concern are air, water, and noise pollution.

The heavy machinery used in construction still leans heavily on fossil fuels. Operational carbon emissions can be reduced gradually over time by increasing energy-efficiency measures and using more renewable power sources [26].

More construction and consulting firms are adopting sustainable construction methods. ISO 14001 certification is considered the gold standard in environmental management and confirms that a business is working to minimize its impact on the environment.

Example: At a quarry near the city of Gothenburg, Volvo provided electric machines and know-how to test what level of emissions could be slashed at the site. After 10 weeks of testing, the results showed a 98% reduction. By using electric loaders, crushers, and excavators, energy costs decreased by 70% and operator costs went down by 40%. Researchers from two of Sweden's universities joined the project at its inception and will look into factors like battery aging and energy management.

15.7.2 Construction Innovations Fighting Pollution: Examples

15.7.2.1 Smog-Eating Photo-Catalysis

The external façade of the Italian pavilion at the Milano Expo 2015 Palazzo Italia was composed of 900 biodynamic concrete panels with TX Active Technology (Italcementi's patent). Photo-catalytic titanium dioxide and cement react with light neutralizing certain pollutants, purifying the atmosphere. This dynamic material permits the creation of complex shapes and helps in air pollution reduction. Furthermore, it is made of 80% recycled aggregates (marble, cement) [35].

15.7.2.2 Smog-Eating Concrete

Research shows that all concrete has some tendency to absorb NO_2. However, this tendency depends on the concrete mix design, and it is also reduced significantly over time by carbonation. The use of small additions of activated carbon can greatly enhance the NO_2 absorption properties of concrete and also reduces the negative influence of carbonation. This technology does not rely on photo-catalysis and therefore can function perfectly well in the dark. It is especially suitable for use in tunnels or parking garages [25].

15.7.2.3 Bio-Carton

This technology converts CO_2 from polluted air into oxygen. A bioplastic prototype creates a thin membrane with artificial habitat for the cultures to grow. The sun shines through the curtain, activates photosynthesis, the polluted air comes in contact with the molecules of the algae, which cleans it, then oxygen is released back into the atmosphere. Biotechnology is integrated with the building design: the building itself becomes a living organism able to interact with change.

15.8 How to Reduce the Environmental Impact of Construction Materials

15.8.1 Life-Cycle Requirements for Sustainability

The construction industry, by its very nature, is a big user of natural resources. But with growing concerns over climate change and the finite nature of these resources, there is increasing pressure on construction firms to reduce their environmental impact.

The life cycle of construction materials includes resource extraction, manufacturing, occupancy and maintenance, demolition, as well as recycling, reuse, and disposal.

Designers must explore possibilities of using products and materials already available within existing buildings and use materials that are easy to recycle or reuse, and which facilitate high-quality waste management.

Mining for raw materials can result in the pollution of local water tables. The fabrication and shipping of materials can have a great impact on carbon emissions.

After water, concrete is the most widely used substance on the planet. The manufacture of concrete has resulted in over 2.8 billion tons of CO_2, a figure that is only going to keep increasing as 4 billion tons of concrete is poured every year [39].

Construction can also result in hazardous waste, and the improper disposal of such waste can result in pollution that affects not just the environment, but also the health of people living in that area.

Regulation No. 305/2011 made sustainability requirements for construction products mandatory for EU Members. They shall

- Not threaten the hygiene or health and safety of people
- Not have an exceedingly high impact, over their entire life cycle, on the environmental quality or the climate
- Keep the noise perceived by the occupants or people nearby to a level that will not threaten their health and will allow them to live in satisfactory conditions
- Be energy-efficient, using as little energy as possible during their construction and dismantling
- Ensure sustainable use of natural resources through reuse or recyclability after demolition, durability of the construction works, and/or use of environmentally compatible raw and secondary materials

Further move toward sustainability of the Construction Products Regulation was already announced in the European Green Deal and in the Circular Economy Action Plan.

Following up on the Construction 2020 Strategy and in line with the new Circular Economy Action Plan, the EC presented the paper "Circular Economy -Principles for Building Design," suggesting an approach to a circular design, focused on durability and adaptability, as well as waste reduction and high-quality waste management.

Nine percent of carbon emissions stems from the production of construction materials. To curb this, the EU has decided that after 2020, 70% of all construction waste must be recycled (Directive 2008/98/EC).

Example: HoHo Vienna is currently considered to be the world's tallest hybrid timber tower (84 m). Its slender profile is made through an innovative combination of concrete, steel, and timber structural components. The use of wood saves 2800 tons of CO_2 compared to a reinforced concrete structure. The construction methods saved 300,000 MWh of energy. Thanks to the level of thermal insulation, resource conservation, energy efficiency, and sound insulation, it has been awarded the certification LEED Gold and ÖGNB gold from the Austrian Society for Sustainable Buildings [4].

15.8.2 Reusing Existing Buildings

Preserving and extending the lifetime of existing buildings normally results in fewer carbon emissions than demolishing and building new ones. To reduce emissions created during the production of building materials, existing buildings should be reused where possible and space efficiency maximized to avoid building more.

15 Construction Industry and Sustainability

The new buildings shall be planned for a longer lifetime with flexibility accommodating different functions.

Example: In Oslo, there is a plan to reuse an 18-floor building from 1970. If the building is demolished and a new one built, the building materials would account for approximately 4000 tons of carbon. But by refurbishing and retaining most of the heavy structure, the emissions will be 55% lower [34].

15.8.3 Designing for Reuse

Solutions exist to make movable buildings. Parts of buildings can be detached without destroying the main structures. Elements of lightweight wooden structures are quick to assemble on-site and can then be moved somewhere else or even implemented as a part of another building. Modular steel structures with sandwich panel roofing and walls also provide similar opportunities.

Example: A Finnish building supplier provides prefabricated learning spaces that can be dissembled and moved to another site when needed. Schools, preschools, gyms, and cantinas are made like this.

15.8.4 Reusing Materials in New Built

When entire structures cannot be retained and reused satisfactorily, reusing existing construction materials makes a good choice, especially for concrete. Global cement production represents about 5% of total carbon emissions, twice as much as emissions from air travel [34].

Researchers present different end-life scenarios to be considered selecting materials: steel is 93% recyclable, but with a big amount of energy and is very heavyweight, so it represents a suitable and sustainable solution only if the transport request is minimal. Otherwise, wood is recyclable or reusable with a smaller percentage, but if the construction is designed to be easily moved, wood can be the most sustainable choice. Increasingly, carbon-negative material such as cross-laminated timber (CLT) or glue-laminated timber, commonly known as glulam, is being used.

Circular economy efforts will be central to reduce the use of virgin resources in construction. Embodied carbon, or the materials that go into a building's construction like steel and cement, makes up 11% of global emissions. According to the nonprofit group Architecture 2030, slashing emissions from this part of the construction process is essential as it is difficult to decarbonize materials once they are locked into a building.

The concrete, structural steel, and copper from demolishment can be recycled and reused, diverting the materials from the landfills and cutting the costs.

Research and development activities help to find new ways to recycle or reuse construction and demolition waste. For example, road rehabilitation can be done by reusing the damaged asphalt layer. Some innovative solutions demonstrate the reuse of tires as geo-grid, to be filled with crushed stone in road construction.

Some of the most interesting innovations are found in the transformation of agricultural and industrial waste into ecological raw building materials. It is possible to refer, for instance, to the steelmaking slag, which is a by-product of steel production, that can be used as a material for road construction, or to the gypsum resulting from the desulphurization of a TPP that can be used in construction. Further, "orb" is a material made out of food or agricultural waste like cocoa husks, dried orange peel, ground blue pea flowers, and mycelium insulation, obtained by feeding waste to the root system of mushrooms. Also, an industrial by-product substituting cement in ultra-low-carbon-dense concrete blocks is becoming a common material.

15.9 From Sustainable Architecture to Carbon-Neutral Countries

15.9.1 Sustainable Architecture

"Sustainable architecture" projects minimize the negative environmental impact of buildings by efficiency and moderation in the use of materials, energy, and development space [30]. At the same time, they secure healthier conditions and are in harmony with the existing landscape.

Sustainable buildings are designed and constructed with (i) environmentally friendly building materials: mostly recycled or renewable, requiring the least energy to be manufactured and transported, often locally obtained woods and stone; (ii) renewable energy sources, energy-efficient lighting and appliances, maximum protection against the loss of warm or cool air, appropriate insulating materials and window glazing, also they are oriented to take full advantage of the sun's position; (iii) adequate ventilation, temperature control, nontoxic materials; and (iv) earth shelters, roof gardens, and extensive planting.

Sustainable architecture should re-embrace local and traditional designs for sustainable solutions. Examples include roof slopes and shapes for optimal protection from the snow and wind, the realization of different walls, depending on the climate and local materials, window design to maximize exposure to light, and the protection from the sun, the shaping of apartments to allow natural ventilation.

Traditions shall be integrated with new technologies and building management systems [6].

Passivhaus buildings are a good model: They provide a high level of occupant comfort while using very little energy for heating and cooling. They are built with meticulous attention to detail and rigorous design and construction according to principles developed by the Passivhaus Institute in Germany and can be certified through an exacting quality assurance process.

New technology comes at a price, but once installed, makes for more cost-effective buildings in the long term, with high energy efficiency and savings in water and gas. Maintenance costs in LEED-certified buildings are thought to be 20% less than conventional commercial buildings. Retrofits are reducing costs by 10% for 1 year.

Example: The renovation of the Empire State building was paid back in 4 years from the savings (CNBC).

15.9.2 Green Structures

The current revival of green architecture began in 1970 from the energy crisis coupled with growing awareness of humankind's impact on the environment. An increasing number of architects began to incorporate green roofs and other energy-saving measures into their projects. Others integrate trees into the buildings. Green space has a substantial physical and psychological impact on our urban environments. Studies have shown that even a small green park can reduce local surface temperature by 7°C, while urban trees planted along streets can reduce temperatures by up to 3.9°C. Green roofs used across an urban area could reduce temperatures by 7°C. Some studies claim that trees can reduce local concentration of NO by up to 57%.

Example: Since 1991, the city of Zürich has made it mandatory to have all flat roofs that are not used as roof terraces to be "greened" when constructing new housing developments or renovating older ones. The main reason for this policy is to increase biodiversity. Of course, these green roofs also play a role in the water cycle as well as reducing heat stress.

The designers and engineers working on the latest generation of tree-scrapers have also decided to tackle the issue of embodied carbon and the considerable amounts of energy consumed when manufacturing steel and concrete.

Examples: Royal Pickering Hotel Singapore and Bosco Verticale Milan.

The carbon-eating 100-m-high skyscraper Tao Zhu Yin Yuan Tower in Taipei, in the form of a DNA double helix, was designed by the Belgian architect Vincent Callebaut. The 42,000 sqm building is covered in approximately 20,000 trees and shrubs, aiming to absorb around 130 tons of CO_2 annually. It will produce more energy than it needs with a solar and wind power system on its roof [23]. The core cylinder creates natural ventilation that can reduce indoor temperature and decrease air conditioning consumption: with its double skin facade, it extracts air at ground level by heating it in a specially adapted glass greenhouse at the base before letting the air pass through a series of filters and releasing clean air at the top. There's a rainwater recycling system installed to feed the ground floor sprinklers. The materials are selected through recycled and/or recyclable labels. Yet, this unique example of modern green architecture gives home to only 80 families and the innovations hardly may balance the CO_2 emissions from its huge construction.

15.9.3 Sustainable Transport and Urban Planning

Sustainable transport and urban planning can be implemented, in varying proportions, depending on the urban development situation and the prevailing land-use patterns. Better outcomes may be achieved moving from isolated investments to integrated approach [11]. Pursuant to the European Green Deal, to achieve climate neutrality, a 90% reduction in transport emissions is needed by 2050 [14]. Construction projects for sustainable transport include:

- Metro lines
- Tram and district light railways
- High speed railways
- Intermodal terminals to shift freight from road onto rail and inland waterways;
- Integrated city transport projects, securing smart zero emissions public transport, safe cycling alleys and digital traffic control
- Re-designing of the existing transport system from "car-oriented" to "mass-transit-oriented", with smart traffic management prioritizing pedestrian, cycling and public transport traffic [36]
- Infrastructure for electric transport
- Ring roads taking the traffic out of the cities

Underground streets and parking areas, reducing pollution and congestion. Improving infrastructure in line with the sustainability goals includes planning and construction ensuring the connectivity of urban areas and cutting down the excess mileage implementing the principles of "mixed-use" urban planning (the 15-minute city). The "15-minute city" is an approach to urban design that aims to improve quality of life by creating cities where everything a resident needs can be reached within 15 minutes by foot, bike or public transit. This concept puts an emphasis on careful planning at the neighbourhood level, giving each district the features it needs to support a full life – including jobs, food, recreation, green space, housing, medical offices, small businesses and more. And importantly, it's a full life that doesn't require a car [32]. Sustainable urban planning and construction are essential to implement this concept, becoming the best solution in the COVID-19 "new normal'. Paris, Melbourne, Detroit, Portland, Ottawa are pursuing the 15- or 20-minute-city concept. C40 Cities, a city-led coalition focused on fighting climate change, elevated the 15-minute city idea as a blueprint for post-Covid economic recovery [5]. Example: Nordhavn was converted from Copenhagen's industrial port to a mixed-use residential and commercial neighborhood, designed as a self-contained community. The 'five-minute city' is planned in a way to take five minutes to walk from an apartment to a kindergarten, shops and public amenities [28].

15.9.4 Sustainable Land-Use Planning

Wise land-use planning shows great potential to reduce negative impacts on biodiversity, animal habitats, soil structures, and other qualities of nature.

Designing good landscape plans at an early stage makes it possible to avoid unnecessary blasting, digging, and transport. For every 5–10 cubic meters of rock blasted, one truck is needed to transport it. When a significant change of terrain is needed, the masses should be used for landscaping on or near the site.

15.9.5 Mega Projects, Mega Impact

The International Olympic Committee President reported in 2018 that its "goal is to make the Olympic Games a catalyst for sustainable development of cities."

Actually, many sport venues are demolished after the games. Past experiences show that outcomes from staging major events are mostly harmful.

Examples: In the founding city of the Olympics, the Olympia in Athens built for the 2004 Games, is a total ruin. Venues like the beach ball volleyball stadium, weightlifting, table tennis, and swimming have all been left to rot. Many Greeks blame the 2004 Olympics for contributing to their country's economic collapse, spending over $11 billion on an Olympic Village and then abandoning it. Only the Olympic Stadium has remained actively used [29]. In 2016, Rio de Janeiro hosted the first Olympic Games in Latin America in nearly 40 years. Brazil had big plans for the recycling and repurposing of the venues, put on hold due to lack of funds. The photograph shows what's left of the demolished Rio Media Center. Many Olympic sites are a health hazard littered with rusty metal.

London 2012 is the first Olympic host to try to deliver a holistic sustainability program from construction and creating legacy for future generations [8]. The Olympics allowed to reclaim 75 ha of polluted and contaminated soil, give a new park and open space to the local communities, reducing the imbalance between the west and east sides of London. An on-site soil-washing center was built to reduce the distance that soil had to travel. Great attention was given to the sustainability of each building. To illustrate, the velodrome was built with 100% sustainably sourced timber, and its resource-efficient approach to construction led to £1.5 million savings from the cable-net roof design alone, requiring about 1000 tons less steel and embodied carbon savings of over 27%. The site is highly accessible by public transport [2].

15.9.6 Smart Carbon-Neutral Cities

Example: The Copenhagen 2025 Climate Plan is a collection of concrete goals and initiatives that Copenhagen must implement in order to become the world's first carbon-neutral capital. Of course, many of the initiatives involve construction activities:

- Renewable energy sources—wind turbines, solar panels, biomass, geothermal energy, waste to energy
- Cogeneration of heat and power

- A central heating distribution system
- Using the distribution system for cooling with cold water from the river
- Energy retrofitting of buildings
- New build constructed for low-energy use
- Green roofs
- Infrastructure for a City of Cyclists, developing public transport and implementing Intelligent Traffic Systems
- Smart-grid integration of electricity, heat, energy-efficient buildings and transport into one intelligent energy system

195 countries have joined the Paris Green Deal to strive for minimum CO_2 emissions. Examples: Sweden pledged to reach carbon neutrality by 2045. UK, France, Denmark, New Zealand, Hungary set into law a target of reaching net-zero emissions by 2050, followed by others [27].

15.10 Sustainable Construction Enablers

15.10.1 Regulations

Regulatory initiatives at an international, EU, and national level provide a strong incentive for the optimization of energy and resource consumption [10]. Recent developments in the international regulatory framework on sustainability and climate change have given a strong impetus for more progress in the construction sector.

At the European level, there are several initiatives and directives focusing on resource and energy performance, such as the Energy Efficiency Directive (EED) (2012/27/EU), the Energy Performance of Buildings (EPBD) (2010/31/EU), the Renewable Energy Sources (RES) Directive (2009/28), and the Waste Framework Directive (2008/98/EC), just to name a few. A longer-term perspective is set out in the Roadmap for moving to a competitive low-carbon economy in 2050 and the Energy Roadmap 2050. Specifically, the climate and energy policy framework for 2030 sets a number of objectives for lowering energy consumption and greenhouse gas emissions, aiming for a 27% improvement in energy efficiency by 2030.

In March 2020, the European Commission adopted a new Circular Economy Action Plan – one of the main building blocks of the European Green Deal. Pursuant to it, the EC is launching a new Strategy for a Sustainable Built Environment, based on learnt lessons. This strategy will ensure coherence across relevant policy areas such as climate, energy and resource efficiency, management of construction and demolition waste, accessibility, digitalization, and skills. It will promote circularity principles throughout the life cycle of buildings [12, 13] by addressing construction products' sustainability in line with the Construction Product Regulation's revision, including recycling requirements, by promoting the durability and adaptability of built assets and by using level(s) to integrate life-cycle assessment in public procurement and the EU sustainable finance framework as well as to explore potential carbon reduction targets and carbon storage. Higher material recovery targets for construction and demolition waste will be also considered, as well as the promotion of initiatives to reduce soil sealing, rehabilitate abandoned or contaminated brownfields, and increase the safe, sustainable, and circular use of excavated soils.

Furthermore, the Green Deal's "renovation wave" initiative can lead to significant improvements in energy efficiency in the EU. The initiative will be implemented in line with circular economy principles, notably optimized life-cycle performance, and longer life expectancy of built assets.

National legislations introduce mandatory requirements for environmental impact assessment, sustainable water, energy, and waste management, energy efficiency, protection of cultural heritage and biodiversity, sustainable planning, and land use.

The introduction of energy-efficiency requirements in EU national building codes resulted in reducing half of the energy consumption of new buildings compared to typical buildings constructed in the 1980s.

National governments, alongside industry stakeholders, launch different initiatives to foster the adoption of BIM in the sector.

Example: The *EU Construction & Demolition Waste Management Protocol and Handbook* shows best practices of waste management in the construction workplace and in concrete production, raising awareness among multilingual construction workers.

15.10.2 *Sustainability Standards and Voluntary Certification*

There is now a proliferation of standards, rating, and certification programs to help guide, demonstrate, and document efforts to deliver sustainable, high-performance buildings. There are hundreds of green product certifications in the world, and the numbers continue to grow. Green building rating programs vary in their approach: some outline prerequisites and optional credits; others take a prescriptive approach; others suggest performance-based requirements that can be met in different ways.

Green building rating and certification systems require an integrated design process to create projects that are environmentally responsible and resource-efficient throughout a building's life cycle: from siting to design, construction, operation, maintenance, renovation, and demolition [38].

Key standards enabling sustainability in the design process are, among others,

- EN and ISO standards for building structures
- Whole-building LCA (EN15978)
- Level(s)
- Reversible Building Design protocol/tool (RBD)15
- Cost–benefit analysis
- ISO standards for DfD/A (ISO 20887)
- BIM-related standards

The incentive for applying for certification is to gain recognition for the sustainability performance of development and gain a competitive advantage in the market.

An increasing number of companies are interested in making their resource-efficiency efforts visible to the outer world by using environmental performance certification schemes such as the EU Eco-Management and Audit Scheme (EMAS).

Demonstration of actual performance in use may be necessary through requirements for Energy Performance Certificates (EPCs).

Example: LEED (Leadership in Energy and Environmental Design) is the most widely used, internationally recognized third-party verification for green buildings. By focusing on resource efficiency, reduction in emissions, health aspects, and a sustainable, cost-efficient building life cycle, LEED considers ecological, economical, and sociocultural factors into consideration with regard to the construction of new buildings. LEED is an open system that adapts to the current state of technical development. In addition to minimum requirements, project-type-specific credits for the following sustainability-related areas of construction can be obtained: "sustainable sites," "water efficiency," "energy and atmosphere," "materials and resources," and "indoor environmental quality [1]."

15.10.3 Financial Instruments

Operating costs of green buildings can be reduced by 8–9% while increasing in value up to 7.5%. Many sustainable buildings have also seen increases of up to 6.6% on return on investment, 3.5% increases in occupancy, and rent increases of 3%.

Sustainable construction brings savings for the whole life of the buildings and facilities. The problem is the higher upfront investment, which is a barrier for many owners and company managers. Financial incentives need to be tailored to each situation, considering both financial aspects through a whole life-cycle perspective, taking into account costs, revenues, and residual value and scenarios in which estimated costs for new materials, furniture, and waste elimination are significantly higher than the actual costs for renovation/refurbishment in which certain elements could be sold for reuse and/or recycling.

Different financial instruments may help to incentivize the stakeholders, including green tariffs to support renewable energy, the volatility of energy and resource costs, the opportunity to sell excess energy from roof or façade solar panels or data centers; reduced taxes for certified energy-efficient buildings, increased taxes for using landfills, reduced waste management costs by decreased reliance on landfill disposal/improved resource efficiency, grants for innovations and sustainable retrofits, credits lines from development and commercial banks with preferential conditions for sustainable projects, ESCO contracts with payments for energy-efficiency outcomes [15], and subsidized schemes for integrated transport and land-use developments.

Energy Performance Contracting (EPC) is a form of "creative financing" for capital improvement that allows funding energy upgrades from cost reductions. Under an EPC arrangement, an external organization (ESCO) implements a project to deliver energy efficiency, or a renewable energy project, and uses the stream of income from the cost savings, or the renewable energy produced, to repay the costs of the project, including the costs of the investment. Essentially the ESCO will not receive its payment unless the project delivers energy savings as expected.

Long-term integrated transport and land-use planning strategies may use tax incentives and subsidized schemes to steer developments in a sustainable direction [31]. It is important to ensure a realistic and detailed budget calculation for each proposed measure/policy/infrastructure and their correlation with the national, regional, and local funding schemes to help prioritize and effectively implement the planned measures. This will be more successful if rooted on a thorough analysis of the funding and financing frameworks and a certain degree of political endorsement from the early steps of the process.

Considering the complexities associated with rate setting and the magnitude of the water challenges, FIDIC recommends that prices for water, wastewater, ecosystems, and stormwater services be established with the assistance of qualified professionals and be sustainable enough to recover operational costs, allow for maintenance, rehabilitation, and expansion, providing reasonable, perhaps subsidized charges for the poorest.

15.10.4 Lean Thinking

The co-evolution of "lean and green thinking" enables sustainable construction.

"Lean thinking" has evolved as a popular strategy in construction, addressing the need for efficiency and waste reduction [3]. Minimizing the interruption of the process, cost, and time overruns in construction projects reduces the negative impact on the environment.

On top of the overall benefits of lean thinking, the following lean practices mostly contribute to the sustainability of the design and construction activities:

- Virtual procurement, design, and construction methods: They reduce the level of paperwork, traveling, and energy consumption
- Prefabrication and modularization: They encourage the manufacturing of construction components off-site, promote recycling and efficient use of materials; they also help to reduce the level of noise and air pollution and other harmful effects on the surrounding environment, resulting from the movement of workers, machines, materials, and erecting of temporary structures and other activities for the production of components on-site

- Just-in-time: Helps to reduce damage, waste of materials, and movement of vehicles and equipment resulting from untimely supplies during the construction phase

15.10.5 Digital Technologies

Digital technologies such as artificial intelligence, 5G, cloud, 3D printing, drones, edge computing, and the Internet of Things can accelerate and maximize the impact of policies to deal with climate change and protect the environment. The digitalization of the construction industry lags behind (only agriculture is less digitalized). Its development has a lot of potential for sustainable outcomes.

The first 3D printed social housing projects are already a reality in France.

In the United Kingdom, the self-repairing cities project is delegating road maintenance to drones that can identify, diagnose, and repair potholes.

Construction machines using digital design data save resources: The digitally managed equipment does not excavate excess quantities; the drones and digital survey devices save time and carbon emissions from the fewer miles of the surveyors' vehicles.

The digital information is easily accessible and reusable. The digital survey data can be integrated into the building information model (BIM), where the design is developed. The clashes between different design disciplines are easily detected in the model, thus saving wasted materials for rework due to design errors. Using value-driven visualizations in BIM, a project can get insights about crucial sustainability aspects, such as net carbon emissions, costs, and the potential for reuse, recycle, and upcycle. The energy efficiency of different solutions for roofing, window shades, wall insulations, and systems may be easily assessed. BIM technology makes it possible to maintain information on the plant and materials, which will be essential for assessing the potential for deconstruction as opposed to classic demolition. BIM and building passports can support the transfer of information, enabling best use, maintenance, repair, or adaptation of the building and its systems.

Producing digital twins of the buildings to learn their impact on the environment can provide crucial information on their life performance, enabling improvements. Leveraging artificial intelligence and machine learning, the digital twins can improve energy resilience, reduce operating costs, increase resource efficiency, and help decarbonize the buildings.

Examples: NTU Singapore's goal was to become the greenest campus in the world, reducing the energy, water, and waste footprints by 35% by 2020. Phase 1 of the project concentrated on creating a master plan model of the EcoCampus with energy signatures for each building, used as a baseline. A selection of new technologies relating to building envelope, lighting and occupancy sensors, plug load management, and optimized chillers were simulated using the calibrated models to determine potential savings. The results are 31% energy-saving potential, 9.6 kt carbon savings.

15.10.6 Education and Innovations

Education and training enabe sustainability in construction by

- Raising awareness about the sustainability agenda and lean thinking tools
- Promoting understanding and use of sustainability standards, schemes, calculators, and examples
- Developing a new design culture preventing premature building demolishment
- Using pre-deconstruction or pre-development audit and appropriate information tools
- Stimulating scenario-thinking during the pre-design phase to get insights into the different possible uses of the building over time
- Providing information on the reuse, recycling, and recovery potential of construction products to encourage a reduction of natural resource depletion
- Creating awareness about the fact that reversible products might use more resources at the start, but make it possible to recover resources and also reuse the product in multiple life cycles
- Training in energy-efficiency skills that build on the current expertise of on-site workers and other construction professionals
- Better knowledge about construction techniques to facilitate deconstruction and to enhance durability and adaptability of a building
- Introducing new technologies and digitalization
- Appropriate maintenance of buildings and components using BIM and passports

Fostering research and development can reduce the cost of high-energy performing projects (through BIM technology, for instance) or in recycling/reusing waste, inventing new materials and new ways to save/reuse water and help biodiversity. More support for research, collaboration, and knowledge transfer between academia and industry actors is essential.

15.10.7 Sustainable Procurement

The procurement of services, works and supplies has a major role in achieving sustainable outcomes. The role of public procurement in advancing sustainable construction is essential [37]. Government and building owners can specify in the tender documents the use of sustainable building rating schemes. The digital procurement and the use of open online registers instead of submitting hard copies as evidence for capability, save paper, time and money. The early selection of consultants enables sustainable planning and holistic approach. After the decision to build or renovate, the tender strategy is developed, considering the whole life cycle. Collaborative procurement may use different instruments like BIM, alliance contracting, frameworks, to achieve more sustainable outcomes by the early involvement of:

- the contractors for value engineering solutions during the design phase
- the supply chain for further sustainable solutions
- the operators/ beneficiaries to focus on their needs, improving sustainability

Alliancing and frameworks bring sustainability to SMEs, local employment and local supply chain, reducing emissions from transportation of resources. Education and implementation of sustainable practices is much easier in long-term contracts.

Example: The Construction Playbook UK 2020 [16] will, by creating the right environment, enable the construction industry to focus on a whole life sustainable outcomes. It brings together commercial best practices and reforms to drive greener delivery through a safer, more innovative, manufacturing-led approach which will increase the end-to-end speed of projects and programmes, advocating:

- Contracting authorities to have a clear understanding of value, desired/required outcomes and how these align to net zero GHG emissions by 2050
- Robust evaluation processes and criteria for better, faster, greener delivery
- Good contract management improving delivery
- Strategic supplier relationship management unlocking additional value and innovation
- Operators' early and continuous engagement

15.11 Conclusion

The construction industry has a huge impact on the sustainability agenda. Its transformation from the biggest polluter and resource consumer toward the sustainability enabler is happening, but rather slowly. Led by the architects and engineers consultants, all stakeholders must spread the knowledge, support innovations, and consider the optimal solutions with less intervention into the natural environment. Every day we make choices: to drive or walk, to build or renovate. Now we have to act with a sense of urgency since obviously we cannot reach by 2030 the UN goals for sustainable urbanization, affordable housing, and basic services for everyone. Robust regulations, financial instruments supporting sustainability, digitalization, and standardization can multiply the green procurement and construction pilot initiatives.

References

1. Alpin Limited. (2016/2017) *LEED costs, benefits and ROI*. Retrieved June 7, 2021, from https://www.alpinme.com/leed-costs-benefits-and-roi/
2. Azzali, S. (2017). Queen Elizabeth Olympic Park: an assessment of the 2012 London Games Legacies. *City, Territory and Architecture, 4*, 11.
3. Babalola, O., Ibem, E., & Ezema, I. (2018). Implementation of lean practices in the construction industry: A systematic review. *Building and Environment, 148*, 34–43.
4. Build Up. (2020). *The European portal for energy efficiency in buildings, HoHo Vienna: The first city office built of timber in Vienna*. Retrieved June 7, 2021, from https://www.buildup.eu/en/practices/cases/hoho-vienna-first-city-office-built-timber-vienna
5. C40 Cities. (2020). *C40 Mayors' Agenda for a green and just recovery*. Retrieved June 7, 2021, from https://www.c40.org/other/agenda-for-a-green-and-just-recovery
6. Dasym. (2019). *Sustainable architecture needs technology and tradition*. Retrieved June 7, 2021, from https://www.dasym.com/sustainable-architecture-needs-technology-and-tradition/
7. Davis S. (2020) What abandoned Olympic venues from around the world look like today. *Insider*. https://www.businessinsider.com/abandoned-olympic-venues-around-the-world-photos-rio-2016-8?r=US&IR=T
8. Dongre A. (2015) *Green game and societal sustenance: A case of London Olympic 2012*. Retrieved June 7, 2021, from https://mpra.ub.uni-muenchen.de/63818/
9. European Commission. (2015). *Savings and benefits of global regulations for energy efficient products*. Retrieved June 7, 2021, from https://ec.europa.eu/energy/studies/savings-and-benefits-global-regulations-energy-efficient-products-%E2%80%98cost-non-world%E2%80%99-study_en
10. European Commission. (2019a). *European construction sector observatory: Improving energy and resource efficiency*. Retrieved June 7, 2021, from https://ec.europa.eu/docsroom/documents/33883/attachments/1/translations/en/renditions/native
11. European Commission. (2019b). *Sustainable urban mobility planning and governance models in EU metropolitan regions*. Retrieved June 7, 2021, from https://sumps-up.eu/fileadmin/user_upload/Tools_and_Resources/Publications_and_reports/Topic_Guides/sump_metropolitan_region_guide_v2.pdf
12. European Commission. (2020a). *Circular economy principles for building design*. Retrieved June 7, 2021, from https://ec.europa.eu/docsroom/documents/39984
13. European Commission. (2020b). *Designing buildings in the context of the circular economy*. Retrieved June 7, 2021, from https://ec.europa.eu/growth/content/designing-buildings-context-circular-economy_en
14. European Commission. (2020c). *European Green Deal and construction: Commission launches open public consultation for the revision of the Construction Products Regulation*. Retrieved June 7, 2021, from https://ec.europa.eu/growth/content/european-green-deal-and-construction-commission-launches-open-public-consultation-revision_en
15. European Energy Efficiency Platform. (2021). *Energy performance contracting*. Retrieved June 7, 2021, from https://e3p.jrc.ec.europa.eu/articles/energy-performance-contracting
16. HM Government. (2020a). *The construction playbook 2020*. Retrieved June 7, 2021, from https://assets.publishing.service.gov.uk/government
17. HM Government. (2020b). *Energy white paper: Powering our net zero future*. Retrieved June 7, 2021, from https://www.gov.uk/government/publications/energy-white-paper-powering-our-net-zero-future
18. International Federation of Consulting Engineers. (2012). *Sustainable infrastructure*. Retrieved June 7, 2021, from https://fidic.org/books/state-world-report-2012-sustainable-infrastructure
19. International Federation of Consulting Engineers. (2015). *Water challenges*. Retrieved June 7, 2021, from https://fidic.org/books/state-world-report-2015-water-challenges
20. International Federation of Consulting Engineers. (2021a). *Time to $Tn-vest*. Retrieved June 7, 2021, from https://fidic.org/events/time-tn-vest-fidic%E2%80%99s-state-world-report-series

21. International Federation of Consulting Engineers. (2021b). *Establishing the value of water—The business case for change*. Retrieved June 7, 2021, from https://fidic.org/events/establishing-value-water-business-case-change-fidic%E2%80%99s-state-world-report-series
22. Hvistendahl M. (2008) China's three gorges dam: An environmental catastrophe? *Scientific American*. https://www.scientificamerican.com/article/chinas-three-gorges-dam-disaster/
23. Karan K. (2017) Tao Zhu Yin Yuan, the DNA look-alike pollution and carbon-eating tower. *The Plunge Daily*. https://mybigplunge.com/news/carbon-eating-tower-tao-zhu-yin-yuan/
24. Katy Bartlett K., Blanco J.L., Rockhill D., G. Strube (2019) *Breaking the mold: The construction players of the future*. Retrieved June 7, 2021, from https://www.mckinsey.com/business-functions/operations/our-insights/breaking-the-mold-the-construction-players-of-the-future#
25. Laurent B., Horgnies M., Dubois-Brugger I., Gartner E. M. (2014) "Smog-eating concrete": A new technology for better cities. *Paper presented at the 9th international concrete sustainability conference, Boston, 12–14 May 2014*
26. Morgan, S. (2019). *Construction clean-up looms as Green Deal approaches*. Retrieved June 7, 2021, from https://www.euractiv.com/section/energy-environment/news/construction-clean-up-looms-as-green-deal-approaches/
27. Murray, J. (2020). *Which countries have legally-binding net-zero emissions targets?* Retrieved June 7, 2021, from https://www.nsenergybusiness.com/news/countries-net-zero-emissions/
28. Nikel, D. (2019). Copenhagen aims to be world's first carbon neutral capital by 2025. *Forbes*. https://www.forbes.com/sites/davidnikel/2019/08/22/copenhagen-aims-to-be-worlds-first-carbon-neutral-capital-by-2025/?sh=e9fb8fe7fb3c
29. Ponic, J. (2020). *Abandoned Olympic venues*. Retrieved June 7, 2021, from https://howtheyplay.com/olympics/AbandonedOlympicVenues
30. RISE. (2015). *Sustainable architecture principles: environment, energy, efficiency*. Retrieved June 7, 2021, from https://risedesignstudio.co.uk/blog/index.php/2015/08/26/sustainable-architecture-principles/
31. European Platform on Sustainable Urban Mobility Plans. (2019). *Guidelines for developing and implementing a sustainable urban mobility plan*. https://www.eltis.org/mobility-plans/sump-guidelines
32. Sisson, P.. (2020). *What is a 15-minute city?* Retrieved June 7, 2021, from https://citymonitor.ai/environment/what-is-a-15-minute-city
33. Stratton, M.. (2020). Carbon-free Copenhagen: How the Danish capital is setting a green standard for cities worldwide. *National Geographic*. https://www.nationalgeographic.co.uk/travel/2020/03/carbon-free-copenhagen-how-danish-capital-setting-green-standard-cities-worldwide
34. SWECO. (2020). *Going circular—A vision for urban transition*. Retrieved June 7, 2021, from https://www.swecourbaninsight.com/climate-action/going-circular-a-vision-for-urban-transition/
35. Tegola, A., Longo, F., & Lanzilotti, A. (2019). The pavilions of Expo 2015 in Milan, as a privileged observatory about the concept of sustainable construction in all languages of the world. *Sustainable Buildings, 4*, 1.
36. UNECE. (2020). *Handbook on sustainable urban mobility and spatial planning promoting active mobility*. Retrieved June 7, 2021, from https://unece.org/transport/publications/handbook-sustainable-urban-mobility-and-spatial-planning
37. United Nations Environment Programme. (2015). *Sustainable consumption and production: A handbook for policymakers*. Retrieved June 7, 2021, from https://webcache.googleusercontent.com/search?q=cache:0WIPIMqgpMYJ:https://sustainabledevelopment.un.org/index.php%3Fpage%3Dview%26type%3D400%26nr%3D1951%26menu%3D35+&cd=2&hl=it&ct=clnk&gl=it
38. Vierra, S. (2019). *Green building standards and certification systems*. Retrieved June 7, 2021, from https://www.wbdg.org/resources/green-building-standards-and-certification-systems
39. Watts, J. (2019). Concrete: The most destructive material on Earth, *The Guardian*. https://www.theguardian.com/cities/2019/feb/25/concrete-the-most-destructive-material-on-earth

Adriana Spassova is a civil engineer with 35 years' professional experience as a designer, construction manager, and consultant with MSc in construction law and dispute resolution at King's College London and a PhD in construction project management.

She managed international construction contracts for 10 years as head of Logistics Department, construction manager, later head of Tender and Contracts Department in the biggest steel structures contractor in Bulgaria. In 2000, she founded the consulting company EQE Control, as part of ABS Group, and continued her career as partner, managing projects in nuclear safety/RES/waste/infrastructure/buildings. She is a FIDIC expert, Dispute Board in Europe and Asia. She has delivered FIDIC Accredited trainings for more than 1000 participants in Europe, Asia, and America.

She has published more than 20 articles. She is a co-author in:

- Mosey, D. (2019) Collaborative Construction Procurement and Improved Value, Chapter 22.
- BSCL Construction Law Guide (2021) Chapters 1 & 10.

In 2020, she published in Minsk "Construction Contract Tools for Risk and Cost Management" (Инструменты строительных контрактов для управления рисками и ценой).

Chapter 16
Impact of Climate Change in Agriculture: Estimation, Adaptation, and Mitigation Issues

Alessandro Olper and Daniele Curzi

Abstract There is an inherent link between how climate change affects different populations and how the burden of environmental policy is unevenly distributed across individuals within a given population. Who pays or benefits for environmental regulations largely depends on how the impact of climate change is heterogeneous across countries and individuals. In this chapter, we first introduce the relationship between climate change, agriculture, and food security, providing also an overview of the main adaptation and mitigation strategies. To follow, we summarize the emerging methods developed in the last two decades to empirically quantify the economic costs of climate change, by presenting a general framework to explain why it is crucial to better understand how and where climate affects economic behavior.

16.1 Introduction

Over the last decades, human welfare has improved considerably worldwide. This is mostly due to important progress in technology and innovations in production systems. However, despite these important improvements, millions of people still live in extreme poverty without adequate food intake or access to essential services (FAO, 2017). It has been estimated that around 800,000 million people worldwide are chronically undernourished and poor. A large share of these people are children and live in the sub-Saharan region, an area where 40% of the population is poor (Fischer et al., 2002). A large proportion of poor people live in rural areas, where agriculture is essential both as a primary source of food and income. Improving agriculture in this region thus represents a means through which to escape from poverty and undernourishment. Evidence from Christiaensen et al. (2006) suggests

A. Olper (✉) · D. Curzi
University of Milan, Milan, Italy
e-mail: alessandro.olper@unimi.it; daniele.curzi@unimi.it

© The Author(s), under exclusive license to Springer Nature Switzerland AG 2022
S. Valaguzza, M. A. Hughes (eds.), *Interdisciplinary Approaches to Climate Change for Sustainable Growth*, Natural Resource Management and Policy 47,
https://doi.org/10.1007/978-3-030-87564-0_16

that agricultural (GDP) growth is more than two times effective in reducing poverty than a growth in GDP driven by other sectors. However, the agricultural system, especially in the poorest tropical areas, is threatened by climate change (Hertel & Rosch, 2010). Climate change is the consequence of most of the human progress achieved in the last two centuries, which has come at a remarkable cost for the environment. The impacts of changes in climatic conditions are already evident worldwide, and if not properly addressed, they may exacerbate shortly (FAO, 2017). All these issues raise the concern toward the possibility of reaching the sustainable development goal (SDG) of eradicating hunger and malnourishment by 2030 (SDG 2) while improving the sustainability of the agricultural and food system (Campanhola & Pandey, 2018).

Within this framework, an additional channel that raises the concern toward achieving this objective is population growth. Despite a decreasing growth rate in worldwide population growth, there are some regions where the population will increase even beyond the end of this century. In terms of food demand, it is of fundamental importance not only to consider the population in absolute number but especially its growth dynamics across regions, age groups, and locations (FAO, 2017). Global trends of population growth suggest an important heterogeneity between high- and low-income countries. Estimates of the United Nations (UN) indicate that the former group will reach their peak of maximum population by 2040, while the latter will continue growing well beyond 2050. Within low-income countries, there are important differences as well. While Latin American countries are expected to reach their zenith before 2060, South Asia around 2070, African countries are likely to continue their growth beyond the end of this century. It seems, therefore, reasonable to expect that the global population will likely exceed 11 billion people in 2100. Such an important dynamic is also associated with a rapid urbanization and income growth process, which leads to a dietary transition toward higher consumption of animal proteins, fruit, and vegetables, whose production is more resource-intensive than other food (e.g., grains).

There is an inherent link between how climate change affects different populations and how the burden of environmental policy is unevenly distributed across individuals within a given population. Who pays or benefits for environmental regulations largely depends on how the impact of climate change is heterogeneous across countries and individuals. In this chapter, we first introduce the relationship between climate change, agriculture, and food security, providing also an overview of the main adaptation and mitigation strategies. To follow, we summarize the emerging methods developed in the last two decades to empirically quantify the economic costs of climate change by presenting a general framework to explain why it is crucial to understand better how and where climate affects economic behavior. Our discussion focuses particularly on the agricultural sector, admittedly the most sensitive sector to climate change, although the econometric methods described below can be applied to any economic activity.

16.2 Impact of Climate Change on Agriculture and Food Security

It is widely acknowledged that climate change will lead in the future to a rise in the carbon dioxide (CO_2) concentration in the atmosphere, which will determine an increase in worldwide temperatures. Current climate change mitigation policies are unlikely to amend this trend over the next 20–30 years (Hertel & Baldos, 2016). Projections of future rise in temperatures depend on many factors, but it seems reasonable to expect an increase of about 0.3–0.4 °C per decade in most agricultural areas by the mid of this century. This increase in temperature will have an important impact on agricultural productivity, although with relevant differences depending on the latitude. Tubiello and Rosenzweig (2008), in reviewing the main literature on the impact of climate change on agriculture, summarize the main findings as follows: an increase in temperature up to 3 °C is likely to have beneficial effects on agricultural and pasture yields in the first part of this century for countries in temperate zones. These effects would turn negative in semiarid and tropical zones, even with a lower increase in temperature (1–2 °C). However, further temperature increases projected by the different climate change scenarios for the second half of this century would certainly lead to negative consequences for crop and pasture yields in worldwide regions. The Fifth Assessment Report of the IPCC reviewed several studies on the impact of rising in temperature on agricultural yield considering different geographical areas (IPCC, 2014). The reports point out that until about 2030 positive and negative effects of higher temperature on agricultural yields will continue counterbalancing each other, but in the long run, negative consequences will outweigh positive ones, especially in low-income countries. In the poorest, tropical regions of the world, which are characterized by a warm climate, a further increase in temperature will be harmful for crop production. For instance, at very high temperature some crops may not grow at all, or soil may become infertile, due also to faster water evaporation, and thus change in the soil moistures, which eventually threatens crops' productivity (Sachs, 2015).

Climate change will also affect other important determinants of agricultural productivity, such as precipitation patterns, water use, and may lead to a more frequent occurrence of extreme weather events (e.g., droughts or floods). Many dry regions in the tropics that are on the very edge of crop growing will experience a further reduction in precipitation, thus leading the production of food almost impossible. Conversely, wet regions are likely to see an increase in precipitation, which could result in a more frequent occurrence of flooding and storms (Sachs, 2015). Extreme weather events will also intensify their prevalence due to climate change in the next years. These weather-related natural disasters represent relevant concerns, especially for crop production, fisheries, livestock, and forestry, as these sectors strongly depend on weather conditions. FAO (2015) estimated that in the period 2003–2013 the overall agricultural sector accounted for about 22% of all the damages caused by natural disasters in developing countries.

Climate change is likely to threaten agricultural and food production through some additional channels, such as the rising sea level, which could lead some coastal lowlands that are currently farmed to be submerged. Ocean acidification, due to the increasing CO_2 concentration in the atmosphere, is likely to threaten the aquatic environment for marine life, thus reducing the availability of fish for human nutrition and loss of biodiversity as well (Sachs, 2015). The rise in temperature and humidity level is likely to increase the population of insects (e.g., mosquitos, midges, and flies), which may be vectors of diseases (FAO, 2017). These insects could also expand over their present geographical area, thus acting as invasive species in a new environment potentially creating several damages to the entire ecosystem (Sachs, 2015). An important issue is also related to the water availability for irrigated lands, responsible for a large proportion of the global grain production. The current increase in temperature is leading to a retreat of glaciers worldwide. Glaciers play an important role in irrigation, as through their melting, they make water available in the river flows, which is then used for irrigation. However, if the rise in temperature will lead glaciers to retreat further or even disappear, this will lead to the loss of a fundamental source of water for crop production based on irrigation and thus to an enormous loss of food availability.

All these elements are likely to affect agriculture negatively and represent an important threat to adequate food production, especially in more vulnerable areas (Tubiello & Rosenzweig, 2008). The impact of climate change on food security is already evident, but if it is not properly addressed, the consequences will be much more severe beyond 2030.

The effects of climate change on agriculture will also affect food security through some indirect channels, especially in developing countries. The access to food will be limited by the reduction of rural incomes and livelihoods. Such a contraction will likely be due to a shrink in agricultural productivity and the higher occurrence of extreme weather events, which will cause damages to important assets and thus preventing future earning capacity (FAO, 2017). The losses in food supply will also lead to an increase in food prices, which again will negatively affect people in developing countries, as they spend a much higher share of their income on food than in developed countries (FAO, 2016).

When dealing with projections for food supply, it is important to consider in addition to the climate change impact on agriculture also other socioeconomic factors, such as population-growth dynamics, or other elements related to the production processes or market mechanisms (Schmidhuber & Tubiello, 2007). When considering these issues, climate change impacts result to be extremely heterogeneous across countries and regions (IPCC, 2007). Developing countries, especially in the sub-Saharan Africa and South Asia regions, will pay the higher consequences as they are much more vulnerable to climate change than developed countries. This is due to their warmer baseline climate, which stresses the environment, limiting food production. The additional challenges posed by climate change on the environment, the lack of capital for development, and the poor diffusion of adaptation measures will likely lead to enormous difficulties in producing food in these regions (Tubiello & Rosenzweig, 2008).

When considering the absolute number of undernourished people worldwide, it emerges that enormous progress was made in the period 2005–2015. During this period, more than twice as many people escaped from undernourishment compared to the previous decade (FAO, 2017). If this trend confirms this pace in the next years, the SDG of zero hunger will not be achieved in 2030 as still more than 630 million people in developing countries would be undernourished by that time (FAO, 2017). However, this *business as usual* scenario does not take into account potential consequences due to climate change and thus should be interpreted with caution. Some estimations suggest that if the negative effects of climate change in developing countries were considered, an additional number of 5–170 million people could be at risk of hunger by 2100 (Schmidhuber & Tubiello, 2007). However, these projections should be taken with caution as, in addition to the uncertainty on the climate change scenario, they do not consider potential mitigation and adaptation options.

The future projections of fast population growth in some of the poorest worldwide regions call for the necessity to expand agricultural and food production. Looking at the growth in agricultural production of the last 50 years, this would not represent a relevant problem. This is because such growth was mostly due to an increase in agricultural productivity due to the green revolution technologies, land expansion, and higher water availability. The contribution of yield increase was particularly relevant as it has been estimated to be responsible for almost 77% of the increase in agricultural output (Alexandratos & Bruinsma, 2012). However, conditions have changed over time, and any increase in the agricultural output in the future should come from different sources (FAO, 2017). This is because, first, agricultural land expansion is limited by several issues, such as water scarcity in the Middle East or sub-Saharan Africa. Such expansion would come at the costs of further deforestation in other regions, which is not a desirable perspective in environmental sustainability. Moreover, the likely consequences of climate change on agriculture will limit growing foods in many regions (FAO, 2017).

One viable way to expand agricultural output is therefore represented by a further yield growth. However, during the last years, the average pace of yield growth has slowed down. This is mainly due to the intensification of some agricultural practices (e.g., fertilizers and pesticides), which have led to soil depletion and, more in general, to important negative environmental consequences (Hertel & Baldos, 2016). Yet, there exist some viable strategies to increase agricultural yield. The reduction of the yield gap represents the first step that should be taken to increase average yield growth globally (Hertel & Baldos, 2016). The yield gap represents the distance between farmers' actual yield and their maximum potential yield in case the production process was to be implemented in the most efficient way. Lobell et al. (2009) argue that, on average, yield gaps for irrigated agriculture are quite close to the maximum potential. Conversely, when considering rain-fed agriculture, which represents about 40% of the total agricultural production, it is estimated that the average yield gap in low-income countries is higher than 50%, and it is especially large in sub-Saharan Africa (about 76%). Licker et al. (2010) suggest that yield gaps are especially large for maize, wheat, soybean, and rice, and that closing such a gap would lead to an increase in production of about 50%. These numbers reflect the suboptimal use of inputs and the lack of adoption of more productive technologies in these regions (Campanhola & Pandey, 2018). The main problem behind such a large yield gap in developing countries is the lack of economic incentives to adopt high-productive technologies. Therefore, with appropriate investments, there would be room for significant improvements in agricultural yield growth in developing countries, which would increase agricultural output in the next decades (Bruinsma, 2009; Hertel & Baldos, 2016).

However, increased agricultural yield depends not only on economic incentives but also on country-specific socioeconomic factors, such as institutional and economic reforms, investments in infrastructure, education, and research and development (Hertel & Baldos, 2016). R&D investments play a particularly important role in increasing yield gaps, although their actual impact is usually visible in two decades or more (Fuglie, 2012). Since the beginning of this century, R&D investments have been increasing in countries where economic growth is more evident (e.g., China, India, and Brazil). However, investments in R&D in the poorest countries are still much lower than in developed countries, and this gap is growing over time (Campanhola & Pandey, 2018).

In sum, investments in reducing the agricultural yield gap in developing countries, through the right economic incentives and by promoting investments in infrastructure and institutions, appear to be the most viable solution to the problem of feeding a growing population without further compromising the environment.

16.2.1 Main Adaptation and Mitigation Strategies

The challenges posed by climate change call for immediate adaptation and mitigation actions in agriculture. When considering adaptation, it is worth noting that their successful adoption may depend on several issues. First, it is important to consider the pace at which climatic conditions change, as, especially in developing countries, short-run climate shocks do not guarantee enough time to adapt to it (Hertel & Rosch, 2010). Adaptation is also a complex process that comes in the different countries together with other endogenous changes, which involve, for instance, population growth, sociopolitical transformations, migration dynamics, or technological progress. Other main determinants may include environmental–geographical elements or the policy environment in which adaptation strategies are implemented, such as market conditions, social safety nets, or political rights (Hertel & Rosch, 2010).

When considering different adaptation options, it is worth mentioning that this should be considered a two-step process (Antle & Capalbo, 2010). First, adaptation based on existing technologies, and second, adaptation based on the development of new technologies (Hertel & Baldos, 2016). There exist several examples of adaptation strategies. In general, adaptation should aim at increasing the resilience of farmers to climatic change. Among the adaptation options, it should be worth mentioning the improvement in resource use efficiency through sustainable intensification of the existing agricultural farming practices (Campanhola & Pandey, 2018). Innovative adaptation options in the near future should also be devoted to improving water resource management, especially in farming systems already equipped with irrigation (Hertel & Baldos, 2016). Market integration, especially in the poorest agricultural regions, can be an effective tool to reduce the negative consequences of climate change. However, this is problematic due to the lack of adequate infrastructure that connects producers and consumers to the main markets in the most remote regions. As shown by Burgess and Donaldson (2010), higher connection through railroads was an effective tool for rural areas in India to develop and address the negative effects caused by rainfalls in those regions. Access to credit is also an important tool to allow farmers to adopt more innovative and effective agricultural technologies (Hertel & Baldos, 2016).

Concerning new technologies, these should go in the direction of reducing the sensitivity of agriculture to change in climatic conditions, such as extreme droughts, and heat conditions. The use of seeds with an improved resistance capacity to pests and disease or the use of cropping systems that are able to survive inundations (e.g., the use of some rice varieties resistant to flooding) are other important adaptation options (Hertel & Baldos, 2016). The use of biotechnologies may also be effective in adapting agriculture to a change in climatic conditions (Campanhola & Pandey, 2018).

It is really complicated to quantify the actual contribution of the different adaptation strategies. For instance, the use of crop varieties more resistant to droughts or heat waves may reduce agricultural yields, which may be not acceptable for some farmers (Sinclair et al., 2004). Moreover, innovative strategies need some time to show their potential effectiveness, and so the return for innovation investments will come with some lags (i.e., 10–20 or even more years) (Alston et al., 2008). The differences in the dissemination of the adoption strategies in developed vs. developing countries, together with the differences in the R&D investments, cast additional concerns toward the global effectiveness of the different adaptation strategies. Perhaps the most important difference between poor and rich countries is in the farmers' willingness to accept the risk of using alternative strategies to cope with the threats by climate change (Hertel & Baldos, 2016). The decision of using an adaptation strategy and the uncertainty that derive from it may be intolerable for farmers in the poorest regions, who instead may prefer to postpone such investment to reduce the associated risks (Antle & Capalbo, 2010).

When considering mitigation strategies, it is worth highlighting, first, that agriculture, forestry, and land used (i.e., AFOLU) are responsible for more than 30% of global greenhouse gas (GHG) emissions (Crippa et al. 2021). Second, emissions from AFOLU are expected to grow shortly, due to the increasing demand for food, which is driven by population growth, and to a change in dietary patterns toward more animal-source foods, due to the rise in incomes (Campanhola & Pandey, 2018). Therefore, agriculture will not only pay some of the highest consequences of changes in climatic conditions, but it will be itself one of the main contributors to climate change. However, AFOLU can play a crucial role in mitigation strategies.

This would be possible by decoupling the production increase from the emissions increased by reducing the emission per unit of production (Campanhola & Pandey, 2018). There exist several viable climate change mitigation strategies that involve the agricultural sector. Perhaps the most important contribution of AFOLU to climate change mitigation comes from the capacity of forests to absorb carbon dioxide. It has been estimated that nowadays around one-third of global carbon dioxide emissions are absorbed by forests (Angelsen et al., 2010). However, once disrupted, such an important storage system cannot play this vital role anymore, and, in contrast, it is transformed into a major source of emission (Campanhola & Pandey, 2018). Mitigation strategies should therefore point to a twofold objective. First, preserving forests to allow their carbon dioxide absorption capacity. Second, avoid forest degradation and disruption to limit the consequent greenhouse gases emissions.

Other important mitigation strategies point especially to better management of carbon and nitrogen in agriculture through an increase in the efficiency of the different farming practices, which would allow achieving the expected results while minimizing the environmental damages. Other strategies involve reducing the energy and limiting food waste and losses along the food chain, which is estimated to be around 30% of the total food production (Campanhola & Pandey, 2018).

16.3 Estimating the Economic Costs of Climate Change

A better understanding of the relationship between climate and economic outcomes is crucial to implement sound climate change mitigation policies (Newell et al., 2021). In the last decade, a growing body of econometric evidence linking random variation of weather to economic outcomes has amassed both at the country and global level (see Dell et al., 2014; Auffhammer & Schlenker, 2014; Kolstad & Moore, 2020 for recent surveys). An important innovation of the emerging literature is the use of modern panel data econometrics (Deschênes & Greenstone, 2007; Dell et al., 2014; Hsiang, 2016; Blanc & Schlenker, 2017; Auffhammer, 2018) together with short-run (i.e., interannual) variation of weather, to better isolate the (causal) temperature–productivity relationship. This approach is different from the so-called Ricardian model, often used to evaluate the impact of climate change in agriculture, that exploits cross-sectional variation and long-run climate variables.

A crucial aspect of this emerging climate econometric literature is to understand the extent of environmental impacts in terms of a function linking climate exposure to the economic cost that can affect different segments of the population, hence translating into heterogeneous impacts. This is because there is an inherent link between how climate change affects different populations and how the burden of environmental policy is unevenly distributed across individuals within a given population (Hsiang et al., 2019). Who pays or benefits for environmental regulations largely depends on how the impact of climate change is heterogeneous across countries, populations, and individuals.

In this section, we summarize the emerging methods developed in the last two decades to quantify the economic impact of climate change empirically. We start by presenting a simple general framework based on Hsiang et al. (2019), which is useful in explaining why it is important to understand how climate affects economic behavior. Next, exploiting the model of Burke and Emerik (2016), we discuss the main methods currently used to estimate the economic damage of climate change. We conclude with a summary of actual empirical evidence and the related policy implications.

16.3.1 Heterogeneity of Climate Impacts and Mitigation Policy

An overwhelming debate between (environmental) economists is how climate change affects economic behavior, particularly across two fundamental dimensions: the people income level and the initial climate conditions. These characteristics represent two important elements of the so-called *vulnerability* to climate change.

Following Hsiang et al. (2019), exposure to climate change can be converted into economic cost through a function that describes the vulnerability of an individual or population, that is, how exposure (treatments) translates into costs (treatment effects). Within this framework, the crucial question is that a policy that would like to reduce the economic damage from climate change, that is, a mitigation policy, will induce distributional effects to the extent to which the marginal damage from climate change is heterogeneous across individuals or populations. For example, suppose that the marginal costs from climate change are positively correlated with income, then a mitigation policy that uniformly reduces emission, and thus exposure, will induce a (regressive) redistribution toward the "rich" because wealthier populations benefit more from that policy. Differently, if the correlation between marginal costs and income is negative, then the same mitigation policy will induce a (progressive) redistribution toward the "poor."

Figure 16.1 illustrates the problem through two different damage functions resulting in (different) heterogeneity of the impact from climate change. The left panel assumes a nonlinear damage function, typically of empirical studies in agriculture, where crop yield reacts nonlinearly to temperature or precipitation levels (*exposure*). With a nonlinear relationship between exposure (temperature) and damages (yield loss), two crops facing different baseline temperatures will experience different marginal damages, even if they are identical in terms of all other factors that determine *vulnerability*.

Alternatively, heterogeneity in marginal damages may result from some socioeconomic characteristics, such as wealth, that affect how exposure to warming translates into damages. This situation is shown in the right panel of Fig. 16.1, where the "rich" population 1 has a damage function different from the "poor" population 2, and so will experience lower damage under the same exposure, for example, because it can put in place alternative coping strategy (irrigation, air condition, etc.).

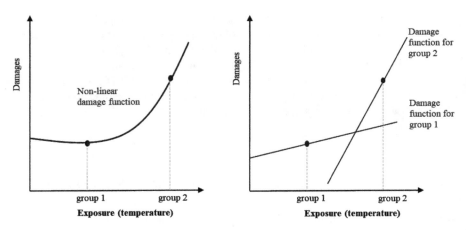

Fig. 16.1 Difference in marginal damages resulting from nonlinear damage functions or differing vulnerability. *Notes*: Different marginal effects are empirically measured (thick black points) for two groups (or populations) at different temperature levels (exposure) because there is a single nonlinear damage function (left panel), or the two groups (populations) exhibit different damage functions, that is, differing vulnerability, based on some socioeconomic attributes (e.g., income) that are correlated with exposure levels (right panel). (*Source:* Hsiang et al., 2019)

Finding evidence of heterogeneity in marginal damages is a sufficient condition to conclude that climate policy may have distributional effects (Hsiang et al., 2019). However, for selecting an efficient climate policy and addressing any distributional effects from mitigation policy, it is necessary to find the source of this heterogeneity. A crucial question is to understand whether the marginal costs differ because baseline *exposure* (e.g., temperature) differs or because *vulnerability* differs, or both. Results from current empirical evidence estimating damage functions are mixed in this respect.

The seminal work of Dell et al. (2012) finds support to the idea that growing temperature harms only poor countries because they are more vulnerable to climate, estimating a linear temperature damage function (right panel of Fig. 16.1). Differently, Burke et al. (2015) find support for a nonlinear damage function (left panel of Fig. 16.1), where poor countries experience more significant damage because they tend to be hotter, a result that holds true both in agriculture and nonagriculture activity. Newell et al. (2021), on the other hand, showed that these nonlinear damage functions are robust only in poor countries, but not in rich ones, a result that holds for both the agriculture and nonagricultural sectors.

Understanding which of the previous explanation holds true is crucial for the design of an efficient climate policy. In fact, if the damage from climate change is mainly due to vulnerability, for instance, because in poor countries the more sensitive agricultural sector is relevant from an economic point of view, then a rational policy prescription could be based on a "development policy strategy," that is, fostering the growth of the economic activities away from, the more vulnerable, agricultural sector. Differently, if nonlinearities induce high marginal damages from higher temperature irrespective of income, then policies should reduce the overall exposure to global warming by implementing mitigating policy. On the other hand, if this nonlinearity is also affected by income level, as the evidence of Newell et al. (2021) shows, then the best policy strategy should be a mix between development policy and mitigation policy, *ceteris paribus*.

The example above gives an idea of how it is important to conduct a sound empirical analysis to understand better the source of heterogeneity in the impacts of climate change.

16.4 Methods to Estimate Damage Function from Climate in Agriculture

There are three main econometric methods to estimate the impact of climate change: cross-sectional or Ricardian models, panel data models, and long-differences approach. The seminal contribution of Mendelsohn et al. (1994) introduced the Ricardian approach to study the impact of seasonal temperature and precipitations on the US counties' land value. This cross-sectional approach has several desirable properties. First, it is derived from sound economic principles because if land market is competitive, the land value should reflect the discounted present value of expected profits for a given parcel of land (Auffhammer, 2018). Second, it identifies long-run economic damage function, that is, the impact of climate is a measured net of how farmers and managers eventually adapt to a persistent change in climate (adaptation). However, the damage functions estimated form Ricardian models do not incorporate the costs of adaptation (Auffhammer, 2018). In addition, being based on cross-sectional data, the model suffers from the standard omitted variables bias, that is, we are not able to estimate the causal effect of climate on agricultural outcomes.

The second method to identify a damage function is using panel data models, first proposed by Deschênes and Greenstone (2007) in a study on US agriculture. Other than the cross-sectional dimension of the observation unit, a panel dataset displays variation for each unit over time. Hence, we can exploit the within-unit variation in outcome and climate variables (e.g., temperature and precipitations) for identification in the estimation. The key advantage of the panel data approach lies in the possibility to include fixed effects in the model specification, allowing the researcher to control for any confounding factor that is time-invariant within each unit of observation. This is crucial because many farms or local characteristics of agricultural activity that strongly affect production outcomes such as soil quality and management ability are simply not observable. In addition, given the exogeneity of weather shocks to farmers' input choice, the research design identifies the causal effect of weather on economic outcomes. However, studies that exploit short-run variations in weather tend to capture limited adaptation as weather changes tend to be unexpected by farmers, consequently having fewer possibilities to adjust their choices (Mendelsohn and Massetti (2017).

More recently, to solve this adaptation problem of panel data models, some papers have introduced the so-called long-difference approach, exploiting a mixture between the cross-sectional and the time-series identification strategies (Dell et al., 2012; Burke & Emerik, 2016). The general idea exploits the fact that, since climate changes are gradual, averaging across long time spans, it should offer the possibility to pick up also farmers' adaptation to climate change.

16.4.1 A Formal Model of Climate Change Impact in Agriculture

To better understand the conceptual differences between the three methods introduced above, we summarized the model of Burke and Emerik (2016). These authors start from a stylized model of farmers' choice among two different crop varieties, the first one adapted to cold climate (V1), the other to a warmer climate (V2). Each farmer i can choose a variety $x_{it} \in \{0,1\}$, with $x_{it} = 1$ when she opts for the V2 heat-tolerant variety. The output of farmer i in period t is represented by the following production function: $y_{it} = f(x_{it}, z_{it})$, where z_{it} is the realized temperature in period t, and it is drawn from a normal distribution with average w and variance σ^2.

It is assumed a quadratic production technology with respect to temperature, so that

$$y_{it} = \beta_0 + \beta_1 z_{it} + \beta_2 z_{it}^2 + x_{it}\left(\alpha_0 + \alpha_1 z_{it} + \alpha_2 z_{it}^2\right)$$

with production for the conventional varieties given by $\beta_0 + \beta_1 z_{it} + \beta_2 z_{it}^2$, and the differential productivity between the conventional and heat-tolerant variety, given by $\alpha_0 + \alpha_1 z_{it} + \alpha_2 z_{it}^2$. Farmer i in year t will choose the variety x_{it} to maximize expected output before she knows the weather realization, z_{it}. The heat-tolerant crop will be chosen if $\alpha_0 + \alpha_1 z_{it} + \alpha_2 z_{it}^2 > 0$, which can be written as

$$\alpha_0 + \alpha_1 w_t + \alpha_2 \left(w_t^2 + \sigma^2\right) > 0.$$

Finally, it is assumed that parameters α and β are known to the farmer but not to the econometrician. Next, we can graphically represent the two production functions for variety V1 and V2, assuming that they are «similar». These are depicted in Fig. 16.2.

An informed farmer will choose variety V2 heat-tolerant if she expects a temperature $> \tilde{w}$ with climate change characterized by a shift in temperature $w \rightarrow w'$, with $w < \tilde{w} < w'$. It is assumed that climate change increases the mean of temperature, but not variance, such that after climate change the farmer experiences $z_{it} \sim N(w', \sigma^2)$ in each year. A fully informed farmer recognizes this increase in temperature and will opt for V2 heat-tolerant variety, which Burke and Emerik (2016) considered a form of "adaptation."

However, in the real world, farmers are likely to learn about changes in climate over time and only adjust behavior after acquiring strong enough information that climate has changed. This framework suggests that farmers should be more likely to recognize changes in climate and thus adapt to those changes in areas where the temperature variance is low and when they are given more time to observe realizations of the new climate.

16.4.2 From Model to Empirical Estimation of Damage Functions

Burke and Emerik (2016) used this model logic to interpret what different empirical models really capture about the impact of climate change. Considering Fig. 16.2, the long-term damages imposed by a shift in climate are represented by $v_0 - v_1$ when adaptation takes place. Now consider a cross-sectional Ricardian approach to infer the damage impose from climate change, that is,

$$y_i = \alpha + \beta_1 w_i + \beta_2 w_i^2 + c_i + \varepsilon_i$$

where y_i is the outcome in the unit of observation i (a country/county/region), w_i is a long-run (30 years) average temperature, and c_i is a time-invariant vector of unit of observation characteristics affecting outcome, such as the average farmers' ability in the unit of observation i or its soil quality.

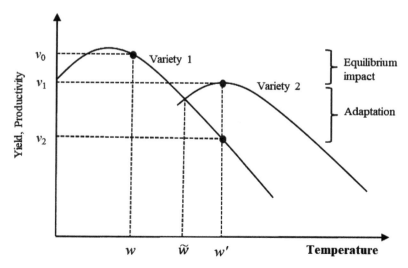

Fig. 16.2 Productivity of two different corn varieties as a function of temperature. (*Source*: based on Burke and Emerik, 2016)

In the Ricardian approach, the dependent variable is normally land values that, if the land market is efficient, embody any possible long-run adaptation to average climate. If this is the case, then a unit of observation with an average temperature of w will achieve ν_0 on average, while the other unit with an average temperature of w' will achieve ν_1. Thus by exploiting this variability across many units of observation, the estimates of β_1 and β_2 under a long-run horizon where mean temperature changed from $w \to w'$ should identify the desired damage effect of $\nu_0 - \nu_1$, that is, the impact of climate change accounting for adaptation, *ceteris paribus*.

The crucial assumption of this cross-sectional research design is that the average climate is not correlated with other unobserved factors, c_i (e.g., soil quality, labor productivity, technology availability, etc.) that also affect outcomes of interest. If this is the case, then the estimated β_1 and β_2 coefficients, by suffering from omitted variables bias, will not be the true ones.

The aim of the panel data approach is precisely to solve this omitted variable bias. Under this research design, the model specification becomes

$$y_{it} = \alpha + \beta_1 z_{it} + \beta_2 z_{it}^2 + c_i + \varepsilon_i$$

where now long-run average temperature w is subsumed by interannual weather variation z_t. Due to the new time dimension of the dataset, the vector c_i can be fully controlled with unit fixed effects, a specific constant for each unit that absorbs all time-invariant unobserved characteristics. In addition, because this year-to-year variation in temperature tends to be exogenous, the fixed-effects model will assure that the effect of temperature on outcomes such as yield or productivity can be interpreted causally, and this is a desired property for any research design. Yet, there could be a downside problem because by using interannual weather variation and not climate for identification, farmers can only make short-run adaptation. That is, with yearly observation, and temperature that grows from $w \to w'$, the model identified a movement along with one of the two functions in Fig. 16.2 so that estimates are a weighted average of the slop of the two curves, with weights related to how many farmers switch from V1 to V2 in the observed period, *ceteris paribus*. Thus, the estimated coefficients β_1 and β_2 will identify a damaging effect in between $\nu_0 - \nu_1$ and $\nu_1 - \nu_2$, only partially accounting for adaptation. Clearly, if the time period is sufficiently long, more farmers will shift to V2 heat-tolerant crop, and the damage function will account also for adaptation.

The problem is that empirical studies used the estimated damage function from panel data models to project the future impact of global warming, so that if the estimated coefficients do not sufficiently incorporate long-run farmers' adaptation, the risk is to overestimate the true impact of climate change. Therefore, panel data models solve identification problems of the cross-sectional approach at the cost of more poorly approximating the idealized climate change experiment (Burke & Emerik, 2016).

For this reason, some recent studies proposed a hybrid approach in between Ricardian and panel data models, called long-difference research design (Dell et al., 2012; Burke & Emerik, 2016). To see how, take two periods «a» and «b» of ~15 years each. Then take the average over the period a, that is, $\overline{y_{ia}} = 1/15 \sum_{t \in a} y_{it}$, and similarly, for $\overline{z_{ia}}$. Defining the variables in period b similarly, then it is possible to differentiate between the obtained quantities ($\Delta \overline{y_i}$, $\Delta \overline{z_i}$) in the two periods, obtaining

$$\Delta \overline{y_i} = \beta_1 \Delta \overline{z_i} + \beta_2 \Delta \left(\overline{z_i} \right)^2 + \Delta \overline{\varepsilon_i}$$

generating ideally unbiased estimates of β_1 and β_2.

The long-difference approach offers advantages over panel data methods because it better approximates the ideal experiment, addresses potential omitted (time-invariant) variable bias, and, at the same time, should capture medium-run to long-run adaptations that farmers put in place against trends in weather, thus estimating something close to $v_0 - v_1$. In addition, as argued by Dell et al. (2014), intensification effects should also be captured with this method, namely situations when climate change may cause damages that are not revealed by small weather changes, but that can be relevant in agriculture. An example are situations where the permanent reduction of precipitations strongly affects the reservoir availability of water for agriculture. However, the downside of the approach is that by averaging across years, we lose a huge variability (and degree of freedom) in the estimation (in comparison with panel data models), which could reduce the power of statistical inference.

16.4.3 Empirical Evidence

The first empirical application of a Ricardian model to estimate the economic impact of climate change in US agriculture has been given by Mendelsohn et al. (1994). They showed that, overall, global warming appears to have had a very low impact on US agriculture. Different conclusions are reported by Schlenker et al. (2006), who focus only on US counties where agriculture is mostly nonirrigated and included climatic variables as degree days, that is, the sum of the days when the daily average temperature is above (or below) a certain threshold, during the growing season. Contrary to Mendelsohn et al. (1994), they found that for the US counties considered, the aggregate impact in the near- to medium term is a 10–25% loss in land value, depending on the climate scenario chosen.

Cross-sectional Ricardian studies of climate impact on agriculture have been carried out for more than 40 countries to date (see Mendelsohn & Massetti, 2017). The emerging results are strongly related to the weather and climatic conditions prevailing in the countries considered. However, since the effect of weather variables is strongly *nonlinear* in agriculture, an increase in temperature due to global warming will become more and more negative and problematic in the medium- to long run. Considering mostly affected areas (Asia, Africa, and South America), current damage loss from Ricardian studies, under mild (2 °C) and more severe (4 °C) temperature increase scenarios, suggests overall farm net revenue reduction going from 8% to 30%, respectively (see Table I in Mendelsohn & Massetti, 2017). Interestingly, similar loss is estimated for European Southern regions (Van Passel et al., 2017; Bozzola et al., 2017), suggesting that the potential damage from global warming displays a pattern correlated with the development level. This result has potential policy implications.

Moving to empirical evidence from panel data studies, it is useful to start from results obtained by global studies that estimated an economy-wide damage function using real GDP per capita growth as the outcome variable. The relationship between weather variables (temperature and precipitation) and GDP per capita, both in levels and in growth rates, has been analyzed by Dell et al. (2012) and Burke et al. (2015). The first study estimated a linear in weather variables damage function showing that weather, notable temperature, is significantly negative only in developing countries and this relationship appears to be mainly driven by the negative effect of warming on the agricultural output and less so from the industrial output.

The study of Burke et al. (2015) gets to different conclusions. First, the impact of temperature on GDP per capita growth is not linear but quadratic and does not display heterogeneity between rich and poor countries. Thus, the damage from warming depends largely on exposure (i.e., level temperature) and not vulnerability. At the aggregated level, they found a maximizing GDP growth temperature of about 13 °C, so that with a quadratic reaction function countries with an average temperature below this optimum will grow faster due to a marginal increase in temperature. The percentage change for a marginal increase in temperature is equal to $(\partial \hat{y} / \partial z) / \hat{y} = \beta_1 + (2 \beta_2 \bar{z})$ with $\bar{z}$ the temperature at which the impact is measured (e.g., average temperature). Differently, countries with an average temperature above this optimum will lose progressively from a warming climate, *ceteris paribus*. Projection estimates at the end of the century, under the worse-scenario representative concentration pathway (RCP) 8.5, showed median world GDP loss of about 20–30%, a loss that rises to 40–60% or more when Latin American, sub-Saharan African, and South Asian countries are considered, respectively. However, estimates for continental Europe displayed important GDP gains.

An important drawback of the Burke et al. (2015) analysis is that they regress GDP per capita growth on temperature and precipitations in level, thus assuming that level temperature affects GDP growth. This is crucial when the researcher used the estimated damage function for future projection because growth effects will accumulate over time, while levels effects implied a recovery.

Newell et al. (2021) conducted an important empirical exercise with the aim to select from many different empirical specifications (linear, nonlinear, in levels, or growth rate, etc.) those that work better in out-of-sample prediction. Overall, results suggest that there is huge model uncertainty, that is, several different models perform similarly in out-of-sample prediction. However, they also showed that models in levels (estimated in first-difference) perform better than growth models and also that nonlinear (quadratic) specifications in temperature are superior to linear ones. These results have important implications for the estimated damage function.

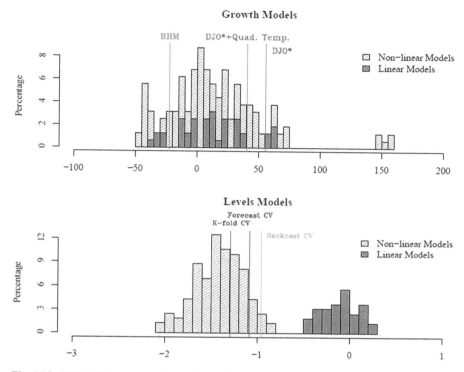

Fig. 16.3 2100 GDP damage estimates in unmitigated warming scenario for GDP growth models (top panel) and GDP levels models (bottom panel). (*Source*: Newell et al., 2021)

Indeed, as emerges clearly from Fig. 16.3, which summarizes their main results, levels models predict future damage in the range of 1–2% of the GDP at the end of the century under the worst-case RCP 8.5 scenario. Instead, growth models display a huge variability switching from positive (+56%) to negative effects (−50%), highlighting important modeling uncertainty. An additional result of Newell et al. (2021) is that they identify statistically significant marginal effects of temperature on poor country GDP and agricultural production, but not rich country GDP, or non-agricultural production. This is an important result because it suggests that poor countries are more vulnerable to climate change, also because a large part of their economic activities is concentrated in the more sensitive agricultural sector.

After the initial contribution of Deschênes and Greenstone (2007) that introduced the panel data approach, an influential study focused on agriculture is the one of Schlenker and Roberts (2009), which studied the impact of temperature and precipitations on corn, soy, and cotton yields at the US counties' level. Results confirm the existence of a solid nonlinear relation between temperature (precipitations) and crop yield, with estimated damage from global warming in the range of 20 ÷ 70% under RCPs 4.5 and 8.5, respectively. All panel data studies applied to developing countries' agriculture confirm the above relationship. Important negative effects of higher temperature on the yield of the main cultivated crops, ranging from −3% to −22%, are reported by Schlenker and Lobell (2010) for sub-Saharan Africa (SSA) and by Lobell et al. (2011) at the global level.

A potential limitation of the above estimates is that by focusing on a specific crop they disregard possible impacts on the livestock sector and labor use. Indeed, farming is an outdoor activity, and thus workers' performance can be affected by weather fluctuations. The alternative is to use as outcome variable sectoral labor productivity. In this respect, Olper et al. (2021) estimated damage functions for both GDP per capita and agricultural labor productivity for Italy, exploiting information at the provincial level. Results confirm the Newell et al. (2021) finding that levels models are superior to growth models. The estimated damage from future warming is close to zero for Italian GDP per capita, though losses displayed a clear geographical pattern. In agriculture, the average loss is important and in the range from 5 to 30% moving from north to south provinces and across different climate scenarios. Thus, results suggest that global warming in Italy disproportionally affects regions that, as an effect of their lower development level, tend to be more vulnerable to weather variability, precisely because their economy is more dependent on the (climate-sensitive) agricultural sector.

Finally, studies that applied the long-difference approach (e.g., Dell et al., 2014; Burke & Emerik, 2016) are somewhat inconclusive, in the sense that the estimated damage function is closed to the one estimated by the panel data model. One possible interpretation of this finding, suggested by the authors, is that there has been only limited long-run adaptation. However, another interpretation could be that under specific conditions—that is, nonlinearity of the damage function—what the panel data model estimates is not just a short-run reaction function. Instead, as recently formally derived by Mérel and Gammans (2021), a (nonlinear) damage function estimated from many (current) panel data systematically incorporated also the long-run impact of climate change. This intuition has been first put forward by McIntosh and Schlenker (2006), who consider the case in which the weather variable on the right-hand side enters as a second-degree polynomial (a quadratic specification). Because the response function is calibrated by two parameters, the coefficient on the higher-order term uses both variations from within units and across units. Hence, it has been argued that studies using this nonlinear specification allow plausibly causal estimates that also incorporate adaptation (e.g., Auffhammer, 2018).

16.5 Conclusion and Policy Implications

The threats posed by changes in climatic conditions will affect the global agricultural system in the near future. Poorest countries, especially in the tropical areas, which are already on the very edge of crop growing, will pay the most severe consequences in food security. Nowadays, more than 800,000 million people worldwide are chronically undernourished. However, the effect of climate change on agriculture in the areas where food-insecure people are concentrated may increase the number of undernourished people in the next years. This explains the diffuse concerns about the capability of the food and agricultural systems to meet in the near future an increasing demand for food for a population that is growing at a fast rate, especially in those areas where chronic hunger and malnutrition are already evident. To address these issues, immediate adaptation and mitigation actions in agriculture are needed. These policies should go in the direction of making agriculture more resilient to changes in the climatic conditions and reducing the impact that agriculture has on the environment.

When considering the incentives toward implementing adaptation and mitigation strategies, an important difference should be noted. While adaptation deals with something that goes in everyone's interest, mitigation is something that concerns global interest (Campanhola & Pandey, 2018). Therefore, GHG mitigation assumes a global public good character, where any improvement can be reached only through a collective effort. This point seems to meet a rather universal consensus, as demonstrated by the United Nations Framework Convention on Climate Change's Paris Agreement in 2015. However, planned mitigation actions set by the signatory countries appear to be still not sufficient to meet the long-term goal of limiting the average increase in temperature to 2°C above the pre-industrial level. One reason for this has to be found in the burden of mitigation policy, particularly on their possible regressive effects.

Clearly, there are important obstacles to a low-carbon world as such transformation should take into consideration different market (and political) forces and require global cooperation. Politics often represent an important obstacle too, especially in those countries where fossil fuel extraction plays an important economic role (Sachs, 2015). In general terms, each region of the world should set its strategy, but with the common goal of important and extensive decarbonization. This should rely on joint R&D programs dealing with low-carbon challenges and a cooperative effort to help the poorest countries reach these goals (Sachs, 2015). An additional problem toward achieving the objectives of the Paris agreement may be represented by the COVID-19 pandemic, which will likely lead to a global economic recession in the next years. Thus, economic recovery plans should not overlook the impact that climate change can have in the next future if unabated. Therefore, policies should take this as an opportunity to elaborate forward-looking strategies to achieve sustainable global economic growth.

A better understanding and a proper quantification of the effect of climate change on agriculture (and elsewhere) is, therefore, of vital importance. Current empirical evidence linking random variation of weather to agricultural (and nonagricultural) economic outcomes suggests that damage functions are strongly nonlinear. Importantly, this evidence seems to hold both across and within countries, mitigating the comparability issue that emerges from global studies (see Auffhammer, 2018). This implies that future climate change could affect economic activities also depending on their initial climate condition. In these contexts, marginal damages tend to increase at higher temperatures. While a future rise in temperature (exposure) is projected to be somewhat larger in cold areas of the rich North than in the poor hot tropics, because of the heterogeneous marginal effects of these changes, it does appear likely that poor populations will face more considerable future damages from climate change, both in agriculture and elsewhere (Hsiang et al., 2019).

A problem linked with this evidence is the potential distributional effects of climate change because the baseline climate of rich and poor populations differs systematically, at both world and country levels. For example, in the United States, the strong nonlinearity in damage functions in different sectors (Hsiang et al., 2017) leads to more serious damages from global warming in the hot and poorer Southern States, relative to the cooler and richer North, exacerbating preexisting patterns of inequality. Similar results are detected for Italy, where the strong nonlinearity of the damage functions, in both agriculture and at the aggregated levels (Olper et al., 2021), predicts that poor and hot South provinces will significantly be more affected by climate change relative to cooler and richer North provinces. In addition, results both at the global (Newell et al., 2021) and at country (Olper et al., 2021) level suggest that the effect of global warming disproportionally affects regions that, as an effect of their lower development level, tend to be more vulnerable to weather variability because their economy is more dependent on the sensitive agricultural sector.

From a policy perspective, this can have some interesting implications. On the one hand, policies could be relevant for mitigating the impact of climate change on some vulnerable agricultural activities, and particularly investments in agricultural infrastructure, such as irrigations, are crucial in many developing countries. On the other hand, an alternative strategy to exploit the potential adaptation of the economy in Southern countries and regions could be to consider a policy mix fostering economic growth away from the agricultural sector.

These policy implications rest on the assumption that the estimated nonlinear damage function from panel data models captures at least partly long-run adaptation. However, both adaptation and nonlinearity in current empirical applications, in some sense, tend to be imposed by the econometrician through specific functional forms assumption (e.g., a quadratic specification), more than from sound theory (Newell et al., 2021). Thus, the great difference in the marginal damage across a population in cold vs. hot locations, which emerged in the literature to date, still needs to be better scrutinized in the applied literature. In fact, some emerging evidence based on more flexible nonlinear damage functions, exploiting dynamic panel models, seems to show that the stark dichotomy between climate change impacts in poor relative to rich countries could be somewhat different (see Kahn et al., 2021). Finally, it is always important to bear in mind that the larger marginal damages of climate in poor populations may still have an economic origin, such as less access to credit or technology and, more in general, low quality of institutions, more than from (estimated) nonlinear damage function. Only future work will better clarify this crucial point.

References

Alexandratos, N., & Bruinsma, J. (2012). *World agriculture towards 2030/2050: the 2012 revision.* ESA Working paper No. 12-03. Rome, FAO. Retrieved Jun 10, 2021, from http://www.fao.org/3/ap106e/ap106e.pdf

Alston, J. M., Pardey, P. G., & Ruttan, V. W. (2008). *Research lags revisited: concepts and evidence from US agriculture.* Staff paper No. 50091. University of Minnesota, Department of Applied Economics.

Angelsen, A., Baur, H., Haskett, J. et al (2010). Forests and climate change: a toolbox. Bogor, Indonesia, Center for International Forestry Research (CIFOR).

Antle, J. M., & Capalbo, S. M. (2010). Adaptation of agricultural and food systems to climate change: An economic and policy perspective. *Applied Economic Perspectives and Policy, 32*(3), 386–416.

Auffhammer, M. (2018). Quantifying economic damages from climate change. *Journal of Economic Perspectives, 32*(4), 33–52.

Auffhammer, M., & Schlenker, W. (2014). Empirical studies on agricultural impacts and adaptation. *Energy Economics, 46*, 555–561.

Bozzola, M., Massetti, E., Mendelsohn, R., & Capitanio, F. (2017). A Ricardian analysis of the impact of climate change on Italian agriculture. *European Review of Agricultural Economics, 43*, 217–235.

Bruinsma, J. (2009). The resource outlook to 2050: by how much do land, water and crop yields need to increase by 2050. Expert meeting on how to feed the world in 2050. Retrieved from http://www.fao.org/wsfs/forum2050/wsfs-background-documents/wsfs-expert-papers/en/

Blanc, E.; Schlenker, W., (2017). The Use of Panel Models in Assessments of Climate Impacts on Agriculture. *Review of Environmental Economics and Policy 11*, 258–279

Burgess, R., & Donaldson, D. (2010). Can openness mitigate the effects of weather shocks? Evidence from India's famine era. *American Economic Review, 100*(2), 449–453.

Burke, M., Hsiang, S., & Miguel, E. (2015). Global non-linear effect of temperature on economic production. *Nature, 527*, 235–239.

Burke, M., Emerick, K., (2016). Adaptation to climate change: evidence from US agriculture. *American Economic Journal: Macroeconomics 8* (3), 106–140

Campanhola, C., & Pandey, S. (Eds.). (2018). *Sustainable food and agriculture: an integrated approach.* Academic Press/FAO.

Crippa, M., Solazzo, E., Guizzardi, D., Monforti-Ferrario, F., Tubiello, F. N., & Leip, A. (2021). Food systems are responsible for a third of global anthropogenic GHG emissions. *Nature Food 2*, 198–209.

Dell, M., Jones, B. F., & Olken, B. A. (2014). What do we learn from the weather? The new climate-economy literature. *Journal of Economic Literature, 52*, 740–798.

Dell, M., Jones, B. F., & Olken, B. A. (2012). Temperature shocks and economic growth: Evidence from the last half century. *American Economic Journal: Macroeconomics, 4*, 66–95.

Deschênes, O., & Greenstone, M. (2007). The economic impacts of climate change: Evidence from agricultural output and random fluctuations in weather. *American Economic Review, 97*(1), 354–385.

FAO. (2017). *The future of food and agriculture—Trends and challenges.* Rome.

FAO. (2015). *The impact of natural hazards and disasters on agriculture and food security and nutrition: a call for action to build resilient livelihoods*, May 2015. Rome.

FAO. (2016). *The state of food and agriculture: Climate change, agriculture and food security.* Rome.

Fischer, G., Shah, M., & van Velthuizen, H. (2002). *Climate change and agricultural vulnerability.* IIASA, Internat. Inst. for Applied Systems Analysis.

Fuglie, K. O. (2012). Productivity Growth and Technology Capital in the Global Agricultural Economy. In K. O. Fuglie, S. L. Wang, & V. E. Ball (Eds.), *Productivity growth in agriculture: an international perspective* (pp. 335–368). MA.

Hertel, T. W., & Baldos, U. L. C. (2016). *Global change and the challenges of sustainably feeding a growing planet*. Springer International Publishing.

Hertel, T. W., & Rosch, S. D. (2010). Climate change, agriculture, and poverty. *Applied Economic Perspectives and Policy, 32*(3), 355–385.

Hsiang, S. (2016). Climate econometrics. *Annual Review of Resource Economics, 8*, 43–75.

Hsiang, S., Oliva, P., & Walker, R. (2019). The distribution of environmental damages. *Review of Environmental Economics and Policy, 13*(1), 83–103.

Hsiang, S., Ro Kopp, R., Jina, A., Rising, J., Delgado, M., Mohan, S., Rasmussen, D. J., Muir-Wood, R., Wilson, P., Oppenheimer, M., Larsen, K., & Houser, T. (2017). Estimating economic damage from climate change in the United States. *Science, 356*, 1362–1369.

Intergovernmental Panel on Climate Change (IPCC). (2007). *Climate change 2007: Impacts, adaptation, and vulnerability. Contribution of working group II to the third assessment report of the intergovernmental panel on climate change*. Cambridge University Press.

Intergovernmental Panel on Climate Change (IPCC). (2014). *Climate change 2014: Impacts, adaptation and vulnerability. Part A: Global and sectoral aspects*. Cambridge University Press, Contribution of working group II to the fifth assessment report of intergovernmental panel on climate change.

Kahn, M. E., Mohaddes, K., Ng, R. N. C. et al. (2021). Long-term macroeconomic effects of climate change: A cross-country analysis. *Energy Economics*, https://doi.org/10.1016/j.eneco.2021.105624.

Kolstad, C. D., & Moore, F. C. (2020). Estimating the impact of climate change using weather observation. *Review of Environmental Economics and Policy, 14*(1), 1–24.

Licker, R., Johnston, M., Foley, J. A., et al. (2010). Mind the gap: how do climate and agricultural management explain the 'yield gap' of croplands around the world? *Global Ecology and Biogeography, 19*(6), 769–782.

Lobell, D. B., Cassman, K. G., & Field, C. B. (2009). Crop yield gaps: their importance, magnitudes, and causes. *Annual Review of Environment and Resources, 34*(1), 179–204.

Lobell, D. B., Schlenker, W., Costa-Roberts, J. (2011). Climate trends and global crop production since 1980. *Science 333*(6042), 616–620.

McIntosh CT, Schlenker W (2006) Identifying non-linearities in fixed effects models. Working Paper, Sch. Int. Relat. Retrieved from https://gps.ucsd.edu/_files/faculty/mcintosh/mcintosh_research_identifying.pdf.

Mendelsohn, R., Nordhaus, W. D., & Shaw, D. (1994). The impact of global warming on agriculture: A Ricardian analysis. *American Economic Review, 84*, 753–771.

Mendelsohn, R., & Massetti, E. (2017). The use of cross-sectional analysis to measure climate impact on agriculture: theory and evidence. *Review of Environmental Economic Policy, 11*, 280–298.

Mérel, P., & Gammans, M. (2021). Climate econometrics: Can the panel approach account for long-run adaptation? *American Journal of Agricultural Economics*. https://doi.org/10.1111/ajae.12200

Newell, G. R., Prest, B. C., Sexton, E. S. (2021). The GDP temperature relationship: implications for climate change damages. *Journal of Environmental Economics and Management 108*, 1–26.

Olper, A., Maugeri, M., Manara, V., Raimondi, V. (2021). Weather climate and economic outcomes: Evidence for Italy. *Ecological Economics 189*, 107156.

Sachs, J. D. (2015). *The age of sustainable development*. Columbia University Press.

Schlenker, W., & Roberts, M. J. (2009). Nonlinear temperature effects indicate severe damages to U.S. crop yields under climate change. *Proceedings of the National Academy of Sciences of the United States of America, 106*(37), 15,594–15,598.

Schlenker, W., & Lobell, D. B. (2010). Robust negative impacts of climate change on African agriculture. *Environmental Research Letters 5*, 014010

Schmidhuber, J., & Tubiello, F. N. (2007). Food security under climate change. *Proceedings of the National Academy of Sciences, 104*, 19,703–19,708.

Sinclair, T. R., Purcell, L. C., & Sneller, C. H. (2004). Crop transformation and the challenge to increase yield potential. *Trends in plant science, 9*(2), 70–75.

Tubiello, F. N., & Rosenzweig, C. (2008). Developing climate change impact metrics for agriculture. *The Integrated Assessment Journal, 8*(1), 165–184.

Van Passel, S., Massetti, E., & Mendelsohn, R. (2017). A Ricardian analysis of the impact of climate change on European agriculture. *Environmental and Resource Economics, 67*, 725–760.

Alessandro Olper is professor in agricultural economics at the Department of Environmental Science and Policy of the University of Milan. His research interests focus on three main topics: international agri-food trade, political economy of trade and environmental policy, and climate econometrics. In the last years, he worked extensively on the economics of climate change, focusing, in particular, on the quantification of the impacts of climate change on different economics outcomes, such as productivity, migration, trade, and food security.

Daniele Curzi is research fellow at the Department of Environmental Science and Policy of the University of Milan. He has a degree in agricultural economics from the University of Milan. His research activity is mainly focused on international trade in the agri-food sector and especially on the role of product quality as determinant of trade patterns as well as on the effect of trade policies in affecting trade flows. During the last years, he extended his research activity on the analysis of the effect of changes in climatic condition on agriculture, as well as on the effect of the adoption of agri-environmental measures on farmers' environmental and economic outcomes.

Part IV
People's Behaviors to Address Climate Change: Proactive and Conflicting Actions

Chapter 17
Climate Change and Consumer Behavior

Elisa De Marchi, Alessia Cavaliere, and Alessandro Banterle

Abstract Fighting climate change and its detrimental effects on the environment and the society has become a key priority for many governments. With the Paris Climate Agreement, many countries worldwide have committed to cut greenhouse gas emission and reach climate neutrality by mid-century, setting the roadmap for a sustainable economic growth. However, this ambitious goal will not be achieved without the contribution of the consumers. Indeed, everyday consumption decisions, especially food-related ones, can play a relevant role in alleviating pressure on the environment. Reducing meat consumption and choosing food products with sustainability labels are just two of many examples of how consumers can become more environmentally friendly. Although people have become increasingly aware and sensitive about sustainability-related issues, they sometimes fail to behave accordingly. This chapter describes some of the key factors underlying consumers' intention–behavior gap and discusses their main implications for designing effective demand-side policies geared at leveraging sustainable consumption decisions.

17.1 Introduction

Since the beginning of the nineteenth century, humans have experienced major scientific progress and technological advances, which led to unprecedented global growth and a substantial improvement of living standards both from a life expectancy and life satisfaction standpoint. The side effect of such increased well-being is that human activities have become dominant on the planet, to the extent that their impact is now posing major risks to the environment. We have reached a critical situation where some planetary boundaries have already been crossed, including the core boundary of climate change (Rockstrom et al., 2009; Steffen et al., 2015). According to recent data of the Intergovernmental Panel on Climate Change (IPCC), human activities have caused an increase in global mean temperature of about 1 °C

E. De Marchi · A. Cavaliere (✉) · A. Banterle
University of Milan, Department of Environmental Science and Policy, Milan, Italy
e-mail: elisa.demarchi@unimi.it; alessia.cavaliere@unimi.it; alessandro.banterle@unimi.it

© The Author(s), under exclusive license to Springer Nature Switzerland AG 2022
S. Valaguzza, M. A. Hughes (eds.), *Interdisciplinary Approaches to Climate Change for Sustainable Growth*, Natural Resource Management and Policy 47,
https://doi.org/10.1007/978-3-030-87564-0_17

compared to preindustrial levels, with severe consequences in terms of ocean acidification, global ice reduction, species extinction, and biodiversity loss (IPCC, 2018).

In 2015, governments worldwide committed to reducing greenhouse gas emissions (GHG) to limit global warming with the Paris Climate Agreement, the first legally binding international treaty on climate change. The treaty represents a milestone for reaching climate neutrality by mid-century and sets the roadmap for more sustainable production and consumption patterns. The ambitious goals of the Paris Agreement have been further extended within the EU borders with the 2020 European Green Deal and the issuing of the first-ever Climate Law. European countries committed to gradually transform the EU into a resource-efficient competitive economy, where economic growth is decoupled from resource use and where zero net GHG emissions will be reached by 2050.

Institutional engagement is essential in the global challenge for environmental sustainability, but the climate neutrality goals will not be successfully attained unless the civil society is actively mobilized. Indeed, evidence indicates that a sizeable and increasing share of GHG emissions derives from private consumption (European Environment Agency, 2010), especially in the most affluent high-consuming classes (Lorek & Fuchs, 2013; Thøgersen, 2021).

Housing- and mobility-related consumption have been recognized as two main contributors to human-generated carbon impact (Ricci & Banterle, 2020). It is estimated that housing is responsible for ~26% of the total carbon footprint, especially due to household electricity use and heat production (IPCC, 2014). Mobility accounts for ~20% of the carbon footprint. A significant part of the mobility-related impact is ascribed to private car use and air transport, which in Western countries account for respectively ~80% and ~10% of the total carbon footprint of mobility (IGES et al., 2019). Essentially, consumers could make a significant contribution to sustainability by reshaping their consumption decisions. For instance, choosing green energy for their domestic needs or preferring public transport and car-sharing to personal car use.

A third key contributor to GHG emissions is represented by food consumption, which probably represents the most challenging target for environmental sustainability. Over the past two decades, most developed countries have started showing food overconsumption trends, which imply multiple negative environmental consequences. Firstly, the increasing demand for food poses high pressure on the production system, and food production is one of the most carbon-intense human activities. It is responsible for land use and soil degradation, biodiversity loss, and considerable freshwater use, such that it has been recognized as the largest cause of environmental change globally (Willett et al., 2019). In addition, overconsumption of food is associated with high waste generation and with high plastic packaging use (mostly single-use) that is frequently dispersed in the environment, thus increasing pollution. Tackling the environmental impact of food production and consumption is urgent but extremely complex. Food is an essential need that people cannot renounce as they could do with air travel or cars, and, importantly, food is intrinsically associated with cultural heritage, it carries local traditions over time, and is strongly related to personal values and individual's health. For all these reasons, changing food-related behaviors might be especially tough.

It is widely recognized that consumers could significantly contribute to alleviating environmental pressure with their daily consumption choices. However, it is anything but obvious that they commit to doing so, even when they are aware of and interested in climate-related issues.

In a recent paper, Vermeir et al. (2020) proposed a theoretical framework to explain sustainable behaviors, which involves the following five components: (1) consumers should positively value the environment; (2) they should be able to recognize the difference between the ideal environment where they would like to live and the actual state of the planet; (3) they should opt for behaviors that can close such gap (i.e., goal intention); (4) they should have the intention to engage in such environmentally friendly behaviors (i.e., behavioral intention); and (5) they should behave coherently with their intention. In fact, the sustainability-related literature demonstrates an intention–behavior gap for which consumers sometimes fail to behave in accordance with their concerns and positive attitudes toward the environment.

Given the critical role that consumers might have in mitigating climate change, demand-side policies have gained increasing attention (Warren, 2018; Creutzig et al., 2018). To date, most demand-side policies targeting sustainable behaviors have been based on classical information instruments such as specific labels and information campaigns. More recently, the interest has shifted to the so-called "behaviorally informed" policies. The latter is mainly based on nudging techniques and seems to have a high potential of application in climate-related policy contexts (Sunstein, 2014).

17.2 Consumer Awareness About Climate Change and Sustainable Behaviors

A recent survey on "Climate change and consumer behavior" conducted across 28 countries worldwide reports that 70% of the total sample states to have made changes to their consumption behaviors out of concern about climate change (IPSOS, 2020). Of these, 17% refers "a lot of changes" and 52% refers "a few changes." Countries where consumers seem to be more prone to modify their behaviors to counteract climate change are India (88% of total interviewed), Mexico (86%), Chile (86%), China (85%), Malaysia (85%), and Peru (84%). Japan is the only surveyed country where almost 50% of people declared that they have not changed their behaviors to impact less on the environment.

As for the European Union, in all surveyed countries, a high percentage of people states to have made at least some changes in their consumption behaviors pushed by environmental concerns. Spain ranks first with 76% of total interviewed, immediately followed by France and Italy (73% respectively), and Poland (72%). In these countries, the percentage of people that report to have made "a lot of changes out of concern about climate change" is around 15%. Countries that report fewer changes in the last few years are Belgium, Sweden, and Germany.

According to the Ipsos survey, the sustainable actions undertaken by consumers are mainly centered around the household and regard especially the amount of water consumed, both the volume and frequency of product recycling, the energy use, the volume and the frequency of reusing goods. Less than half of the total sample reported changes in the amount and type of food purchased.

Overall, the survey results indicate that consumers have become increasingly engaged in sustainability and, in some cases, they are willing to take action to fight climate change. However, these data tell only one side of the story.

In fact, climate change is a complex matter, and sometimes consumers find it difficult to fully understand its causes and implications and behave accordingly. For instance, it is challenging to concretely predict the effects of a 2 °C increase in temperatures starting from scientific forecasts. It is also difficult to foresee what will be the effects for human beings and when the singles will experience these on a day-to-day basis. Climate change is mostly seen as an issue for future generations, which implies that some people tend to procrastinate choices (Hansen et al., 2013). Additionally, facing climate change requires a common worldwide commitment (Walker & King, 2008), and this can cause a "social loafing" across all involved players, such that, essentially, the responsibility is left to others (Beattie et al., 2011).

There is also evidence that consumers might behave sustainably in some contexts (e.g., energy or water use) but not in others (e.g., food consumption). This is referred to as the "domain dependence" issue.

Furthermore, a considerable body of sustainability-related literature has highlighted the existence of an intention–behavior gap. This concept refers to the fact that often consumers who declare to be interested in climate change issues and to be intentioned to act accordingly, fail to do so. In other words, they face the challenge of translating concerns and positive attitudes toward the environment into concrete behaviors (Thøgersen, 2005; Grunert, 2011; Beattie et al., 2016). The reasons for the intention–behavior gap have gained increasing attention over the past decade. Indeed, an in-depth understanding of why people fail to behave sustainably would be of great value for formulating policy measures that can effectively redirect consumer consumption choices toward a more climate-friendly trajectory.

17.3 Understanding the Gap Between Intentions and Behaviors

The intention–behavior gap has been analyzed from different angles, but it is hard to find a unique and exclusive explanation. Several endogenous and exogenous factors contribute to determining consumer climate-related decisions, and these factors are often interrelated. Some of the most important are discussed in detail in the following subsections.

17.3.1 *The Role of Individual Time Preferences*

The concept of time preferences relates to how individuals value present over future events. People with high time preferences tend to outweigh immediate utility over future utility. Their decisions mostly aim at maximizing present gains despite possible consequences that might occur later in time. On the opposite, people with low time preference attach greater importance to future outcomes and are more prone to defer gratification to obtain benefits in the future (Strathman et al., 1994; Chisholm, 1998; Frederick et al., 2002; Joireman et al., 2012). Time preferences have a key role in all intertemporal decisions, namely, every time individuals are called to make tradeoffs between costs and benefits occurring at different points in time (Frederick et al., 2002). Time preferences affect human behaviors in a multiplicity of different contexts, such as health (e.g., eat well for being healthy in the future), finance (e.g., long-term investments), education (e.g., graduate to have a more remunerative job), as well as in the sustainability domain. Sustainability-related decisions are intrinsically intertemporal. Indeed, present generations with their actions can affect both the environment and the society in the future. If they succeed in attaining the climate goals, future generations will live on a better planet (McCollough, 2010; Allcott & Greenstone, 2012; Carmi & Arnon, 2014; Cavaliere et al., 2014; Cavaliere et al., 2016; De Marchi et al., 2020).

However, as mentioned above, individual decisions to engage in sustainability-oriented behavior can be influenced (among others) by their time preference. Past studies have demonstrated that people with low time preferences are more likely to recycle, save energy, and invest in energy-efficient technologies (McCollough, 2010; Allcott & Greenstone, 2012; Carmi & Arnon, 2014). Low time preferences are also associated with higher preferences for food products that carry sustainable characteristics, such as organic products (De Marchi et al., 2016). On the other hand, people with strong preferences for immediate utility are more likely to make inefficient decisions regarding the environment (Cairns, 1992; Allcott & Greenstone, 2012; Bradford et al., 2017). Such inefficient decisions may negatively impact society in the long run, generating negative externalities (Bradford et al., 2017).

17.3.2 *The Double Reasoning System*

Consumer choices are always the result of a decision-making process. The Neoclassical economic theory assumes that such decision-making is based on consumers carefully evaluating all possible alternatives and then choosing the one that maximizes their utility within specific budget constraints. We now know that these assumptions poorly explain what happens in real life. In fact, most of the times, consumers do not act rationally, and their decision-making processes are guided by limited, automatic, and effortless reasoning that leads to suboptimal decisions. These key insights come from the studies of Daniel Kahneman (2011). Kahneman theorized that people have two cognitive systems acting independently from one another. They are called System I and System II. System I is automatic, very fast, effortless, mainly works with associative memory (i.e., recalling familiar concepts in mind), and is largely unconscious. System II, instead, is the problem solver. Under System II, decisions are taken because of a reflective, skilled, and conscious process, which is as efficient as effortful. As explained by Kahneman, System I is the director of most of our everyday life decisions. It quickly jumps to conclusions, saving us time and cognitive efforts. The problem solver, instead, is lazy and comes into play only when we deliberately activate it to solve difficult tasks. Such uneven workload between the two reasoning systems implies that in many cases our decisions are based on intuitive thinking rather than on careful reasoning.

How does this relate to climate-friendly behaviors? Moreover, how does it explain the intention–behavior gap?

As anticipated, climate change and related sustainable behaviors are complex, and evaluating their effects requires attentive reasoning and tradeoffs between present and future outcomes. We would be able to make such decisions efficiently under System II, but evidence indicates that we mostly rely on System I. Moreover, System I and System II are strictly linked to individual attitudes. Specifically, System I is associated with "implicit" attitudes generated through fast-learning process, whereas System II to "explicit" attitudes formed through a slow-learning system. Both types of attitudes influence behaviors, but they come into play in different decision-making contexts. Explicit attitudes are good predictors of behaviors when people are highly motivated and have the time to deliberate about a choice (Beattie, 2010; Beattie & Sale, 2011). Implicit attitudes, instead, are much better in predicting behavioral outcomes in situations where individuals experience time pressure, the cognitive load is high, and/or they are emotionally involved (Gibson, 2008; Hofmann et al., 2007).

Unfortunately, most of our everyday choices—including those related to household energy and water use, recycling, and food shopping—belong to the latter category and are therefore better predicted by implicit attitudes.

This is one of the reasons behind the intention–behavior gap. People might be interested in climate-related issues, but when it comes to choices, interest is not sufficient because decisions are the result of more complex mechanisms.

17.3.3 Exogenous Factors

In addition to time preferences and our reasoning processes, some exogenous factors come into play when we are choosing across different products or services and affect our choice behavior. Price is probably one of the most important. Past studies reported that when consumers are asked to indicate the main obstacles to purchasing and using sustainable products, their high prices are among the top answers (Grunert, 2011). In a very recent study on solar energy adoption, Colasante et al. (2021) found that the monetary aspect was relevant for both strongly sustainability-involved consumers and those whose interest in the environment was less marked. In both cases, their willingness to adopt solar energy was mainly guided by savings in the energy bill and the possibility to receive subsidies. Prices are also relevant in the food context. Food purchasing represents an almost daily action and largely impacts consumer finances. Therefore, the higher prices of environmentally friendly food products might represent a significant discouraging factor. Furthermore, the results of the Eurobarometer survey (European Commission, 2013) showed that even when consumers are willing to pay a premium price for sustainable alternatives, their willingness is dependent on whether they trust the sustainability-related information provided by the producers. Only half of the European consumers involved in the Eurobarometer survey reported trusting producers' environmental commitments, while a considerable part of them does not trust what firms state to do about their environmental performances.

17.4 Food Production and Climate Change: How Can Consumers Contribute with Their Day-to-Day Behaviors?

When dealing with sustainability issues and climate change, a specific focus on the food system is unavoidable. In fact, there are very strong connections between food production and consumption and the environment.

First, every stage of the food system—particularly the agricultural sector—depends on environmental inputs, namely the availability of seeds, land, water, soil, nutrients, energy, etc. Second, the outputs of the production and consumption activities, as well as the outputs of the disposal, are introduced in the natural environment in the form of GHG emissions, residuals, waste, etc. Additionally, the food system relies on ecosystem services, which are several benefits to humans provided by a healthy environment and ecosystems. Such services include aquatic and forest ecosystems, biocapacity, agroecosystems, pollination of crops, clean air, weather control, etc.

Among all food-system activities, food production is one of the most impacting ones (Meyer-Hofer, 2014). Agricultural activities consume more than 70% of the world's freshwater, and the food system uses almost 30% of the total energy worldwide (FAO, 2011; Jeswani et al., 2015). The environmental impact of food production is not limited to resource use. Intensive agriculture is related to soil depletion due to the high use of chemicals, biodiversity loss, and high GHG emissions. The latter mostly derive from livestock farming, which is the most carbon-intensive activity in food production (Darre et al., 2019; McLaughlin & Kinzelbach, 2015).

The increasing demand for food in the past two decades was due to both the overconsumption trends that characterized developed countries and the population growth in developing countries. The latter phenomenon is expected to further increase, and food demand is projected to rise from 59 to 98% by 2050. Moreover, the growing number of people living in urban areas and the consequent increase in average income levels will contribute to change dietary habits determining high consumption of protein-based foods (especially meat and dairy), as well as highly processed convenience food.

To reverse these projected trends, the concept of sustainable diet has gained increasing attention in the last decade. Sustainable diets have been firstly conceptualized in the early 1980s (Gussow & Clancy, 1986) to generically describe dietary patterns that can be beneficial both for human health and environmental preservation. This concept has been modified and extended by the Food and Agriculture Organization (FAO) that reports, "Sustainable diets are those diets with low environmental impacts that contribute to food and nutrition security and healthy life for present and future generations." Moreover, sustainable diets are "nutritionally adequate, safe and healthy, while optimizing natural and human resources." A valuable example of a sustainable diet is represented by the Mediterranean diet, which is based on high consumption of low-processed plant-based food products and by moderate to low consumption of dairy foods, fish, and meat (Burlingame & Dernini, 2011; Schröder et al., 2016). In fact, it is amply recognized that plant-based diets are more climate-friendly than animal protein-based ones, such as the so-called "Western" diet highly diffused in Western countries, such as the United States and United Kingdom.

Consumers have the power to mitigate food-related climate impacts with their day-to-day consumption decisions. Certainly, a remarkable contribution would derive from a shift from meat-based diets to mainly plant-based nutritional patterns. However, this represents a remarkable change in consumer habits, which requires time and might be hard to reach due to cultural and identity-related issues. However, consumers could reduce their carbon footprint also in other, less intrusive ways. For instance, choosing more sustainable alternatives at the supermarket, avoiding buying food products with multiple packaging, and managing waste production at the household levels.

17.4.1 Changing Diets: The Unsustainable (M)eat

While the scientific evidence demonstrates that livestock production is highly detrimental in terms of environmental impact, the demand for meat and other animal-based food is increasing worldwide. In Western industrialized countries, meat consumption has always been the norm. According to OECD and FAO, European citizens consume on average 70 kg per capita a year of meat (OECD/FAO, 2019), but this quantity is much higher in the United States, where yearly individual consumption reaches almost 100 kg. However, the greatest change in terms of increased meat consumption is happing in eastern developing countries where the traditional, mainly plant-based patterns are gradually abandoned in favor of more "Westernized," meat and fast-food-based eating. Recent OECD data (OECD, 2020) indicate that meat consumption and the demand for other animal-based protein food are dramatically increasing in China and India, where the increasingly affluent middle class drives the increase in animal-based foods and, therefore, it is reasonable to expect that the trend will increase further in the next years. As a result, the growing demand for meat in the most populated countries worldwide will remarkably worsen GHG emissions (Reisch et al., 2021).

This trend cannot be reversed unless consumers start being actively involved in the process by reducing meat consumption, and this is not trivial. Multiple factors influence how consumers relate to this issue.

For instance, a recent review (Sanchez-Sabate & Sabaté, 2019) reported that (i) most consumers are not aware of the environmental impact of meat production, while only a minority knows about this issue; (ii) consumers often underestimate or ignore the positive effects that could derive from either stopping or reducing meat consumption; and (iii) they do not fully understand that plant-based diets are more climate-friendly.

Furthermore, other factors influence consumer willingness to renounce or reduce meat consumption. Past evidence suggested that meat-eaters tend to believe that meat is an essential and necessary component of a healthy diet and tend to associate meat consumption with strength and masculinity concepts (Allen et al., 2000; Joy, 2009; Kildal & Syse, 2017; Rothgerber, 2013).

For other consumers, eating meat represents a deep-rooted habit, and for this reason, it is hardly abandoned (Graça et al., 2015). In fact, this would imply modifying usual purchasing behaviors, preparing meals, and would also require improved cooking skills for introducing new food categories into the diet. This aspect is especially problematic for consumers that experience high time constraints in their everyday life (Buckley et al., 2005).

Furthermore, food cannot be simply considered as a primary need for individuals. Consumption habits relate to a wider variety of abstract concepts, like conviviality, personal gratification, sharing, hedonism, religion, and all these affect individual willingness to change dietary patterns.

17.4.2 Choosing Sustainable Product Alternatives

As explained, it might be challenging to reorient consumers toward mainly plant-based diets, or at least this will require a long transition process. In the meantime, there are several other ways in which consumers can lower their carbon footprint.

One of these involves choosing more sustainable product alternatives at the point of purchase.

For instance, consumers can purchase seasonal and/or local products and prefer low-processed foods instead of convenience/ready-to-eat foods, whose production is typically more carbon-intense. Another option is to purchase products that carry sustainability-related certifications. The latter can refer to environmental sustainability, as well as to animal welfare or ethical aspects. It must be considered that sustainability attributes belong to the so-called "credence attributes." This means that such product characteristics cannot be concretely experienced or tasted by consumers, not even after consumption. In some ways, we could say that purchasing sustainable products is based on trust.

In such situation, sustainability labeling plays a crucial role as it is the instrument that enables consumers to identify sustainability-related characteristics at the point of purchase and to attach them a specific value, both monetary and abstract (Grunert, 2011, 2014). In other words, sustainability labels translate credence attributes in search attributes (i.e., attributes that can be recognized by looking at the product packaging), making consumers able to distinguish environmentally friendly alternatives from less sustainable ones.

Over the past decades, the use of sustainability labels and logos has considerably grown. The organic logo is probably the most popular example. Its presence on food products has become prominent over time and was followed by the Fair-Trade symbol, the Rainforest Alliance logo, several carbon footprint schemes, as well as numerous welfare-related symbols. To date, the European cataloguer ecolabelindex.com includes more than 150 sustainability-related labeling standards for the food and beverage product categories (Grunert et al., 2014).

While, as said, such labels are essential for consumers to identify sustainable products, the presence of so many different schemes might have some drawbacks. Consumers might experience difficulties in understanding the specific meaning of the labels, Moreover, this might generate confusion while comparing different alternatives of the same product category, especially when the underlying concepts of sustainability labels overlap (e.g., when comparing different carbon footprint logos or certifications). In this regard, the EU Commission is committed to take concrete action to overcome these issues and facilitate the shift toward more sustainable diets. Within the Farm-2 Fork Strategy, the Commission will explore ways to harmonize the so-called "voluntary green claims." In coherence with the initiatives proposed for providing food-related information to consumers and in line with the legislative framework on sustainable food systems, the Commission will promote the creation of an EU-harmonized sustainable food labeling framework and support enforcement of rules on misleading information.

17.5 Demand-Side Policies for Promoting Climate-Friendly Behaviors

From an economic standpoint, unsustainable behaviors cause negative externalities that negatively impact society and call for public interventions to assure a socially optimal level of environmental protection.

Public interventions can take different forms and can be directed to economic activities or consumers. The latter are referred to as demand-side policies and have the objective to enable, support, and guide consumers in their everyday life toward more sustainable choices and behaviors.

Demand-side policies have been typically based on various forms of information provision. More recently, the adoption of alternative approaches based on behaviorally informed measures has catalyzed the attention of both researchers and policy-makers. These approaches mainly rely on nudging techniques that seem to have great potential in large-scale implementation and effectiveness.

17.5.1 Information and Education-Based Policies

Information provision has always been considered a valuable instrument for demand-side policies. Among the possible ways of providing information, education campaigns for the broad public and, above all, education at school should represent the grounding for reaching behavioral changes. In fact, education can increase consumer awareness and knowledge about specific topics, which in turn can affect their perception of the problem and their decisions. For instance, being aware and having specific knowledge about the concrete consequences of GHG emissions on the environment might lead consumers to carefully reconsider their meat consumption, possibly leading to changes in everyday dietary habits. Education could also represent an effective intervention for shifting (in a long-run perspective) consumer time preferences from present orientation to future orientation, at least in the sustainability domain.

However, educating people is a long process and the results of such policy interventions can only be seen in years. For this reason, there is a need for alternative and complementary policies that can have more immediate effects. The latter include labeling, which aims to guide consumers in their decision-making processes and purchasing decisions across various products and services. Labeling has the key advantage of being available at the point of purchase, and providing context-specific information that is directly usable by consumers. Labels can enable consumers to evaluate alternative products or services better, compare their characteristics, and, ultimately, choose the best (i.e., more sustainable) option.

Moreover, when information is provided in the form of labeling, it increases market transparency and reduces asymmetric information between producers and consumers. Asymmetric information represents a market failure that can cause opportunistic behaviors. Indeed, at the point of purchase, consumers always have less information about products than producers, which implies that they might not be able to properly evaluate product characteristics and attach them a value (monetary or abstract). In such a situation, producers might fix prices that are higher than the real value of the good, which negatively impacts on consumers.

For all these reasons, information provision through labeling is overall considered a viable policy instrument for demand-side policies. However, past evidence demonstrated that it might prove ineffective in changing consumer behaviors (Staats et al., 1996; Abrahamse et al., 2005). With specific regard to sustainability labeling, recent literature highlighted several limiting factors (Torma and Thøgersen 2021). For instance, consumers often experience high time pressure at the point of purchase and do not have time to look for specific indications or read the labeled information. Furthermore, the high variety of labels and logos in a limited space makes the reading difficult and information processing might be complex and time-consuming.

Another line of research explained information ineffectiveness within the framework of System I and System II reasoning system. Most of the time, the information requires a high cognitive effort to be processed and understood and should be handled by System II. However, as explained in Sect. 17.3.2, System II is lazy and most of our decisions result from the effortless and intuitive reasoning of System I, which does not have the skills to manage complex tasks.

17.5.2 Nudging-Based Policies

Given the sometimes-limited effectiveness of information provision, in recent years other instruments have been investigated to achieve behavioral changes. These alternative approaches are based on the so-called "nudges" (Thaler & Sunstein, 2008). Thaler and Sunstein (2008) first defined nudges as "any aspect of the choice architecture that alters people's behavior in a predictable way without forbidding any options or significantly changing their economic incentives." The main characteristics of nudges are that they are typically very easy to be implemented in specific decision-making contexts, they are cheap, and importantly, they do not mandate people to act in a certain way (Thaler & Sunstein, 2008). Nudges aim at changing people's behavior playing with their known cognitive limitations. To explain, information provision gears at enhancing consumer ability to make rational decisions and requires the effortful involvement of System II. On the contrary, nudges interventions specifically target System I. Nudges activate very fast, intuitive, and associative responses that guide-not force-behaviors toward a specific outcome without consumers being fully aware.

These stimuli are "behaviorally informed" as they systematically consider the underlying decision-making processes of consumers and how they empirically choose. They can be designed and adopted to nudge (i.e., "gently push") healthy, social, as well as sustainable choices (Hansen et al., 2019; Jachimowicz et al., 2019; Sunstein, 2014; Weber, 2017).

Thaler and Sunstein (2008) argue that in some situations nudging-based tools are needed. For example, when choices have a delayed effect, when choices are difficult, when choices are infrequent (and learning not possible), when feedback is poor, and when the relationship between the choice and its outcome is ambiguous. Most environment-related choice situations have several of these characteristics. Therefore, nudges can be considered as a viable intrument to tool to promote pro-environmental behavior.

Some examples of nudges are reminders, warnings, salient disclosures such as labels, social norms, and defaults of choice situations. All these nudges can allow consumers to make green choices, such as food choice with low GHG emissions, reducing domestic food waste, or investing in a new energy system for the house.

17.5.3 Product Labeling for Both Informing and Nudging

As explained before, nudges are based on modifying the "choice architecture" to redirect consumer choices so that they can be beneficial for the society in the long run (Thaler & Sunstein, 2009).

With specific regard to sustainability labels, the debate on whether they should be considered as information provision tools or as nudges is still open.

Sustainability labels can affect consumers' choices in two different ways. Sustainability-involved consumers can actively and consciously look for sustainability-related information. When so, they deliberately and effortfully process information and make their purchasing decision accordingly. This happens when the "problem solver" (System II) is at work. As anticipated, however, not all consumers actively engage in sustainability-related behaviors and, more in general, purchasing decisions are often guided by the "automatic mind" (i.e., System 1) because consumers have limited knowledge, time, and attention. When choices are made under System I, sustainability labels can act as a nudge, especially when they are easily recognizable (e.g., when they are provided as logos or color schemes). In fact, Some studies argue that sustainability labels alter the choice architechture in which consumers are used to make prchasing decisions by providing cues about the most environmentally friendly option (Johnson et al. 2012; Ölander & Thøgersen, 2014). Therefore, sustainability labels can activate associative memory (e.g., green = good/red = bad) and effortlessly guide decision-making (Hollands et al., 2017; Ölander & Thøgersen, 2014; Thøgersen, 2005; Thøgersen et al., 2012). The latter mechanism explains why in some cases nudges can prove more effective than conventional information provision in affecting consumer behaviors.

References

Abrahamse, W., Steg, L., Vlek, C., et al. (2005). A review of intervention studies aimed at household energy conservation. *Journal of Environmental Psychology, 25*, 273–291.

Allcott, H., & Greenstone, M. (2012). Is there an energy efficiency gap? *The Journal of Economic Perspectives, 26*(1), 3–28.

Allen, M. W., Wilson, M., Ng, S. H., et al. (2000). Values and beliefs of vegetarians and omnivores. *The Journal of Social Psychology, 140*, 405–422.

Beattie, G. (2010). *Why aren't we saving the planet? A psychologist's perspective*. Routledge.

Beattie, G., & Sale, L. (2011). Shopping to save the planet? Implicit rather than explicit attitudes predict low carbon footprint consumer choice. *International Journal of Environmental, Cultural, Economic and Social Sustainability, 7*(4), 211–232.

Beattie, S., Lief, D., Adamoulas, M., et al. (2011). Investigating the possible negative effects of self-efficacy upon golf putting performance. *Psychology of Sport and Exercise, 12*, 434–441.

Beattie, S., Woodman, T., Fakehy, M., & Dempsey, C. (2016). The role of performance feedback on the self-efficacy performance relationship. *Sport, Exercise, and Performance Psychology, 5*, 1–13.

Bradford, D., Courtemanche, C., Heutel, G., et al. (2017). Time preferences and consumer behavior. *Journal of Risk and Uncertainty, 55*(2–3), 119–145.

Buckley, M., Cowan, C., McCarthy, M., et al. (2005). The convenience consumer and foodrelated lifestyles in Great Britain. *Journal of Food Products Marketing, 11*(3), 3–25.

Burlingame, B., & Dernini, S. (2011). Sustainable diets: The Mediterranean diet as an example. *Public Health Nutrition, 14*, 2285–2287.

Cairns, J. A. (1992). Health, wealth and time preference. *Project Appraisal, 7*(1), 31–40.

Carmi, N., & Arnon, S. (2014). The role of future orientation in environmental behavior: Analyzing the relationship on the individual and cultural levels. *Society and Natural Resources, 27*(12), 1304–1320.

Cavaliere, A., De Marchi, E., & Banterle, A. (2014). Healthy–unhealthy weight and time preference. Is there an association? An analysis through a consumer survey. *Appetite, 83*, 135–143.

Cavaliere, A., De Marchi, E., & Banterle, A. (2016). Does consumer health-orientation affect the use of nutrition facts panel and claims? An empirical analysis in Italy. *Food Quality and Preference, 54*, 110–116.

Chisholm, S. J. (1998). Attachment and time preference—Relations between early stress and sexual behavior in a sample of American University women. *Human Nature, 10*(1), 51–83.

Colasante, A., D'Adamo, I., & Morone, P. (2021). Nudging for the increased adoption of solar energy? Evidence from a survey in Italy. *Energy Research and Social Science, 74*(101978), 2214–6296.

Creutzig, F., Roy, J., Lamb, W. F., et al. (2018). Towards demand-side solutions for mitigating climate change. *Nature Climate Change, 8*, 260–263.

Darre, E., Cadenazzi, M., Mazzilli, S. R., et al. (2019). Environmental impacts on water resources from summer crops in rainfed and irrigated systems. *Journal of Environmental Management, 232*, 514–522.

De Marchi, E., Caputo, V., Nayga, R. M., Jr., et al. (2016). Time preferences and food choices: Evidence from a choice experiment. *Food Policy, 62*, 99–109.

De Marchi, E., Cavaliere, A., & Banterle, A. (2020). Consumers' choice behavior for cisgenic food: Exploring the role of time preferences. *Applied Economic Perspectives and Policy, 43*(2), 866–891.

European Commission. (2013). Flash Eurobarometer 367. Attitudes of Europeans towards building the single market for green products. Retrieved from https://data.europa.eu/data/datasets/s1048_367?locale=it

European Environment Agency. (2010). *The European environment—state and outlook 2010*. European Environment Agency.

FAO. (2011) The state of the world's land and water resources for food and agriculture: managing systems at risk. In: O. Dubois (Ed.) Earthscan, London, UK

Frederick, S., Loewenstein, G., & O'Donoghue, T. (2002). Time discounting and time preference: A critical review. *Journal of Economic Literature, 40*(2), 351–401.

Gibson, B. (2008). Can evaluative conditioning change attitudes toward mature brands? New evidence from the implicit association test. *Journal of Consumer Research, 35*(1), 178–188.

Graça, J., Calheiros, M. M., & Oliveira, A. (2015). Attached to meat? (Un) Willingness and intentions to adopt a more plant-based diet. *Appetite, 95*, 113–125.

Grunert, K. G. (2011). Sustainability in the food sector: A consumer behaviour perspective. *International Journal on Food System Dynamics, 2*(3), 207–218.

Grunert, K. G., Hieke, S., & Wills, J. (2014). Sustainability labels on food products: Consumer motivation, understanding and use. *Food Policy, 44*, 177–189.

Gussow, J. D., & Clancy, K. (1986). Dietary guidelines for sustainability. *Journal of Nutrition Education, 18*(1), 1–5.

Hansen, J., Kharecha, P., Sato, M., et al. (2013). Assessing "Dangerous Climate Change": Required reduction of carbon emissions to protect young people, future generations and nature. *PLoS One, 8*(12), e81648.

Hansen, P. G., Schilling, M., & Malthesen, M. S. (2019). Nudging healthy and sustainable food choices: three randomized controlled field experiments using a vegetarian lunch-default as a normative signal. *Journal of Public Health, 43*(2), 392–397.

Hofmann, W., Rauch, W., & Gawronski, B. (2007). And deplete us not into temptation: Automatic attitudes, dietary restraint, and self-regulatory resources as determinants of eating behavior. *Journal of Experimental Social Psychology, 43*(3), 497–504.

Hollands, G. J., Bignardi, G., Johnston, M., et al. (2017). The TIPPME intervention typology for changing environments to change behaviour. *Nature Human Behaviour, 1*(8), 0140.

Institute for Global Environmental Strategies (IGES), Aalto University, D-mat ltd. (2019). *1.5-Degree lifestyles: Targets and options for reducing lifestyle carbon footprints.* Technical report. Institute for Global Environmental Strategies, Hayama, Japan

IPCC. (2014). Climate change 2014: Synthesis report. In R. K. Pachauri & L. A. Meyer (Eds.), *Contribution of working groups I, II and III to the fifth assessment report of the intergovernmental panel on climate change.* Core Writing Team. IPCC, 151 pp.

IPCC. (2018). Summary for policymakers. In V. Masson-Delmotte, P. Zhai, Portner, et al. (Eds.), *Global Warming of 1.5°C. An IPCC Special Report on the impacts of global warming of 1.5°C above pre-industrial levels and related global greenhouse gas emission pathways, in the context of strengthening the global response to the threat of climate change, sustainable development, and efforts to eradicate poverty.* World Meteorological Organization, 32 pp.

IPSOS. (2020). *Climate change and consumer behavior.* Global changes in consumer behavior in response to climate change. An Ipsos survey for the World Economic Forum, Jan 2020

Jachimowicz, J. M., Duncan, S., Weber, E. U., & Johnson, E. J. (2019). When and why defaults influence decisions: a meta-analysis of default effects. *Behavioural Public Policy, 3*(2), 159–186.

Jeswani, H. K., Burkinshaw, R., & Azapagic, A. (2015). Environmental sustainability issues in the food–energy–water nexus: Breakfast cereals and snacks. *Sustainable Production and Consumption, 2*, 17–28.

Johnson, E. J., Shu, S., Dellaert, B., Fox, C., Goldstein, D., Häubl, G., & Weber, E. (2012). Beyond nudges: Tools of a choice architecture. *Marketing Letters, 23*, 487–504.

Joireman, J., Shaffer, M. J., Balliet, D., & Strathman, A. (2012). Promotion orientation explains why future-oriented people exercise and eat healthy: Evidence from the two-factor consideration of future consequences-14 scale. *Personality and Social Psychology Bulletin, 38*(10), 1272–1287.

Joy, M. (2009). *Why we love dogs, eat pigs and wear cows: An introduction to carnism.* Conari Press.

Kahneman, D. (2011). *Thinking, fast and slow.* Farrar, Straus and Giroux.

Kildal, C. L., & Syse, K. L. (2017). Meat and masculinity in the Norwegian Armed Forces. *Appetite, 112*, 69–77.

Lorek, S., & Fuchs, D. (2013). Strong sustainable consumption governance—Precondition for a degrowth path? *Journal of Cleaner Production, 38*, 36–43.

McCollough, J. (2010). Consumer discount rates and the decision to repair or replace a durable product: A sustainable consumption issue. *Journal of Economic Issues, 44*(1), 183–205.

McLaughlin, D., & Kinzelbach, W. (2015). Food security and sustainable resource management. *Water Resources Research, 51*(7), 4966–4985.

Meyer-Hofer, M. V. (2014). *Product differentiation and consumer preferences for sustainable food*. Georg-August-Università at Göttingen.

OECD. (2020). Meat consumption (indicator). Retrieved from https://data.oecd.org/agroutput/meat-consumption.htm

OECD, FAO. (2019). *OECD-FAO agricultural outlook 2018–2027*. OECD Publishing/Food and Agriculture Organization of the United Nations.

Ölander, F., & Thøgersen, J. (2014). Informing versus nudging in environmental policy. *Journal of Consumer Policy, 37*(3), 341–356.

Reisch, L. A., Sunstein, C. R., Andor, M. A., Doebbe, F. C., Meier, J., & Haddaway, N. R. (2021). Mitigating climate change via food consumption and food waste: A systematic map of behavioral interventions. *Journal of Cleaner Production, 123717*.

Ricci, E. C., & Banterle, A. (2020). Do major climate change-related public events have an impact on consumer choices? *Renewable and Sustainable Energy Reviews, 126*, 109,793.

Rockstrom, J., Steffen, W., Noone, K., et al. (2009). Planetary boundaries: Exploring the safe operating space for humanity. *Ecology and Society, 14*(2), 32.

Rothgerber, H. (2013). Real men don't eat (vegetable) Quiche: Masculinity and the justification of meat consumption. *Psychology of Men & Masculinity, 14*(4), 363–375.

Sanchez-Sabate, R., & Sabaté, J. (2019). Consumer attitudes towards environmental concerns of meat consumption: A systematic review. *International Journal of Environmental Research and Public Health, 16*(7), 1220.

Schröder, H., Gomez, S. F., Ribas-Barba, L., et al. (2016). Monetary diet cost, diet quality, and parental socioeconomic status in Spanish youth. *PLoS One, 11*, e0161422.

Staats, H. J., Wit, A. P., & Midden, C. Y. H. (1996). Communicating the greenhouse effect to the public: Evaluation of a mass media campaign from a social dilemma perspective. *Journal of Environmental Management, 45*, 189–203.

Steffen, W., Richardson, K., Rockstrom, J., et al. (2015). Planetary boundaries: Guiding human development on a changing planet. *Science, 347*, 1259855e1–1259855e10.

Strathman, A., Gleicher, F., Boninger, D. S., et al. (1994). The consideration of future consequences: Weighing immediate and distant outcomes of behavior. *Journal of Personality and Social Psychology, 66*, 742–752.

Sunstein, C. R. (2014). *Why nudge? The politics of libertarian paternalism*. Yale University Press.

Thaler, R. H., & Sunstein, C. R. (2008). *Nudge. Improving decisions about health, wealth, and happiness*. Penguin.

Thaler, R. H., & Sunstein, C. R. (2009). *Nudge: improving decisions about health, wealth, and happiness*. Penguin.

Thøgersen, J. (2005). How may consumer policy empower consumers for sustainable lifestyles? *Journal of Consumer Policy, 28*, 143–178.

Thøgersen, J. (2021). Consumer behavior and climate change: Consumers need considerable assistance. *Current Opinion in Behavioral Sciences, 42*, 9–14.

Thøgersen, J., Jørgensen, A.-K., & Sandager, S. (2012). Consumer decision making regarding a "green" everyday product. *Psychology and Marketing, 29*, 187–197.

Torma, G., Thøgersen, J. (2021). A systematic literature review on meta sustainability labeling e What do we (not) know? *Journal of Cleaner Production, 293*, 126–194.

Vermeir, I., Weijters, B., De Houwer, J., et al. (2020). Environmentally sustainable food consumption: a review and research agenda from a goal-directed perspective. *Frontiers in Psychology, 11*, 1603.

Walker, G., & King, D. (2008). *The hot topic: What we can do about global warming*. Harcourt.
Warren, P. (2018). Demand-side policy: Global evidence base and implementation patterns. *Energy & Environment, 29*, 706–731.
Weber, E. U. (2017). Breaking cognitive barriers to a sustainable future. *Nature Human Behaviour, 1*(1), 0013.
Willett, W., Rockstrom, J., Loken, B., et al. (2019). Food in the anthropocene: The EATeLancet Commission on healthy diets from sustainable food systems. *Lancet, 393*, 447–492.

Elisa De Marchi is assistant professor at the Department of Environmental Science and Policy of the University of Milan. She is teaching assistant for the course "Agri-food Economics" of the master degree in "Environmental and Food Economics" and lecturer for the course "Agricultural and Natural Resources Economics and Policy" for the master degree in "Environmental Change and Global Sustainability" at the University of Milan. She got her PhD in "Technological Innovation for Agricultural, Food and Environmental Sciences" in 2015 at the University of Milan. During her PhD, she was visiting scholar at the University of Arkansas, at Korea University, and at Michigan State University. She has been involved in several national and international research projects. Her main research interests include behavioral and experimental economics applied to the study of sustainable diets and the role of food labeling in shaping consumer decisions, the economics of obesity, and the role of time preferences on human behaviors. She is author of more than 20 scientific publication with more than 200 citations.

Alessia Cavaliere is associate professor at the Department of Environmental Science and Policy of the University of Milan, where she teaches courses in Food Economics, European Food Legislation and Food Policy. She received her PhD in agricultural economics at the University of Milan. During the PhD she was visiting researcher at "Centre de recerca en economia i desenvolupament agroalimentari" CREDA UPC-IRTA of Barcelona (Spain). She is member of the doctoral committee of the PhD in environmental science. She participated as researcher to numerous national and international research projects such as TRUEFOOD (2006–2010), CAPINFOOD (2011–2014), MOSE4AGROFOOD (2010–2012), and Long Life–High Sustainability (2014–2017). She delivered more than 30 oral presentations at national and international congresses. Her main research interests and fields of specialization are related to the study of consumer behavior, the economic aspects of health and environmental-sustainable food consumption, and information-based policy. She was author of more than 60 international and national publications, which include 30 full papers, 17 chapters in books of international interest, 6 educational materials, and others proceedings. She has the following bibliometric indicators: 30 publications, 366 citations, and 11 H-index.

Alessandro Banterle is full professor at the University of Milan. He graduated at University of Milan (110/110 magna cum laude) and he got his PhD in agri-food economics and policy at the University of Padua. He has been visiting researcher at University of Reading (UK). He is the head of the Department of Environmental Science and Policy (ESP—www.esp.unimi.it/ecm/home) at the University of Milan and professor of agri-food economics and policy. He is member of the doctoral committee of the PhD in economics. He was founder and editor in chief of the international scientific journal Agricultural and Food Economics, published by Springer. He was member of the International Jury for the "EXPO Awards" of EXPO Milano 2015. He participated in the coordination of several research projects, such as TRUEFOOD (2006–2010), CAPINFOOD (2011–2014), SOSTARE (2009–2011), MOSE4AGROFOOD (2010–2012), Long Life–High Sustainability (2014-2016), IPCC MOUPA (1918–2020), LEGERETE (2017–2021), DIVERFARMING (2017–2022), and BIOSURF (2021–2023). His research interests include economic analysis of food supply chain, consumer behavior, food policy, environmental sustainability, and circular and bio-economy. He has the following bibliometric indicators: Scopus 54 publications, 904 citations, and 20 H-index; Google-Scholar 166 publications, 2.051 citations, and 27 H-index.

Chapter 18
Climate Change Litigation: Losing the Political Dimension of Sustainable Development

Sara Valaguzza

Abstract Climate change litigation has been spreading worldwide at an impressive rate in the last decade. These types of lawsuits often take the form of strategic litigation that, based on scientific knowledge, aim to reach rulings that extend far beyond the interests of the parties involved in the case. What seems to be emerging in this trend is the advancing of an argument of a subjective "right to climate," invoked to solicit public policies and interventions to protect the lives of present and future generations. The current article aims at illustrating some reflections while also highlighting shortcomings of climate change litigation lawsuits. In particular, considering the relevance of climate change consequences, it will be stressed how climate disputes risk undermining the democratic structure of our societies. Effectively, the author argues, the processes involved in forging policies, naturally characterized by checks and balances, seem to be weakened in climate dispute decisions. Therefore, the judicial decisions are left to influence and shape development models and, conversely, take away responsibility of public actors, whose action should be precisely managing and caring for the environment and society's well-being at large.

18.1 Climate Change Litigation: Context and Purpose

Climate change litigation is often a form of strategic litigation, or a series of lawsuits filed to seek an advantageous ruling that extends far beyond the interests of the parties involved in the case. If we compare climate disputes to other types of strategic litigations—for example, civil and political rights—some interesting characteristics immediately emerge.

Strategic litigation regarding human rights requires, on the one hand, a higher level of abstraction. On the flipside, it raises divisive ethical issues that are rarely

S. Valaguzza (✉)
University of Milan, Milan, Italy
e-mail: sara.valaguzza@unimi.it

decided based on common feeling. For example, the decision whether to allow all couples to make use of fertility treatments, or only those experiencing certain pathologies, depends on evaluations concerning the desire to assign a privileged position to the family, and the meaning that one may wish to attribute to the right of parenthood. Furthermore, disputes regarding civil and political rights often do not rely on objective, specific, and scientific data that courts can take and reference as demonstrably true facts when deciding cases.

Instead, climate litigation—those cases in which states' policies to reduce global warming are challenged—are built on common scientific grounds, and on a widespread understanding that global warming is harming both human beings and the environment.

Generally speaking, climate litigation refers to cases in which the judgment's subject matter is directly or indirectly connected to climate change. Climate litigation includes actions against states, government agencies and administrative authorities, local entities, industries, and private citizens (Lord et al., 2012). Normally, governments are sued in order to challenge the policies adopted to reduce emissions and combat rising temperatures. Lawsuits against government agencies and administrative authorities typically challenge decisions that appear illegitimate or disproportionate, challenging the balance between environmental interests and economic freedoms. Actions against local entities tend to request a review of urban redevelopment programs or building permits that have been approved without sufficiently considering the carbon footprint of what has been approved. In the compensatory realm, some climate litigation also involves business activities and resolves disputes between private parties.

Currently, the most interesting types of climate lawsuits, in terms of the field's growth and extension, are those where private parties bring legal actions against public bodies in response to ineffective mitigation and adaptation policies. In this type of dispute, public bodies are criticized for not implementing adequate measures to protect communities from the devastating effects of global warming. Thus, underlying climate litigation is a dissatisfaction with the ability of law, understood as a normative and regulatory framework, to cope with the problems of modern society. The complaint is one against the lack of efficiency and effectiveness of public intervention. A closer look at the phenomenon reveals it is characterized by the opening of a rift between legality and justice. This means neither climate policies nor the law are not viewed as capable of solving the problem, and therefore, a gap emerges between the protections offered by the legal framework and that which would be perceived as "the right thing" to do.

As a reaction to the law's inability to be "just," the world is witnessing a rise in climate justice claims that are currently outside the regulatory parameters. These demands call for a greater consideration of the environmental, economic, and social effects of climate change on the part of national governments. There exist various types of protection requests related to climate issues based on the inherent responsibility of public actors, with management and care for the environment as the primary concerns.

Through both climate litigation and movements for climate justice—including street demonstrations—climate activists are challenging a development model that exploits territories, leaving behind deep wounds that harm the relationship between humanity and nature, to the detriment of current and future generations. Columbia University's Sabin Centre for Climate Change Law, which monitors climate disputes, confirms their impressive growth all around the world due to the presence of common and widespread contributing causes.

The goal of this chapter is not to analyze in detail the various and ongoing climate disputes around the world. Rather, it will try, given the book's interdisciplinary nature, to offer a systematic overview of the cases, highlighting their limits, criticalities, and virtuosity. In particular, this chapter will critically analyze how climate disputes impact the role of jurisdiction and environmental protection policies.

It is worth noting at the outset that an excessively personalized and selfish connotation of environmentalists' claims creates a risk of decreasing understanding of the environment as a common good. Consequently, it might also divert attention from the need to negotiate, with the community, protection policies that are the result of a thoughtful mediation between economic, social, and environmental needs.

18.2 Climate Change Litigation Fever: Reasons and Extension

The spread of intergenerational movements demanding respect for their own rights and for those of future generations, in the form of concrete and immediate action from national governments that protect the planet and its resources, can be identified as the main reason for the rise in climate disputes.

Promoted by activists and environmental associations, climate litigation appears as the result of a declared antagonism that pits people and the civil community against those who exercise public powers. Groups, associations, foundations, individuals, and local communities all advocate for a new right: the right to climate, which has been denied by the inability of public policies to identify effective resilience and sustainability actions in response to rising global temperatures.

A common feature of climate litigation is that plaintiffs do not act only for their own good, but for an altruistic purpose that benefits future generations (Ogletree & Hertz, 1986, 23).

In brief, the climate disputes that are spreading everywhere at an impressive rate are a form of strategic litigation with a strong ideological thrust. Here, militant environmental activism is ignited and, promoted by ethical and ideal purposes, calls on the courts to express moral judgments for the national governments.

Another contributor to the spread of climate litigation is the increase of so-called "science for all"—scientific evidence translated into understandable language that is accessible to everyone. Communities are able to understand the risks and be aware of the causes of the phenomenon.

The growth of awareness regarding climate issues is certainly the basis of what animates civil society to promote strategic climate disputes.

In fact, climate disputes are built by using scientific data. The most used are the International Panel on Climate Change (IPCC) reports, even if the IPCC seems to reduce the scope of its reports, and certainly excludes the possibility that its findings be binding for national governments.

"The IPCC's reports are neutral, policy-relevant but not policy-prescriptive. The assessment reports are a key input into the international negotiations to tackle climate change," according to its website (www.ipcc.ch). Still, these reports are used to prove the truthfulness of the assumptions on which climate disputes are developed, and the seriousness of the problem that litigation intends to address.

The established parameters could condition the scientific evaluations, and the continuous movement of the reference data makes the science less mathematical than one might think. However, it is undeniable that general agreement exists with regard to the well-known and alarming data demonstrating the need to reduce global warming.

Moreover, the importance of science in climate disputes effectively reduces the uncertainty of judicial decisions, at least in relation to the assessment over the dangers brought by global warming.

The issue's interdisciplinary nature leads to a broader recognition of the problems that afflict the world and increases critical thinking within communities. Consequently, the opportunity arises to open a clear, decisive, and sometimes even critical dialogue with the relevant institutions.

Moreover, this awareness has certainly increased as a result of the Covid-19 pandemic that impacted the entire planet in 2020.

Indeed, scientific data is emerging that shows how the virus's negative effects are exacerbated in areas with high air pollution.

In particular, a 2020 study by the Euro-Mediterranean Centre on Climate Change found a potential correlation between exposure to the most dangerous air pollutants and Covid-19 incidence and mortality levels (Granella et al., 2021).

All of this inevitably leads to a search for ways, including judicial means, to solicit public policies and interventions that protect the life of present and future generations.

Moreover, it is fully understandable that, in a manner directly proportional to the areas where there is greater attention for the environment, strategic climate disputes increase.

Starting from *Leghari v. Federation of Pakistan*, which is considered the forerunner of the lawsuits on climate change and was inspired by the protection of human rights, climate litigation is now progressively increasing both in terms of numbers and strength of demands (Peel & Osofsky, 2018). The increase is significant enough that the current reality must be subject to legal study and modeling.

The exceptional increase in climate litigation over the past few years is significant from different perspectives, and especially so if we consider that these are strategic disputes. This type of legal action is difficult to initiate; requires a particularly long preliminary phase; presupposes the acquisition of technical and scientific data that are not easy to formulate; and is the result of a social awareness tactic that requires time and a good dose of premonition (regarding the sensibilities of the judge deciding the case, the setting of judicial requests, the "evidence" produced in court, etc.).

In recent years, activists in Europe have achieved ground-breaking judgments in which national governments were convicted for not having adopted adequate internal policies to fight climate change.

In 2019 in the Netherlands, in a famous case brought by Urgenda Foundation et al. (Fasoli, 2018, 90; Scovazzi, 2019, 619; Maxwell, 2020, 620; Antonopoulos, 2020, 119), the Supreme Court found the State guilty of failing to adopt adequate measures to reduce greenhouse gas emissions to combat climate change, reasoning that "it is an essential feature of the rule of law that the actions of (independent, legitimised, and controlled) political bodies, such as the government and parliament can—and sometimes must—be assessed by an independent court" (Supreme Court of the Netherlands, *State of the Netherlands v. Urgenda Foundation*, Judgement, 20 December 2019).

In 2021 in France, the *Tribunal Administratif* de Paris ruled in favor of an appeal brought by four environmental associations—Oxfam France, *Notre Affaire à tous*, *Fondation pour la Nature et l'Homme*, and *Greenpeace France*—supported by more than 2.3 million people who signed the "*Affaire du siècle*" petition condemning the state for ecological damage.

The judges observed that, "In the period from 2015 to 2018, France exceeded its first carbon budget by 3.5% (…) achieving an average decrease in emissions of 1.1% per year, while the budget required a reduction of around 1.9% per year (…). For the year 2019, the decrease in emissions compared to 2018 was 0.9%, while the second carbon budget, set for the period 2019–2023, foresees a decrease of 1.5% per year."

Consequently, the State was held liable for not having complied with the commitments it had undertaken to reduce greenhouse gas emissions.

In Italy, in late 2020, a group of organizations, associations, committees, and individual citizens launched a campaign called the "Last Judgment." The campaign's goal was to induce the Italian government—including by threat of legal action—to implement stringent measures to tackle climate change. A lawsuit is expected to be filed in the coming months.

Leaving aside decisions in individual cases, it should be noted that the large number of climate disputes is a symptom of a serious problem, and a rift between those in charge of public powers compared to a particularly active share of community members.

This is how a collective fever starts—aggravated by a phenomenon covered in mass media that psychologists have termed "eco-anxiety"—that identifies global warming as the main issue that needs to be solved at all costs.

The problem is certainly known and defined, but its contextualization is sometimes not very realistic, and the desired solutions are not seen on the horizon.

How can global warming be solved? Where and what is the effect of climate policies? What and how many implications exist between climate problems and economic and social problems?

If it were possible to turn off a switch or adjust a thermostat in our collective home, planet Earth, the problem would be solved. But this is not the case. The planet needs adaptation policies that cannot and must not only consider the environment; industrial, labor, housing policies, the relationship between costs and benefits, and so on also require a deep analysis.

In connection with the abovementioned points, a few limits, from a legal perspective, arise with regard to climate litigation. Despite increasing government attention on environmental policies, climate lawsuits have the downside of attempting to allocate political or administrative powers to those exercising jurisdiction. The pitfall of this process is that it risks losing sight of the complexity and multidimensionality of the concept of sustainability, which carries with it three different dimensions of development: economic, environmental, and social.

Among other things, a conviction cannot determine the content of the mitigation policies and, above all, the adaptation policies that are indispensable for the right to enjoy a healthy environment. Therefore, climate litigation does not solve the problem from which it starts.

Accordingly, in addition to analyzing the case law and acknowledging the enthusiasm surrounding it, jurists should also highlight the risk that climate litigation could overshadow the three-dimensionality that in the past has been difficult to achieve when passing from a focus on the environment in absolute terms to valuing sustainable development policies.

18.3 The Subject Matter of Climate Litigation: The Right Sought

Climate litigation in Europe typically frames the subject matter of the dispute—namely, the legal theory under which an action is brought—as "the right to climate." As already mentioned, in some instances, the lawsuits are promoted not only for the benefit of the plaintiff, but also to protect the interests of future generations.

Climate litigation can be linked to human rights disputes (OHCHR) and, in particular, the right to health. The right to climate is reinterpreted as the right to a healthy climate, with legal action taken to protect one's individual sphere from the negative effects of climate change on the person's psychophysical condition.

At the center of the European climate legal proceedings are humans, not nature. Those who bring legal actions believe they have the right to first, protect their health, and second, the right to a climate that allows life to exist on earth in present and future generations (Bogojevic & Rayfuse, 2018; Shelton, 2012; Boyle & Anderson, 1996).

In climate litigation based on human rights, nature—and the ecosystem in general—is functional to humanity's enjoyment of the planet.

In other words, according to this logical approach, aggression against the environment is judicially relevant when it also becomes aggression against human beings, who end up undergoing, as some have tried to argue, a form of torture caused by environmental pollution.

In one dispute based on the European Convention on Human Rights, the court rejected the argument that Article 3 of the Convention, the prohibition against torture, was violated by the state's failure to repair the environmental damage. The plaintiff had argued unsuccessfully that this failure was the equivalent of forcing the population to undergo inhumane and degrading treatment (ECHR, *López Ostra v. Spain*, Application no. 16798/90, Judgment, 9 December 1994).

While in Europe the right to climate focuses on the rights of individuals, in the United States and other parts of the world, the strategy is to identify principles, norms, and obligations—for example, the public trust doctrine—that render public authorities liable for embracing policies deemed ineffective (Blumm & Wood, 2015; Frank, 2012; Wood, 2007). The consequence of this approach brings about a discussion on the legal value of a lawsuit based on legal provisions resulting from international agreements.

Briefly, the public trust doctrine identifies public entities as those responsible for protecting and managing natural assets (waterways, water resources, and wildlife) in the best interest of the administered community.

The subject of what has been defined as "Nature's Trust" are natural resources that are vital for society's well-being.

According to this theory, the trustee—the government—must take care of nature by representing the interests of the trust beneficiaries, which is the population.

The public trust doctrine provides a conceptual reconstruction that is abstractly useful for the courts to consider in climate lawsuits, as well as for repairing the damage caused to the environment as an ecosystem.

The two approaches to climate litigation—the first based on individual rights and the second based on Nature's Trust—express two different identities. The European legal culture focuses on individuals and can be seen as both enlightened and romantic at the same time; the American framework entrusts the legal system's progress with the creative rulemaking of courts.

In countries where the bond with nature is stronger, due to geographical reasons or conformity with urban contexts, one sees a sort of new pantheism of nature (Cullinan, 2002), as if the courts were witness to a search for tools to imbue the environment with human qualities.

Indigenous communities in many parts of the world, from New Zealand to the Philippines, have brought legal actions claiming the "right of nature," which is not understood as their own right to enjoy nature, but rather as the right of nature itself to prosper without being assaulted by humans.

Some regulations also give legal relevance to natural "things," such as the *Te Awa Tapua Act* of 2017. In order to settle a dispute with the indigenous Maori community, the government of New Zealand gave legal status to Whanganui River, considering it an "indivisible and living entity" that included the river and the indigenous groups living in symbiotic relationship with it (Collins & Esterling, 2019; Argyrou & Hummels, 2019, 752). The goal was to protect, with a bundle of rights and duties, both the human and the natural environment as a whole.

In 1993, the Supreme Court of the Philippines, referred by a group of minors represented by their parents, as well as by some nongovernmental organizations, affirmed—contradicting the assessments of the court of first instance—that the right to the full use of natural resources, is a fundamental right: "These basic rights need not even be written in the Constitution, for they are assumed to exist from the inception of humankind. If they are now explicitly mentioned in the fundamental charter, it is because of the well-founded fear of its framers that unless the rights to a balanced and healthful ecology and to health are mandated as state policies by the Constitution itself, thereby highlighting their continuing importance and imposing upon the state a solemn obligation to preserve the first and protect and advance the second, the day would not be too far when all else would be lost not only for the present generation, but also for those to come—generations which stand to inherit nothing but parched earth incapable of sustaining life."

In order to affirm the State's liability, the court defined the national government as *parens patriae*, and thus applied the country's environmental code, which recognized a right to the environment for present and future generations, and laid out the following specific obligations for the public authorities: "(a) to create, develop, maintain, and improve conditions under which man and nature can thrive in productive and enjoyable harmony with each other, (b) to fulfill the social, economic, and other requirements of present and future generations of Filipinos, and (c) to insure the attainment of an environmental quality that is conducive to a life of dignity and wellbeing" (Supreme Court of the Republic of the Philippines, *Oposa v Factorian*, Case No. 101083, 30 July 1993).

18.4 The Undefined Content of the "Right to the Climate" and the Environment as a Duty

Whatever the techniques and instruments through which a case is brought before a court, from a legal theory perspective, the "right to climate" is difficult to define. It could be assumed that it is the right to a "just" climate, that is, a climate that preserves life and is compatible with ecosystems development and protection of biodiversity.

If this were the case, however, it would be difficult to identify a specific individual right based on a specific subjective situation. This interpretation of the right to a just climate would mean that each person could have a generic, indefinite claim against public authorities, which could infringe on the economic or social rights of other actors within the legal system. Some have also observed, provocatively and paradoxically, that acknowledging a right to the climate based on a qualified subjective position could lead, for instance, to a claim of a right to sea breeze or mountain air.

Of course, fighting to mitigate climate change means defending a bundle of rights: the right to climate, the right to nature, the right to food, the right to family and domestic life, and the right to health.

The difficulty in defining the "right to climate" makes theories such as Nature's Trust, which is based on the concept of responsibility rather than a right, more convincing.

The theory could receive further support from analysis carried out in civil law countries in the context of public law studies that understand the environment as a common collective asset.

This concept provides a common ground between the right *to* nature and the right *of* nature, in the sense that recognizing the environment as a common good makes it both subject to protection and an area of responsibility, as well as a potential area of rights. But the rights involved are not individual; rather, they are common public rights.

Judicial protection, therefore, acts as key to unlock substantial protection for mankind's common heritage, based on a subjective situation. Within these widespread interests, there "appears to be a psycho-social state of tension between a need and a good, repeated indefinitely so as to affect the masses, which are by definition undifferentiated" as Mario Nigro (1987, 8).

Even more clearly, some authors have recently identified a path based on a theory of duties (Fracchia, 2018a, Fracchia, 2018b; Cafagno et al., 2019; Martines, 1996, 23). In addition to making the idea of a subjective right to the environment or to climate appear illusory, the theory systematically identifies a web of obligations that clarify the relationship with the commons and with future generations, thus indirectly strengthening the public dimension of the environmental problem. If the environmental discipline aims to guarantee current and future living conditions, then "the environment is not a right, but a duty."

This duty, while primarily ethical, belongs not only to institutions, but to everyone, based on the principle of solidarity referred to in Article 2 of the Italian Constitution. This includes moral and legal obligations of an intergenerational and *intragenerational* nature, since it is not logical to take care of those who will come later but not those who are already suffering now.

The discussion described above raises serious doubts about the fitness of an environmental protection system based on rights, rather than a hierarchy of duties.

18.5 The Limits of Climate Litigation and the Arising of a Related Problem

American scholars define the climate crisis as a "super wicked problem." Local communities in the United States have, indeed, experienced catastrophic events that have been attributed at least in part to rising temperatures, especially an increase in hurricanes and severe weather (Partenteau, 2006).

Europeans, which typically take a policy-based approach, might add that the climate crisis is a relationally complex issue with multiple contributing causes. It interferes with different levels of government and policies, and influences various dimensions of the economy and society.

Such a complex issue should be tackled with an approach where science, policies, rights, and individual versus collective considerations generate a series of reflections based on a dynamic exchange of opinions and evaluations across networks.

The way in which this wicked and relationally complex problem is tackled is so decisive that it becomes, as previously mentioned, a question of identity for who governs us.

Indeed, the European Green Deal will be remembered as the most significant policy choice initiated by the President of the European Commission, Ursula Von der Leyen (Sola, 2020, 462; Durante, 2020; Rifkin, 2020; Chomsky & Pollin, 2020). Its impact had a cascading effect on US President Joe Biden, who immediately declared his plans to rejoin the Paris Agreement and adopt certain Green New Deal policies.

It should be noted, however, that it is not just rising temperatures that generate concern. At the heart of the problem is the choice by national governments about how to confront these changes with future economic development strategies.

Massive use of scarce resources; an unsustainable consumption of raw materials; a linear economy that pays little attention to recycling; a competitive framework that leaves no room for the circulation of value, undercutting the competition with the lowest price; a society based on individualism and on the dominance of the strongest. These are all aspects that, upon closer inspection, must be addressed in order to identify the economic and social development model for our future, including strategies for tackling global warming.

The claim that reducing global temperature is the sole end means neglecting other important factors. Doing so would subject us to new paradigms of global crisis, such as mass migrations due to poverty (Mann & Wainwright, 2019, 131), killer viruses, the loss of democracy, and renewed illiteracy.

There are several limits to an approach that seeks answers only by tackling environmental issues and not the many problems associated with the development model.

Other limits derive from the attempt to seek, through the exercise of legal jurisdiction, answers to questions that are essentially political. On one hand, the theoretical connotation of judicial protection as the forum for guaranteeing individual rights is lost. On the other hand, political debate could diminish, along with the many different perspectives that constitute sustainable development (Giovannini, 2018; Scotti, 2019, 493; Salvemini, 2020; Sachs, 2014; Valaguzza, 2016, 95), which make social, economic, and environmental judgments.

In other words, especially with regard to adaptation policies, the complex processes involved in conceiving, adopting, and implementing public policies cannot realistically be censored, resolved, and redesigned in the exercise of judicial functions.

The fact that a complicated political issue lies at the root of climate disputes is clear. Yet climate litigation looks to the courts, entrusting them with a task that belongs to the Legislature, or at most to the executive power: to express political evaluations on the relationship between scientific evaluations and policies. This is evidence of a significant overturning of the traditional order of legal structures and of the relationship between the State's powers.

18.6 Judicial Decisions as *Extrema Ratio*

The judicial decisions that have found fault with climate policies around the world have shown that the principle of separation of powers, the so-called political triad, does not prevent government bodies from being convicted for not adopting effective policies to combat climate change. This result, however, highlights another, potentially critical issue.

In 2020, in the *Juliana* case, the U.S. Court of Appeals for the 9th Circuit, Oregon District, held that, "We reluctantly conclude, however, that the plaintiffs' case must be made to the political branches or to the electorate at large, the latter of which can change the composition of the political branches through the ballot box. That the other branches may have abdicated their responsibility to remediate the problem does not confer on Article III courts, no matter how well-intentioned, the ability to step into their shoes" (United States Court of Appeals for the Ninth Circuit, *Juliana v. United States*, Appeal from the District of Oregon, filed January 17, 2020).

However, in Europe, the extreme gravity of the issue brought before the courts—the irreparability of the damage being done and the absence of ethical conflicts—is pushing judges to issue decisions based on evidence found in scientific reports, despite a lack of regulations to apply.

One example is the conviction of the Dutch government, where an adverse verdict was reached even though the country's national rules respected European targets.

What is clear is that many of the complexities and facets of environmental policies have been overlooked. These policies require a multistage development process made up of a plurality of elements that diverge in nature and content. Their combination must be used to assess whether or not they achieve the expected public-interest goals.

Within the system of checks and balances of liberal democracies, using judicial evaluation to verify public policies' (in)effectiveness does not appear to be the proper tool. Instead, the task should fall mainly to electoral mechanisms or in any case to the representative democratic systems.

Indeed, it must be kept in mind that the end result of the political process is compromise, in the sense that it includes differing choices, mediations, and priorities.

It seems incoherent, therefore, to examine this process through the lenses of individual rights before a court. The natural place would be within public and institutional confrontation, in the context of political and administrative debate.

It is certainly positive that judicial decisions can act as a force for policy change. But it must be emphasized that judicial intervention is, in the case of climate litigation, an *extrema ratio*, or an extreme solution. No judicial decision can ever lay down policies; if anything, they can only waive them.

Ultimately, case law increases institutional awareness on the climate problem and serves to modernize and make legal systems responsive to the complexities of social needs—and it does so more quickly than the normal legislation course. Judicial decisions, however, cannot forge entirely new models for future development. Courts are not entitled to do so, and at least in Europe, such a practice would fall completely outside the democratic system.

18.7 The Climate Problem as a Question of Identity and Return to the Principle of Sustainability

The irreversibility of climate change, the number of contributing causes, and above all, the different approaches to environmental policies among national governments make an international joint political action an unrealistic solution to the climate issue.

On the other hand, local policies cannot ignore the reconfiguration of a new global order based on issues that affect the entire world. This reconfiguration necessarily requires a rethinking of the mechanisms of democracy and representation traditionally built on a territorial basis.

It is increasingly clear that the weakness of international policies, excluding at least macro targets, is the fact that mitigation policies aimed at reducing global warming are not the priority any longer. As testified by the European Green Deal, adaptation tools cannot ignore local governments' choices and must be based on awareness of the coexistence between socioeconomic development and environmental protection. In other words, the same original compromise that is internal to the mechanisms of sustainable development.

These are mechanisms that try to mediate between economic needs and environmental protection. This mediation is evident in the different international declarations on the issue.

The 1972 Stockholm Declaration emphasized the centrality of the economic and social development found in principle No. 8: "Economic and social development is essential for ensuring a favorable living and working environment for man and for creating conditions on Earth that are necessary for the improvement of quality of life." Principle No. 11 further states: "The environmental policies of all States should enhance and not adversely affect the present or future development potential of developing countries, nor should they hamper the attainment of better living conditions for all."

Later, the 1992 Rio Declaration reaffirmed the same concept, stating the right to (economic and social) development guided by the principle of sustainability and enshrined in principle no. 3—"The right to development must be fulfilled so as to equitably meet developmental and environmental needs of present and future generations"—and principle no. 4: "In order to achieve sustainable development, environmental protection shall constitute an integral part of the development process and cannot be considered in isolation from it."

Not surprisingly, the action program included in the 2030 Agenda for Sustainable Development focuses on the prosperity of people and the planet, providing for common goals that go beyond the relationship with nature and the environment.

The aforementioned compromise is implemented through specific forecasts that depend on local actions since the objective and subjective indicators of sustainability and well-being are measured, weighed, and corrected at the local level.

It is not a coincidence that the issue of sustainability centers on community well-being and ecological awareness in cities. Cities have acquired a central role within the parameters of the Sustainable Development Goals and the 2030 Agenda.

The European Commission supported the launch of the Covenant of Mayors for climate and energy, which today has amassed nearly 10,000 signatories.

Given that adaptation policies require a compromise based on a complex process of investigation, consultation, and balancing of interests, and that consequent resilience actions move through verification, trial, and error based on the scientific method, jurisprudence cannot be the right place to shape them.

The risk is that a purely subjective connotation based on the environmentalists' claim—which views the ecosystem as a good to which human beings should have rights—obscures the fact that we are faced with a common heritage that, above all, requires protection policies.

The self-centered "subjectification" of climate litigation risks undermining the complexity of relationships based on different values and perspectives, all of which must be integrated in search of a sustainable development strategy.

It would be a serious planning error to abandon this complexity and reduce, once again, the ecosystem to a single human being—or at most, to a group of human beings—who acts for himself and others based on his individual needs and his prophecies.

If, as sustainability theories maintain, ecosystems must be respected and protected because it is man who belongs to nature, and not vice versa, the principal road is outside the courts. The correct path is a renewed political debate that targets sustainable development, obtained through a mapping of needs and objectives that are balanced and ordered within a global network of strategic priorities.

As noted in the preceding pages, climate litigation reveals that the issue of climate justice is a question of identity, both for those who take legal action based on their own vision for the future, and for those who adopt resilience and adaptation or mitigation strategies.

The climate agenda is certainly a crucial political topic that belongs neither to the right nor to the political left. It is a wide-ranging macro-theme that attracts consensus or dissent regardless of political affiliation.

Even more so, tackling the climate problem is a question of identity where its social implications are understood; the link between climate problems and social issues is unquestionably on the minds of all.

The spread of climate justice principles and values, such as social and intergenerational equity, must be developed within a scenario that builds a new public ethics based not only on the principle of the rule of law, but also on the promotion of equality and of shared public values.

From this perspective, fighting climate change is part of the public's universal duty of solidarity, which modern democratic states must pursue.

Our future will depend on this: a future that is not up to the courts to design.

References

Antonopoulos, I. (2020). The future of climate policymaking in light of Urgenda Foundation v. the Netherlands. *Environmental Law Review, 22*(2), 119–124.

Argyrou, A., & Hummels, H. (2019). Legal personality and economic livelihood of the Whanganui River: A call for community entrepreneurship. *Water International, 44*(6-7), 752–768.

Blumm, M. C., & Wood, M. C. (2015). *The public trust doctrine in environmental and natural resources law*. Carolina Academic Press.

Bogojevic, S., & Rayfuse, R. (2018). *Environmental rights in Europe and beyond*. Hart.

Boyle, A. E., & Anderson, M. R. (1996). *Human rights approaches to environmental protection*. Clarendon Paperbacks.

Cafagno, M., et al. (2019). The legal concept of the environment and systemic vision. In M. Cafagno et al. (Eds.), *The legal concept of the environment and systemic vision: A rock in the pond* (p. 141). Springer.

Chomsky, N., & Pollin, R. (2020). *Climate crisis and the global green new deal*. Verso.

Collins, T., & Esterling, S. (2019). Fluid personality: indigenous rights and the Te Awa Tupua (Whanganui River Claims settlement) Act 2017 in Aotearoa New Zeland. *Melbourne Journal of International Law, 20*(1), 197–202.

Cullinan, C. (2002). *Wild law. A manifesto for earth justice*. Cormac Cullinan.

Durante, A. (2020). *Il Green New Deal. Rischi e vantaggi di un nuovo interventismo pubblico in economia*. Albatros.

Fasoli, E. (2018). State responsibility and the reparation of non-economic losses related to climate change under the Paris Agreement. *Riv dir internaz, 1*, 90–130.

Fracchia, F. (2018a). L'ambiente nella prospettiva giuridica. In *Diritto amministrativo e società civile. Vol. III – Problemi e prospettive* (p. 619). Bononia University Press.

Fracchia, F. (2018b). *Environmental law: Principles, definitions and protection models*. Editoriale Scientifica.

Frank, R. M. (2012). The public trust doctrine: assessing its recent past & charting its future. *University of California Davis Law Review, 45*, 665–691.

Giovannini E., *Utopia sostenibile,*. Bari-Rome, 2018.

Granella, F., et al. (2021). COVID-19 lockdown only partially alleviates health impacts of air pollution in Northern Italy. *Environmental Research Letters, 16*, 1–9.

Lord, R., et al. (2012). *Climate change liability*. Cambridge University Press.

Mann, G., & Wainwright, J. (2019). *Il nuovo leviatano. Una filosofia politica del cambiamento climatico*. Treccani.

Martines, T. (1996). L'ambiente come oggetto di diritti e di doveri. In V. Pepe (Ed.), *Politica e legislazione ambientale*. Edizioni Scientifiche Italiane.

Maxwell, T. (2020). (Not) going Dutch: Compelling states to reduce greenhouse gas emissions through positive human rights. *Public Law, 4*, 620–631.

Nigro, M. (1987). Le due facce dell'interesse diffuso: ambiguità di una formula e mediazioni della giurisprudenza. *Il Foro it, 110*, 7–20.

Ogletree, C., & Hertz, R. (1986). The ethical dilemmas of public defenders in impact litigation. *NYU Review of Law & Social Change, 14*(1), 23–42.

Partenteau, P. (2006). Come hell and high water: coping with the unavoidable consequences of climate disruption. *Vermont Law Review, 34*, 957–974.

Peel, J., & Osofsky, M. H. (2018). A right turn in climate change litigation. *Transnational Environmental Law, 7*(1), 37–67.

Rifkin, J. (2020). Un Green Deal globale. Il crollo della civiltà dei combustibili fossili entro il 2028 e l'audace piano economico per salvare la Terra. (trad. Parizzi M) Mondadori, Milano

Sachs, J. D. (2014). *The age of sustainable development*. Columbia University Press.

Salvemini, L. (2020). Lo sviluppo sostenibile: l'evoluzione di un obiettivo imperituro. *AmbenteDiritto.it, 2*(20), 1–35.

Shelton, D. (2012). What happened in Rio to human rights? *Yearbook of International Environmental Law, 3*, 75–93.

Scotti, E. (2019). Poteri pubblici, sviluppo sostenibile ed economia circolare. *Dir Econ, 65*(98), 493–529.

Scovazzi, T. (2019). L'interpretazione e l'applicazione ambientalista della Convenzione europea dei diritti umani, con particolare riguardo al caso Urgenda. *Riv giur amb, 4*, 619–633.

Sola, A. (2020). Sostenibilità ambientale e Green New Deal: prime analisi in commento alla legge di bilancio 2020. *Federalismi.it, 10*, 462–478.

Valaguzza, S. (2016). *Sustainable development in public contracts. An example of strategic regulation*. Editoriale Scientifica.

Wood, M. C. (2007). Nature's trust, VA. *Environmental Law Journal, 24*(431), 448–449.

Sara Valaguzza is a full professor of administrative law at the University of Milan, where she teaches administrative and environmental law, sustainable development, green procurement, and public–private partnership. She also leads a research group on environmental issues and policies.

Professor Valaguzza is an expert on public contracts and environmental law and policy. Her publications on governing by public contracts and collaborative agreements are an important reference for national and international scholarship. Her researches on climate litigations, on the juridical dimension of sustainability, and on the issues of resilience and government of the environment are often discussed in academic and institutional seats.

She is the promoter of the global partnership among universities, centers of research, public authorities, and private operators that in 2020 launched the first edition of the Interdisciplinary Approaches to Climate Change (IACC) in the University of Milan, which she coordinates.

She is also the founding president of the European Association of Public-Private Partnership—EAPPP and the scientific director of the Italian Centre of Construction Law & Management.

She wrote six monographs, both in Italian and in English, and more than seventy essays on administrative law topics. She has written extensively on public–private partnership, public contracts policy, European multilevel governance, sustainable development, *res iudicata,* and judicial review.

She also practices as an attorney for Italian and international clients and leads a research group to promote collaborative contracts in the public and private sectors in Italy, to reach environmental and social targets, and to improve public value.

Chapter 19
The Judicial Review of Administrative Decisions with Environmental Consequences

Eduardo Parisi

Abstract The issues of climate change and the need to shape a new model of sustainable development strongly rely on the exercise of administrative functions by public authorities. The application of public powers in food and agriculture, energy, transportation, urban planning, construction, public procurement, landscape, and cultural heritage shapes the human activities that are capable of affecting the environment, both in everyday lives and in the production and consumption of materials.

Undoubtedly, the public–private dynamic that is generated by the execution of the described administrative functions provokes hard conflicts of interests, given that it implies the application of unilateral powers in economically and ethically sensitive matters. This being the case, juridical disputes aimed at assessing the legitimacy of administrative decisions with environmental consequences are an almost unavoidable character of the legal system transitioning toward a greener and healthier dimension.

Detecting the transversal features of this dynamic by analyzing the case law drawn from the US, the EU, Italy, and the UK, the chapter conceptualizes judicial review as a fundamental instrument to guarantee a balanced application of a national model of growth.

19.1 Introduction

The issues of climate change and the need to shape a new model of sustainable development strongly rely on the exercise of administrative functions by public authorities.

E. Parisi (✉)
University of Milan, Milan, Italy
e-mail: eduardo.parisi@unimi.it

The mentioned functions cover the whole spectrum of public matters, including food and agriculture, energy, transportation, urban planning, construction, public procurement, landscape and cultural heritage, soil, water, and air pollution. They consist of a rich array of activities, encompassing, for example, regulating, planning, standard setting, monitoring, and sanctioning, concession and authorization issuing, as well as the management of market-based environmental protection mechanisms, such as fiscal policies, cap-and-trade systems, and incentives.

The enforcement of public powers in the described areas of the law shapes the human activities that are capable of affecting the environment, both in everyday lives and in the production and consumption of materials. It is possible to affirm that the exercise of administrative functions implements the relationship designed by the legal framework between human activities and nature.

In one of the most convincing accounts of the juridical literature on this subject, the concept of *environment* itself is intended as a human problem, pertaining to the ethic of the "respect of the other" (Fracchia, 2006; Cafagno et al., 2019). In this conception, environmental law can be explained as a set of duties imposed on humankind, which assumes the responsibility of the care and protection of the surrounding ecosystem in order to survive.

The specification and application of the duties imposed by the law within the framework of a specific relationship between economic growth, society, and the environment is the essence of the modern administrative functions and the key for the realization of sustainable development. In turn, sustainable development becomes a guide to any administrative decision (Valaguzza, 2016).

Undoubtedly, the public–private dynamic that is generated by the execution of the described administrative functions provokes hard conflicts of interests, given that it implies the application of unilateral powers in economically and ethically sensitive matters. This being the case, juridical disputes aimed at assessing the legitimacy of administrative decisions with environmental consequences are an almost unavoidable character of the legal system transitioning toward a greener and healthier dimension (Macroy, 2008).

This is not a pathological but rather a physiological feature of the States that are based on the rule of law, in which direct appeal to courts against the exercise of public power is granted. As will be explained in the following sections, in these contexts, judicial review is a fundamental subjective guarantee against arbitrary and abusive exercise of public power and an instrument capable of contributing to shape the model of development selected at a political level.

It must be underlined that the characteristics of judicial review systems differ from jurisdiction to jurisdiction because they depend both on procedural law and on the shaping of the substantial judicial positions that can be presented in court. Furthermore, the form of judicial review of administrative decisions strongly relies on the constitutional interpretation of the relationship between the judiciary, the legislative, and the executive powers.

However, as it is true that "all systems of administrative law resolve similar issues" (Craig, 2017, 389), all systems of judicial review deal with the same complexities, which are all related to the way in which the expression of public power can be contested in courts of law. In this regard, it is possible to consider issues concerning (i) access to courts, (ii) the presence of a judicial power to rule over the filed case, and (iii) grounds for review.

The analysis of how the mentioned issues are resolved in different legal frameworks can provide a sufficiently defined picture of the main transnational characters of environmental judicial review. In order to embrace the widest possible number of legal scenarios, it is helpful to focus on jurisdictions that present different traditions of administrative law, such as the British and the Italian ones. Moreover, the case law of the Supreme Court of the United States and of the European Court of Justice can be considered for their capacity of setting forth the political and juridical path toward a more sustainable future at a global level. By estimating the ways in which the distinctive issues of environmental judicial review are dealt with in different courts of law, it will be possible to clarify and reconceptualize the role of judicial review in the ambit of transnationally acknowledged trajectories of sustainable development.

19.2 Environmental Judicial Review and the Sustainable Development Principle

A specification is required to correctly set up the analysis: to affirm that administrative decisions should be guided by the principle of sustainable development does not automatically imply the possibility of using the said principle as a parameter of the legitimacy of the administrative actions.

Since its conception, the idea of sustainable development was intended in general and philosophical (more than juridical) terms. The concept combines the values of intragenerational equity, solidarity, and ecology to indicate the need to frame a new balance between the economic, the social, and the environmental dimensions or, more broadly, between men and the "others," meaning other men and nature (Boyle & Freestone, 1999; Schrijver, 2008; Barral, 2012).

The principle implies an assumption of responsibility that infers the performance of a series of duties that all contribute to the framing of a better world, both for present and future generations. The specification of said duties is contained in other norms and principles that constitute its immediate application.

It is then possible to affirm that nowadays the enforcement of the principle of sustainable development passes through the application of specific rules and principles that can be seen as the direct sources of both the duties of care imposed on private citizens and companies and on the public administrations. In this context, administrative discretion operates as an instrument of selection and balancing of conflicting interest in solving specific issues, which should all be considered and reconciled in a new and more virtuous model of growth (Breyer, 2009). In this regard, it is striking to notice that some legislative expressions of the principle of sustainable development (such as the Italian one) contain a reference to the complexity of the relationship and interferences between nature and human activities (see Legislative Decree n. 152/2006, art. 3-*quarter*, par. 3).

The interpretation of the laws and regulations that compose and implement the sustainable development principle by the administrative authorities is a fundamental exercise in the progress of the environmental transition as it contributes to the concrete realization of the model of growth designed at a political level.

To exemplify, the setting of administrative standards (e.g., in relation to water, air, or noise pollution levels), the provision of measures protecting determined species or habitats (e.g., the management of a national reserve or hunting prohibitions), the permitting or licensing of commercial or productive activities (e.g., town planning, the modification of a stream of water), the environmental impact assessment of risky activities (e.g., the realization of a new airport or the adoption of a municipal plan to reduce greenhouse gases' emissions), the issuing of incentives supporting green conducts (e.g., for the installation of renewable power plants or for the refurbishment of existing building stocks), and the subdivision of competences regarding activities of pollution prevention and solution (e.g., waste management and remediation procedures) are all incisive means of application of the principle of sustainable development, as well as the most direct and concrete expression of progress toward a better future. The erroneous application of these norms may imply the violation of the principle as it is currently intended.

Differently, the principle of sustainable development is not still used as a parameter of legitimation of administrative actions. In this regard, it must be underlined that in the democratic systems of law the framing of the correct relationship between the economic, the social, and the environmental factors pertains to the political level. Consequently and, for instance, a judgment imposing a restriction on vehicle circulation in the area of a historic monument in the absence of any norm or regulation because of the reaching of extremely high levels of air pollution would have to be considered as questionable as far as the constitutional balance of power is concerned (Moules, 2011).

Intended in its correct function, the judicial review of administrative activities with environmental consequences has to be conceived as the institutional mean to control the compliance of the exercise of public power with specific normative expressions of the model of sustainable development relating to a specific legal system.

In other words, judicial review should not auto-generate and substitute a model of sustainable development to the one designed by the norms and principles of environmental law but rather contribute to its application, specification, and execution, in the monitoring of the duties of care imposed by the legal system on public and private actors.

19.3 Environmental Claims for Judicial Review

The judicial review of environmental administrative decisions can be thus defined as the revision by a court of acts that have been implemented by administrative authorities in the exercise of a public power pertaining to the set of duties of care imposed by the political framing of a specific sustainable development system under the parameters of legality and the principles that curtain their action.

It is now relevant to understand how the mentioned instrument can be triggered and by whom. This feature is particularly relevant as it conditions both the kind of claim that it is possible to bring in court, the judicial answer to the party's demand, and the consequences of enforcement of the judicial decision.

The selection and definition of the juridical positions entitled to bring legal action against an administrative exercise of power falls within the concept of "standing." At an irreducible minimum, standing requires that the party who invokes the court's authority to restore legality must show that it suffered an actual or threatened injury as a result of the conduct of the defendant which, in the case of judicial review, is the public authority in the exercise of a public power, that can be redressed by a favorable judicial decision (Schwartz et al., 2014, 681).

The mentioned perspective is particularly helpful to organize and analyze the possible claims that can be brought in court against administrative decisions with an environmental character. Elaborating on the categories proposed by the literature (Moules, 2011, 8), it is possible to distinguish among (i) disputes between disappointed applicants and decision makers, (ii) challenges aiming to quash restrictive regulations or administrative decisions, and (iii) challenges by interest groups or individual members of the public.

First of all, harm may come to a party in case of administrative denial. The rejection of a party's request aiming at acquiring a certain improvement in their legal sphere (as it happens for authorization, concessions, permits, and licenses) may provoke the reaction of the disappointed applicant and may cause a dispute regarding the expression of power of the competent administrative authority. In these situations, the environmental concern is generally the ground for the authority's denial of the applied act. The harm is connected to the missed acquisition of the intended value from the competent authority.

To exemplify, an authority may withhold the issuing of a building permit considering the project to be noncompliant with the beauty of the landscape, or it may deny an authorization to open a landfill to favor recycling sites, or else it may issue a negative environmental impact assessment based on the serious impacts of the intended program on the surrounding habitat.

Secondly, the environmental claim may regard a restrictive administrative regulation. Considering that nearly any human activity is capable of affecting the surrounding environment, administrative regulations aiming at protecting the environment almost always limit freedom of action. Restrictions may impose limitations on industrial productions, commercial practices, transportation services, light and gas provision, with related technical difficulties and costs in the transition toward compliance. For this reason, an impressive number of environmental claims is brought by physical or juridical entities affected by administrative regulations.

Given that regulations are acts of general application directed at a multiplicity of indefinite recipients, in these kinds of disputes, access to court is generally granted only in case of ability to demonstrate a direct impact on a personal sphere. On this basis, several legal systems (such as the European one, see art. 263.4 of the Treaty on the Functioning of the European Union) categorize classes of regulations based on the content of their prescriptions. Should the regulation be able to immediately infringe the juridical sphere of the recipients, the regulations may be directly appealed. Differently, should the regulation contain programmatic and abstract norms that are incapable of producing any, it shall only be annulled in relation to an applicative act.

The harmfulness of an administrative action may be also relevant in regard to administrative decisions capable of unilaterally affecting another public or private party's juridical position. Public entities may be granted standing to quash other administrative decisions with environmental consequences that either affect their sphere of competence or are considered to harm the environmental interest falling in their area of domain (Delsignore, 2020). Likewise, private entities may put in question the legitimacy of an administrative exercise of power, regardless from their intimate motive for trial, as exemplified in the following case. In the United Kingdom, in *R (Feakins) v Secretary of State for the Environment, Food and Rural Affairs* [2003] EWCA Civ 1546, it was significantly held that standing may be questioned in those cases in which the claimant is only moved by a selfish desire to seek compensation from the issuing administrative authority, rather than an environmentally related concern. The case regarded a judicial review of a decision of the environmental department to dispose of farm animals diagnosed with foot-and-mouth disease, slaughtering and incineration of the carcasses. The claimant held that the ashes should have been disposed of in a landfill but the defendant held that the real motive for the claim was the objective to seek for monetary compensation. The court ruled in favor of the claimant, affirming that the role of the court in the evaluation of standing should merely be the one to verify the absence of an ill-motive for trial (in the words of *R v Somerset County Council and ARC Southern Ltd, ex p Dixon* [1997] Env LR 111, the claimant should not be a "busybody" or a "trouble-maker").

The mentioned judgment highlights that the intimate reasons for judicial review should not be relevant in a trial as long as the procedural norms provide an instrument for self-defense toward a decision that is generally affecting one's own juridical positions. By granting standing to the recipient of an administrative decision with environmental consequences, the legal system allows the verification of the correctness of the administrative evaluation on a specific duty of care toward the environment and of the correct interpretation of the sustainable development model in a specific context. Therefore, the value of the controversy transcends the personal reasons on which the claim is based, in light of the function of judicial review to redress a harm caused by the exercise of a public power and to clarify the correct set of public and private duties that characterize a model of sustainable development.

The described function of judicial review in environmental matters also provides the conceptual basis for the recognition of standing to legal or physical persons that are not the recipient of the administrative decision. This may be the case of a person holding the illegality of a building permit issued to a neighbor on the grounds of the impossibility to construct in the area because of the presence of a course of water whose habitat needs to be preserved or of a company challenging an incentive granted to a competitor. Here, the matter of standing is more complex as the applicant is not a direct recipient of the administrative action. To use Justice Scalias' words, "When the suit is one challenging the legality of government action or inaction, the nature and extent of facts that must be averred (...) or proved (...) in order to establish standing depends considerably upon whether the plaintiff is himself an object of the action (or foregone action) at issue. If he is, there is ordinarily little question that the action or inaction has caused him injury, and that a judgment preventing or requiring the action will redress it. When, however, (...) a plaintiff's asserted injury arises from the government's allegedly unlawful regulation (or lack of regulation) of someone else, much more is needed." See *Lujan v Defenders of the Wildlife* (504 U.S. 555 (1992)).

Referring to the conceptual framework that we selected in this reconstruction, it could be affirmed that in these circumstances, to obtain standing the claimant should demonstrate that the same set of duties of the recipient of the administrative decision also applies to its part. In these terms, it is possible to explain both the cases of the claim toward a license granted to a competitor—see, *inter alia*, FCC v Sanders Brothers Radio Station, 309 U.S. 470 (1940) in the United States system, *R (on the Application of Rockware Glass Ltd) v Chester City Council and Quinn Glass Ltd* [2005] EWHC 2250) in the United Kingdom and Cons. St., sec. V, January 3rd, 2002, n. 11 for the Italian one)—and the claim brought by a neighbor in relation to a construction that is potentially harmful to the environment.

In such cases, the equal burden imposed by the legal system on both the claimant and the defendant toward the environment enables them to judicially react to the unfair advantage that the other operators would gain in case of breach of the legal standards. This expedient allows occasional environmental judicial protections by actors who are motivated by a personal interest toward the claim.

Lastly, environmental claims may be presented by claimants who are motivated by purely ideological concerns.

The characterization of the environment as a juridical good (Fisher et al., 2013, 5) and as a widespread interest belonging to the public at large and not to single persons provokes the need for legal systems to elaborate mechanisms to challenge in court the nonalignment or the nonsufficient alignment of administrative actions with the environmental protection value (Hilson and Cram, 1996).

Probably the most relevant norm in this sense is art. 9, par. 2 of the Aarhus Convention of 1998, which grants automatic standing to nongovernmental organizations promoting environmental protection. The Convention thus embraces the logic of the necessity to detect organizations that are considered by the legal system as responsible for specific areas of the environment, thus enabling them to promote legal action in court for any suffered or threatened harm.

The said limitations may be connected to the actual capability of the association to represent the specific environmental interest at stake (Goisis, 2012; Eliantonio, 2012). In this regard, the judicial verification usually falls on the *vicinitas* of the organization to the interest that they represent in court, which is assumed to be harmed by an administrative action. Consistently with this idea, the clauses of the organization's by-laws, the previous actions carried out to defend that same interest in the past and the eventual participation to the proceedings, and the consistency in the protection of the environmental interests have been considered by the case law as indexes to proof standing. On these grounds, standing has been generally denied to committees of citizens occasionally getting together to appeal single administrative measures (see TAR Lombardia, Brescia, sez. I, December 22nd, 2017, n. 1478 and, for the British system, *R v Darlington BC and Darlington Transport Company, ex p the Association of Darlington Taxi Owners* [1994] COD 424).

Contrarily, other features that do not relate to the position of the organization toward the environmental ambit at stake, such as the number of inscribed associates or the geographic location of the headquarters of the associations, have been generally discharged (see ECJ, judgment *Bund für Umwelt und Naturschutz Deutschland, Landesverband Nordrhein-Westfalen*, in C-115/09 and ECJ, sec. II, October 15th, 2009, in C-263/08, *Djurgården-Lilla Värtans Miljöskyddsförening* and ECJ, *Greenpeace and Others v Commission*, C-321/95 P).

This distinction can be considered as an expression of the need of the legal system to detect a group of individuals that are entrusted with similar duties toward the same portion of the environment. Such entities are the only ones allowed to revise the administrative actions that specify and enforce the mentioned legal duties.

19.4 Transversal Characters of Environmental Judicial Review

It is now possible to advance the reasoning on how administrative law claims pertaining to environmental matters are treated in courts of law.

If read with a high-level view, the description of the possible ways in which environmental claims may be presented in court brings to affirm that judicial review in this specific ambit

1. Is triggered by a reaction to a violation caused by the exercise of a public power related to the application of a duty toward the environment
2. Implies the interpretation of the set of norms that describe the system of sustainable development
3. Expresses a secondary judgment on the application of the sustainable development model, which revises the assessment operated by public authorities on specific matters

The described characteristics are common to any analyzed system despite their degree of openness toward environmental claims and deserve further specification as they define the way in which judicial review contributes to the shaping of a model of sustainable development.

19.4.1 Judicial Review on Environmental Matters as a Reaction to a Harm

The characterization of the claim as a reaction to a harm allegedly provoked by a public authority in the shaping of a duty toward the environment brings to underline the necessity to find boundaries between the juridical and political actions in the definition of the sustainable development system (Appel, 2011).

The questions of the legitimacy of administrative decisions like the granting of incentives for extensive renewable energy parks or of the setting of fabrication standards for motor vehicles pose a common issue related to the political value of the judicial decision toward the designed asset of sustainable development. In this regard, environmental judicial review is mightily shaped by the constitutional design of the relationship between the legislative, administrative, and judicial powers.

Firstly, the separation of power principle, "namely the view that the judiciary espouses that the legislative and executive branches should make policy decision" (Appel, 2011, 279), forbids to promote the judicial review of powers that have not been exercised yet. The rule is clearly intended to prevent the judges from becoming administrators by interfering with the exercise of powers that constitutionally are entrusted with the executive branch (Mazzamuto, 2018, 82).

According to the mentioned rule, a court of law may not—for instance—evaluate the possible self-annulment of an act stopping the exercise of an economic activity for environmental reasons as this power would not have been exercised yet. In the mentioned hypothesis, judicial power finds a limit in the need to review a concrete expression of power. Any indication regarding a possible future administrative action, environmentally driven as it may be, is prohibited since the task of the judiciary in this regard is the restoration of the legality whenever an administrative action departed from it.

Secondly, the so-called *trias politica* determines how rigorously the administrative judge can examine exercised administrative powers. Indeed, the separation of power principle justifies the deference that the administrative courts should have toward the discretionary decisions of administrative agencies.

The principle has wide transnational relevance: both the deference doctrine applied in the United States since *Chevron U.S.A., Inc. v Natural Resources Defence Concil, Inc.* (467 U.S. 837 (1984)) and the discretionary margin left by the systems of *droit administratif* to the public agencies are based on the acknowledgment of an unavoidable margin of statutory interpretation left to administrative authorities in the execution of their public powers. Administrative courts must act within the boundaries designed by the law and by the legal principles that curb their actions. If the administration acted *ultra vires* (namely, they breached the boundaries of their capacity of action), courts may intervene in re-establishing legality; however, should this not be the case, the judges should refrain from reviewing their action.

Consequently, judicial review is a matter of legality and not of opportunity of the administrative action. Whether or not administrative authorities had a power to act and if that power was exercised in conformity to the administrative law and to principles are the main concerns of the courts of law in a judicial review (Moules, 2011, 169).

This subjective characterization makes the judicial review system tendencially protective and not proactive as it is intended as a warranty of the application of the correct balance between the social, economic, and environmental spheres of sustainable development. This further implies that judicial review should (i) be limited by the party's claim and (ii) aim toward the correct implementation of the set of duties normatively imposed for the protection of the environment.

19.4.2 Judicial Review on Environmental Matters as an Interpretation of the Law

The exercise of the mentioned role by the courts of law is not always an easy task because especially in the environmental matters their judgment needs to be made in a highly technical and scientifically complex ambit of the law.

More specifically, environmental law judgments are characterized by the presence of a highly detailed normative that often lingers behind the technical advancements of science.

This factor is intrinsically related to the high technical level that characterizes environmental regulation, which needs to be constantly adjourned on the basis of the ever-evolving parameters of science.

The level of air pollution in a city, the chemicals to be classified as pollutants in the waters, the techniques of oil extraction, and the instruments of soils remediation are mutating factors that pose serious interpreting issues whenever the norm that functions as a parameter for administrative action does not explicitly consider a method of its exercise.

In this context, in which the legal framework appears to be fragmented and not adjourned to the technical state of the art, a flexible interpretation is pivotal to ensure the implementation and the advancement of the law.

Flexibility is here ensured by principles such as the ones of precaution, prevention, integration, polluter pays, deriving from the international declarations that helped to shape, per objectives, the positions of national governments, and public administration toward the environment.

Looking at case law, it is possible to notice that an astonishing number of environmental law controversies arise from the silence of the norms on specific matters and are resolved on the basis of an open interpretation of the statutory norms based on the flexible principles of environmental law.

Massachusetts v. Environmental Protection Agency, 549 U.S. 497 (2007) itself—one of the most crucial cases in the US Supreme Court history—was generated by the absence in the US Environmental Protection Act of any reference to carbon dioxide and other greenhouse gases as air pollutants. This lack was not consistent with the progress of the scientific evidence on the capability of those substances to harm the environment by causing global warming.

The same issue can be found in the far more recent case *County of Maui, Hawaii v. Hawaii Wildlife Fund et al.*, No. 18.260, 590 U.S., which regarded the possibility to consider an effluent from a sewage treatment plant that is injected into groundwater wells but ultimately migrates into navigable waters as a discharge of pollutants requiring a permit from federal or state authorities, pursuant to the US Clean Water Act. In this case, the US Supreme Court recognized that the silence of the federal norm was intended to leave space to national legislatures to rule; however, the judges ruled that the intention of the legislator was not to leave a "large and obvious loophole" by allowing polluters to escape the permitting system by discharging in groundwater that could be only a short distance from oceans and rivers. Therefore, the court held that the permit is required not only when there is a direct discharge into navigable waters but also in cases of "functional equivalent of a direct discharge," to be detected from indexes such as the transit time, the traveled distance, and the chemical composition of the material.

Similarly, in the European context, the issue of detailed but incomplete normative to be filled up with principles of environmental law has been evident in the last years with regard to the interpretation of the concept of "waste," defined by Directive 2006/12/EC as "any substance or object in the categories set out in Annex I which the holder discards or intends or is required to discard." The unavoidable incompleteness of the annex that is supposed to contain all possible materials that can be considered as waste provoked numerous controversies in relation to the juridical extent of the term.

In all the mentioned cases, the silence of the norms was filled up by the courts of law on the basis of general principles of environmental law. It is, therefore, possible to underline a second distinctive function of judicial review in the environmental law ambit: the specification of the duties of care normatively imposed by the legal framework on both public and private actors.

The technical character of the environmental regulation imposes the need to continuously adjust the framework of duties to the varying relationship between the spheres of sustainable development. In this regard, the courts of law play the crucial role of filters between the decision makers and the holders of duties of care positions toward the environment, in the specification of the rules descending from a normatively fragmented design of the sustainable development model.

19.4.3 Judicial Review on Environmental Matters as a Secondary Judgment

The mentioned flexibility in the interpretation of the applicable legal framework needs to be analyzed in regard to the characterization of environmental judicial review as a secondary judgment on the one carried out by the public authority on a specific matter.

Because of the mentioned high technical level of the environmental normative, in this ambit judicial review often implies the evaluation by the courts of specific analysis conducted by the administration. Hence, science can be used to challenge administrative regulation (as it is the case of emission tests to challenge a regulation certain construction trademarks for vehicles) or to highlight the causal relationship between a conduct and its outcome on the environment (as it happens for the remedial procedure for soil pollution or in the ban on chemical products for agriculture).

In both cases, the scientific outline that supported the administrative decision is questioned. The discrepancy between the administrative and the party's reconstruction may regard both the treatment of the evidence brought during the procedure or a different perception of the same risk (Beck, 1992; Sunstein, 2002).

The problem is that no scientific evidence is objective. Indeed, the scientific evolution process itself is philosophically founded on questioning the pieces of knowledge that function as a premise of the reasoning (Wynne, 1992), so that scientists themselves differ on the content of technical concepts. This further complicates the matter of adjudication of scientifically intensive legal disputes.

In any case, a more precise trajectory can be fixed by considering that in the case of judicial review of administrative decisions (i) the scientific knowledge enters into a trial only as established by the law (Sulyok, 2020) and (ii) the scientific knowledge is filtered by an administrative evaluation.

This consideration helps to understand that environmental judicial review plays a role in selecting valid data and sound reasoning as declared admissible by the legal framework. Given that the design of a sustainable development model relies on the assumption of scientific data, it is particularly relevant to assess whether that knowledge was assumed and applied correctly by the administration in specific cases.

The parameters of this secondary judgment rely on the perception of risks deriving from the legal framework and on the administrative normative imposing prescriptions on the duties of the public authorities to recollect and evaluate facts.

The thoroughness of the evaluation depends on the level of scientific certainty that is considered sufficient by a legal framework to allow an administrative decision with environmental consequences. Consistently, the courts of law should show deference toward precautionary decisions of the public administration in the context of legal frameworks that have a high perception of the possible risks toward the environment and human health of dangerous conducts and products. Conversely, judicial review should require higher standards in the most regulated sectors that are characterized by abundance of scientific data and preciseness in the tests to be conducted to support administrative decisions.

The European case law provides a rich set of examples of the mentioned statement. In the landmark case T-13/99 *Pfizer Animal Health SA v Council* [2002] ECR II-3305, the applicants sought annulment of a regulation banning the use of certain antibiotics and additives in feeding stuffs. It was held that in a situation in which the precautionary principle is applied, "A risk assessment cannot be required to provide the Community institutions with conclusive scientific evidence on the reality of the risk and the seriousness of the potential adverse effects were that risk to become reality. (…) Rather, (…) a preventive measure may be taken only if the risk, although the reality and extent thereof have not been fully demonstrated by conclusive scientific evidence, appears nevertheless to be adequately backed up by the scientific data available at the time when the measure was taken" (142, 144).

This is coherent with the acknowledgment of the political value of the precautionary decision assumed by the administration for the protection of the environment and of human health. As it was affirmed in the quoted *Pfizer* case, "the contested regulation is founded on a political choice" (468).

A different attitude should characterize those scenarios in which the administrative decisions are carried out in conditions of scientific certainty. The duty of courts of law to intensify their scrutiny when reviewing administrative decisions involving scientifically complex assessments was recently affirmed by the First Section of the European Court of Justice in the judgment June 26th, 2019, C-723/17, *Craeynest et al*. The dispute was brought by a few citizens and a nongovernmental organization who challenged the air quality plan of the Brussels Capital Region in regard to the position of the sampling points for measuring the concentration of relevant air pollutants. The court acknowledged that the selection of the site for the measurement of air pollution is a matter that falls within the administrative discretion as it is "the responsibility of the competent national authorities to choose, within the limits of their discretionary powers, the actual location of the sampling points" (para 44). However, it was affirmed that "the existence of such discretionary powers does not in any way mean that the decisions taken by those authorities in that connection are exempt from judicial review, in particular in order to verify whether they have exceeded the limits set for the exercise of those powers" (para 45 and, to that effect, judgments of October 24, 1996, *Kraaijeveld and Others*, C-72/95, EU:C:1996:404, paragraph 59, and of July 25, 2008, *Janecek*, C-237/07, EU:C:2008:447, paragraph 46). Therefore, it was stated that the discretion of public authorities in carrying out complex assessments based on specific technical considerations may be challenged with scientific evidence. In such a dynamic, the courts of law should verify whether or not the technical analysis was carried out in a way to guarantee the satisfaction of the objective of the norm. In the referred case, the legislative objective consisted in the sound and effective monitoring of air pollution in the urban area, so that the measuring devices should have been placed in highly trafficked zones.

In such instances, the reasons of the administrative decisions lose their strong connection with the standard of risk aversion imposed by the legal framework, thus opening the way to technical considerations, which may be contested in court, at least on the ground of the scientific correctness of the applied method and of the retrieved results of the analysis.

19.5 Environmental Judicial Review as an Instrument of Sustainable Development

Some of the most distinctive characters of judicial review in environmental matters have been conceptualized in the previous pages, thanks to the identification of transversal lines of case law.

In particular, the reasoning carried out on how administrative law claims pertaining to environmental matters are brought into courts has led to underline the imporance of the connection between the normative duties of environmental protection and the sphere of actions of the recipients of said normative burdens.

This has caused to consider judicial review as an instrument of reaction to a harm suffered by the recipients of specific normative duties imposed by the model of sustainable development designed by the legal framework.

In the performance of this activity, courts conform the relationships of all the relevant public and private actors in a specific ambit of environmental protection, indicating the correct weight and width of the norms that are a direct application of the sustainable development principle.

In this regard, judicial review also operates a definition of the normative parameters of sustainable development, in a deeply fragmented and highly technical normative context. It specifies the regulatory parameters of the human behaviors toward the environment, following the evolution of scientific progress and making up for the delays of the legislators in keeping track of the advancements in the human–ecosystem relationships.

Lastly, it selects and promotes sound reasoning in administrative decisions by relying on a normatively predetermined disposition toward risk and setting forward basis and boundaries of the human dangerous activities, thus playing a significant role in the conformation of the market relationships toward sustainable production and consumption.

The effects of this judicial activity are particularly relevant (Shapin, 2000). To exemplify, the annulment of a denial of incentives issued by a public agency on the basis of a restrictive interpretation of regulation may reshape the allocation of public funds for green projects at a national level. The appeal of a building permit on the basis of the impossibility to realize real estate intervention in an area considered to be a protected habitat may induce the municipal authority to redesign its town planning. The rejection of an appeal by a private operator of an administrative decision not to open a landfill based on the argument that waste discharge should be considered as the last possible resource may have an impact also on administrative decisions taken by other equivalent administrative authorities in the national territory.

As the mentioned examples demonstrate, the possible consequences of judicial review in the shaping of the world of tomorrow are wider than other environmental protection tools such as private law and criminal law litigations, which can only resort in the assessment of personal liabilities in relation to harmful behaviors (Wald, 1992; Kotzé & Paterson, 2009; Voigt & Makuch, 2018). Firstly, an administrative judgment may produce the annulment of the administrative act and command to a public authority to shape the future exercise of that same power. Secondly, administrative actions are often composed of a series of acts that are strictly intertwined, so that the revision of a single element may have an effect on the connected components of the chain. Thirdly, administrative case law may constitute the basis for aligning the interpretation of norms by other public offices with similar competences.

It is so explained the significance of judicial review in the environmental sector. The application of a sustainable development model implies the enforcement of administrative powers in almost any sector of human activity. A consequence of this dynamic is the necessity to verify that the detection and application of the normative duties of the members of a community living in a specific ambit by the authorities that are vested with the exercise of public powers are carried out in a correct form. While reacting to the diversions from the path set forth by the legislator, judicial review disentangles the normative loopholes and provides significant indications of conduct to both public and private actors, thus becoming a primary factor in the achievement of a better tomorrow.

References

Appel, P. (2011). Wilderness, the courts, and the effect of politics on judicial decisionmaking. *Harv. Envtl. L. Rev., 32*, 275–312.

Barral, V. (2012). Sustainable development in international law: Nature and operation of an evolutive legal norm. *The European Journal of International law, 23*(2), 377–400.

Beck, U. (1992). *Risk society: Towards a new modernity*. Sage.

Boyle, A., & Freestone, D. (1999). *International law and sustainable development: Past achievements and future challenges*. Oxford University Press.

Breyer, S. (2009). Economic reasoning and judicial review. *The Economic Journal, 119*(535), F123–F135.

Cafagno, M., et al. (2019). The legal concept of the environment and systemic vision. In M. Cafagno et al. (Eds.), *The legal concept of the environment and systemic vision: A rock in the pond* (p. 141). Springer.

Craig, P. (2017). Judicial review of questions: a comparative perspective. In S. Rose-Ackerman et al. (Eds.), *Comparative administrative law* (2nd ed., p. 389). Edward Elgar Publishing.

Delsignore, M. (2020). *L'amministrazione ricorrente. considerazioni in tema di legittimazione nel giudizio amministrativo*. Giapplichelli.

Eliantonio, M. (2012). Towards an ever dirtier Europe? The restrictive standing of environmental NGOs before the European Courts and the Aarhus convention. *CYELP, 7*, 69–85.

Fisher, E., et al. (2013). *Environmental law. Text, cases and materials*. Oxford University Press.

Fracchia, F. (2006). The legal definition of environment: from right to duty. *IJEL*, 17–36.

Goisis, F. (2012). Legittimazione al ricorso delle associazioni ambientali ed obblighi discendenti dalla Convenzione di Aarhus e dall'ordinamento dell'Unione Europea. Note to ECJ, 12/05/2011 n. 115, sez. IV and ECJ, 08/03/2011 n. 240, Grand Chamber. *DPA, 1*, 91–101.

Hilson, C., & Cram, I. (1996). Judicial review and environmental law. Is there a coherent view of standing. *Legal Studies, 16*(1), 1–26.

Kotzé, L. J., & Paterson, A. R. (Eds.). (2009). *The role of the judiciary in environmental governance. Comparative perspectives*. Wolters Kluwer.

Macroy, R. (2008). Environmental public law and judicial review. *Judicial Review, 13*, 115–125.

Mazzamuto. (2018). Il principio del divieto di pronuncia con riferimento a poteri amministrativi non ancora esercitati. *DPA, 1*, 82.

Moules, R. (2011). *Environmental judicial review*. Oxford University Press.

Schrijver, N. (2008). *The evolution of sustainable development in international law: inception, meaning and status*. Martinus Nijhoff Publishers.

Schwartz, B., et al. (2014). *Administrative law: A casebook*. Wolters Kluwer.

Shapin, C. R. (2000). The legislative design of judicial review. *Journal of Theoretical Politics, 12*, 269–304.

Sulyok, K. (2020). *Science and judicial reasoning*. Cambridge University Press.

Sunstein, C. R. (2002). *Risk and reason: Safety, law and the environment*. Cambridge University Press.

Valaguzza, S. (2016). *Sustainable development in public contracts. An example of strategic regulation*. Editoriale Scientifica.

Voigt, C., & Makuch, Z. (Eds.). (2018). *Courts and the environment*. Edward Elgar Publishing.

Wald, P. M. (1992). The role of the judiciary in environmental protection. *Environmental Affairs, 19*, 519–546.

Wynne, B. (1992). Uncertainty and environmental learning: Reconceiving science and policy in the preventive paradigm. *Global Environmental Challenge, 2*(2), 111–127.

Eduardo Parisi is a researcher in administrative law at the University of Milan. Qualified to carry out the functions of associate professor, he teaches administrative and sustainable development law and environmental law in the same Athenaeum. His research interests involve judicial review, environmental law, public contracts, and public private partnership. He holds a Ph.D. in administrative law (University of Milan, 2017) and an LLM in Legal Theory (NYU School of Law, 2014). As a Responsible for the International Relationships of the European Association of Public Private Partnership and a founding member of the Transnational Alliancing Group, he cooperates with several academics and experts in a global perspective to promote environmental and social values in public contracts.

Chapter 20
Mediation in Environmental Disputes

Roberta Regazzoni

Abstract This chapter focuses on the importance of an integrated approach of the various dispute avoidance and/or resolution tools for environmental issues. Environmental conflicts are, in fact, complex and multifaceted; therefore, the synergy for a long-standing and efficient coexistence of all stakeholders has to be ensured also by the adoption of dialogue-based tools and solutions. Specifically, the author reports the experience of the environmental mediation project carried out by the Mediation Service by the Milan Chamber of Arbitration and some public and private stakeholders.

20.1 Introduction

The intrinsic traits of environmental disputes, which are specified in the following paragraphs, in many cases seem to call for a consensual, dialogical approach, represented by mediation and/or facilitation.

Litigation and mediation are both dispute resolution systems. Approaches are different though: the case brought to court "ends" with a decision taken by a third person (the judge), whereas in mediations a third neutral person promotes the dialogue between the parties.

This is based on the idea that only the parties that are directly affected by a dispute can find the best suitable solution. The immediate inference of this reasoning is that a decision imposed is rarely final, while a consensual settlement is long-lasting and fits better the circumstances. Moreover, mediation is a dispute prevention and a damage mitigation tool.

R. Regazzoni (✉)
Milan Chamber of Arbitration, Milan, Italy
e-mail: roberta.regazzoni@mi.camcom.it

© The Author(s), under exclusive license to Springer Nature Switzerland AG 2022
S. Valagussa, M. A. Hughes (eds.), *Interdisciplinary Approaches to Climate Change for Sustainable Growth*, Natural Resource Management and Policy 47,
https://doi.org/10.1007/978-3-030-87564-0_20

20.2 Environmental Conflicts' Features

The definition of environmental conflict proposed by the United Nations Environment Programme (UNEP) is straightforward and comprehensive: conflict is a dispute or incompatibility caused by the actual or perceived opposition of needs, values, and interests (UNEP, 2009).

However, the definition of environmental conflicts takes different nuances depending on the context. Nowadays disputes are intrinsically hybrids: environment blur into social, economic, and cultural.

This feature enhances their complexity: stakeholders are generally numerous as they include public administration (for actions or omissions affecting the environment), companies (public or private), local communities and private citizens, committees, environment protection movements and associations (i.e., the civil society), shareholders, and insurances.

This multitude of potentially interested actors play different roles, have different powers, responsibilities, and use different "languages" as well. To this end, it is worth pointing out how all the experiences that have been carried out in the field presented a common issue regarding the inadequate involvement of the general public in the process, a lack of transparency and the complexity of the information shared with the public about projects and its actual or perceived impact on people's lives. In fact, almost all environmental conflicts are (also) a consequence of gaps in communication and information. The language of the public administration, as well as the technical reports, seems not to be successful in reaching the general public.

The following list, which is nonexhaustive, can give an idea of the complex dynamics originating from environmental conflicts:

- They impact on fundamental interests (perceived as opposite) such as health, safe environment, economy, and social improvement
- They are likely to involve a multitude of stakeholders with a considerable unbalance of power
- They are highly localized (management of the "Not In My Back Yard"—NIMBY phenomenon)
- They focus on complex matters and about technical aspects
- They are often cross-border and even cross-cultural
- They revolve around damages that are difficult to assess
- They reveal transparency issues and information asymmetries
- They need to be solved promptly in order to avoid or contain possible threats to life and the ecosystem

Several scholars argue that the environmental conflict represents a demand for active participation in building/managing a territory in a consultative way (Castiglioni and De Marchi 2009). In another perspective, the environmental conflict is a claim for a lack of involvement of citizens in the decision-making processes affecting the environment (and their own lives).

20.3 Mediation: An Overview

As mentioned in the opening, mediation is a problem-solving tool based on dialogue, which is facilitated by a third neutral party, namely the mediator. Participation is voluntary and consensual in principle, meaning that the actors (the parties to a dispute) should agree to participate in the process and can leave the discussion table at any time. There is no obligation to mediate and no obligation to settle. Parties in the process are empowered and assisted to design their own path toward sustainable solutions and preserve (or improve) their relationship, if they want or need to.

Mediation is a multifaceted tool. There are as many mediation approaches as mediators. Every mediator has her/his own style, which can be more purely facilitative to more directive and often needs to be further adapted to the case depending on the disputed matter, the personality of the disputants, and the context. Simplifying, we can say that mediation is a structured process (but still) flexible and informal, where a third subject does not make a decision on the matter but assists the parties in designing their own solution.

Confidentiality covers the whole proceeding, including all the information shared and the statements made, unless the parties expressly renounce to it in full or partially.

The most important feature of mediation, which makes it different from a common negotiation, is the presence of the third neutral facilitator of the dialogue. One (or more) third neutral person governs the process and assists the parties to negotiate a fully satisfying solution to a problem or the overcoming of an impasse in their relationship. This latter is an important aspect to stress about mediation, given that it works well also as a conflict prevention mechanism.

The abovementioned keystone of mediation is often perceived as a "lack" of power of the third neutral, a flaw of this dispute resolution mechanism. On the contrary, it is crucial when thinking about the huge potential of listening to all the parties involved in a conflict, exploring in depth their real needs and interests so as to detect multiple common grounds upon which to build trust, rapport, and consent. These goals can be achieved only if the participants in the process feel themselves free to speak openly, with no fear that somebody will make judgments on them, on how they behaved, on their responsibilities.

Mediation is much about the relationship between the parties and less about the merit of the dispute; therefore, the primary role of the mediator is to activate (or reactivate) a dialogue that went missing for some reason using her/his skills and mediation techniques.

The process unfolds along the following phases:

- Opening: The mediator informs the parties about the feature of the proceeding and states basic rules (voluntariness, confidentiality, the possibility to talk to each of them also in private sessions if needed, etc.). The parties present their respective points of view on the situation
- Exploration: The parties work with the mediator to identify and generate multiple settlement options and to narrow the issues

- Negotiation: This is the space for the evaluation of options generated in the previous step—parties select the viable one that best meets their needs
- Conclusion/implementation: Drafting of the settlement agreement or implementation program (not always necessary)

The activities of each singular phase can be carried out with all parties in plenary session or in private session (just the party and the mediator) depending on the pathway agreed at the beginning of the process.

Settlements resulting from mediation are likely to meet the parties' needs much more than a court decision for the simple reason they have built them. For the same reason, the settlement is likely to be long lasting. If the parties deem it necessary, the implementation can be monitored, rediscussed, and readjusted over time.

Lastly, it is important to highlight that, despite being a very flexible tool that can be adapted to almost any kind of conflict, the choice on whether or not to use mediation must be carried out on the basis of an accurate strategic assessment of the situation. Indeed, especially in the environmental context, it should be evaluated if the intention of the parties is to solve the dispute or to affirm a more general juridical principle. In the latter cases, traditional litigation (in the form of strategic litigation) may be more effective.

20.4 Collaborative Approaches to Environmental Conflict: The "Decide, Announce, and Defend" (DAD) Approach

In situations of conflict, people interact through perceptions and cognitions and are influenced by their expectations. In any field, and at any level, conflict has much to do with the relationship between the disputing parties, their history, and the context.

As mentioned, many models of mediation are available, and a flexible approach is recommended when considering the concrete situation. The flexibility of mediation consists in the actual possibility to draw a dispute resolution process that perfectly adapts to the context and the specific features of the case but also, and more efficiently, in preventing a dispute.

When a conflict occurs, an assessment should be made on its causes and dynamics. The said analysis should identify all the stakeholders who should be involved in the proceedings and possible independent technical advisors.

Some situations may require the statement of a judge in order to initiate a process of legislative reformation; many others may benefit from a dialogue of all the stakeholders to find a (final or transitory) solution to the issues. Also, most of the situations may be dealt with through a blended approach.

Judicial proceeding is fundamental but limited: It allows a limited reading of a fact, through the lenses of the law, which is taken into consideration under the aspect of liability. The judicial decision investigates the past and states who is responsible for what but it rarely offers a concrete solution of the problem.

Mediation, on the contrary, projects toward the future and to what can be done to prevent or mitigate the damage or to avoid the realization of a risk—more broadly, to resolve the problem now and for the future.

In very broad terms, we can therefore say that when a conflict involves different stakeholders, each with different roles, powers, and competences, in a context where all must somehow interconnect, mediation is appropriate. It is more useful than a judgment as it affects all the relevant aspects (legal and nonlegal) of a problem. Many consider a judicial decision the final word on a dispute. Unfortunately, this is not always the case, especially when we talk about environment, natural resources protection, and economic growth.

In our world and daily experience, we can identify environmental conflicts in contexts such as infrastructure construction projects, waste management and disposal, remediation of polluted areas, energy production, plants producing chemicals or other dangerous material, management of conservation areas, and urban development projects. In most of these situations, a top-down approach by the proponents (public administrations, private operators) is especially useful (according to the "decide, announce, and defend"—DAD approach).

This kind of method is believed to be a quick way to get things done quickly; however, it implies several drawbacks. Not involving citizens into decisions and activities carried out in their (supposed) favor, it implies the risk of protracted and image-damaging legal actions and a disruptive loss of trust and relationship among administrators and citizens.

If local actors were legitimized as interlocutors in the various stages of the decisional process, proponents could improve their planning ability and avoid or reduce conflicts.

20.5 Supporting Dialogue and Mediation in Environmental Matters

Supranational entities paved the way by embedding collaborative dispute resolution mechanisms in their policies, recommendations, and treaties (OECD, 2011; CEPEJ, 2019; UN, 2019). The United Nations Environmental Programme has developed an impressive work about conflict management through mediation and participative tools. In particular, it is worth mentioning here the Guide for Mediation Practitioners (UNDPA & UNEP, 2015) the "toolkit and guidance for preventing and managing land and natural resources conflict" (UN, 2012) and "From conflict to peacebuilding. The role of natural resources and the environment" (UN, 2009).

Moreover, in the 1990s, the International financial institutions created IAMnet (Independent Accountability Mechanisms), which is a virtual network of practitioners who contribute to the exchange of ideas and assist with institutional capacity building in citizen-led accountability and compliance. Many multilateral or bilateral development financial institutions have been joining the network so far, such as the African, Asian, Inter-American and Caribbean Development Banks, the European Bank for Reconstruction and Development, United Nation Development Programme, and the World Bank, insofar introducing dialogue mechanisms within projects they fund, including mediation.

Subsequently, in 1998, the United Nations Economic Commission for Europe (UNECE) launched the Convention on Access to Information, Public Participation in Decision-Making and Access to Justice in Environmental Matters.

As for the European Union, the Regulation (EC) n. 1367/2006 gave life to the principle stated in the Aarhus convention, but even before two important directives (n. 35/2003 and n. 35/2004) tackled the important issue of giving access to environmental decision-making processes to citizens. Particularly meaningful to our purposes is Decision No 1386/2013/EU of the European Parliament and of the Council of November 20, 2013, on a General Union Environment Action Programme to 2020 "Living well, within the limits of our planet." The Decision mentions, both in "whereas," Art. 3 n. 2, the need for connection with citizens ("2. Public authorities at all levels shall work with businesses and social partners, civil society and individual citizens in implementing the 7th EAP" and point n. 15 of the Annex (a program for action to 2020), where it is affirmed that "the public should also play an active role and should be properly informed about environment policy. Since environment policy is a sphere of shared competence in the Union, one of the purposes of the 7th EAP is to create common ownership of shared goals and objectives and ensure a level playing field for businesses and public authorities. Clear goals and objectives also provide policy makers and other stakeholders, including regions and cities, businesses and social partners, and individual citizens, with a sense of direction and a predictable framework for action."

Today, national laws contain provisions for citizens' participation in environmental decision-making processes (see, for the Italian context, Legislative Decree no. 152/2006, especially in relation to the environmental impact assessment).

Despite the huge potential recognized at any level of governance, the use of mediation and participatory processes to address environmental issues is still limited and scattered.

Compensation is not the solution per se, but it has to be part of a strategic and shared framework to avoid increasing the cost of work and leave environmental issues unresolved.

Trust is a critical issue in most environmental conflicts. The relationship between citizens, public administrations, and companies is more and more at risk as citizens feel their interests are not represented. On top of that, administrative proceedings are complex and lack transparency. Offering independent, accessible, and understandable information, as well as a shared evaluation and monitoring of the projects, is crucial in avoiding and mitigating conflicts.

20.6 Mediation and Facilitation in Environmental Matters in Italy: The Milan Chamber of Arbitration

Since 1986, the Milan Chamber of Arbitration (CAM) has been providing the business community with an array of services and tools in the ambit of alternative dispute resolution, which allow for a time-effective resolution of commercial disputes through extrajudicial methods such as arbitration and mediation.

In 2015, CAM seized the opportunity offered by an innovative (for the time) mediation legislation resulting from the implementation at a national level of Directive n. 52/2008 and took it as a starting point for its pilot project on environmental mediation.

Together with Fondazione Cariplo and the Milan Chamber of Commerce, Regione Lombardia and Comune di Milano, CAM coordinated and led a project aiming at testing in vivo the efficiency of mediation for the management of environmental conflicts.

The organizing committee put together experts with different backgrounds spanning from mediation and facilitation, law, environmental matters, to entities directly involved in environmental conflicts such as public administration bodies, companies, and environmental associations for an in-depth interdisciplinary discussion aiming at building a conflict management model.

The working group made an assessment on the legislative and regulatory "state of the art" at a national, European, and international level.

The pilot lasted 1 year and 3 months, after which CAM issued a report with the findings and the results, including

1. The reluctance of several public administrations toward the participation in the mediation proceedings
2. The high success rate (around 75%) of the agreement achieved by putting the parties in condition to sit together at a negotiation table

The described results were particularly evident in cases introducing collaborative approaches in situations characterized by an ongoing trial. One of the cases mentioned in the report regarded a 15-year-long controversy between a producer of raw materials and three different public authorities for the aftermath of a sandpit dismissal and its future use.

Over time, public administrations had accrued a credit worth several million euros for environmental damage (partially recognized by a court decision) as well as administrative fines for irregular digging activities.

The case was particularly complex, both from a legal and technical point of view, and it involved the local community that claimed damage to their landscape.

Several legal actions had been initiated in courts and no solution was on the horizon. The parties had never met in person to discuss the problem.

The company was broken and could not satisfy the creditors, but it was open to alternative solutions.

During the mediation sessions, the parties needed technical help, which proved to be key for all parties involved, to understand the very complex issues on the table and why certain possible solutions were not feasible.

After 12 months of discussion and site visits, an agreement was reached.

The company took on the dismantling of industrial ruins and further remediation activities on the areas (15 years of abandonment played in favor of damage mitigation) and sold part of them to cover a portion of the debt. The slot transferred to the public administrations was totally cleaned up and made available to the local community.

A situation that had been pending in front of the state courts for years has eventually ended. Public money was saved; part of the area was made available to the local community for recreation activities creating value.

Over the years, the attitude of the public administration toward mediation has improved greatly, and nowadays, environmental conflict management involving the public sector is a common practice at CAM.

However, the prevention of environmental conflicts, which should be the most important aspect to address, marks time. Communication and information gaps, and in general a certain reluctance at involving local communities and stakeholders in the decision-making process, are still present. Many public administrators worry that consultations cannot be conducted in a profitable way due to the various educational and cultural backgrounds of the public or that the public meetings can be manipulated and distorted by some antagonist group. More on the project can be found at http://www.ius-publicum.com/repository/uploads/30_04_2019_18_41-Ius_Publicum_ENVIRONMENTALDISPUTES.pdf ("Encouraging Public Entities to settle environmental disputes through mediation in Italy, theoretical analysis, practical evidences, possible solutions" - Lea di Salvatore) and on the CAM web site "*La mediazione dei conflitti ambientali: linee guida operative e testimonianze degli esperti.*"

This is the reason why the Chamber of Commerce of Milan, Monza Brianza and Lodi, together with CAM, started a second project on environmental facilitation called "*FacilitAmbiente.*"

Facilitation shares with mediation some principles and techniques and sometimes the lines between the two tend to blur. However, facilitation is a method for stimulating productive and collaborative discussion and problem solving. In most cases, there is no conflict to deal with (as it happens in mediation).

FacilitaAmbiente aims at designing a structured model to be used for environmental conflict prevention using methodologies that are already available and widely used, but they are not formalized at a legislative level yet.

The "Rules" have been approved on March 29, 2021, and currently there is an open tender to recruit the professionals acting as facilitators in the project.

Any interested person, association, company, and public administration (stakeholders) can submit a request to start a dialogue and a facilitation process. The Secretariat of the service will conduct a preliminary interview with the proponent to get more details on the case, and it will choose the facilitator who best fits in the case. The facilitator will meet with the proponent to gather more information and will submit a "facilitation path." Should the proponent agree on the proposed activity plan, the facilitator will conduct the activities agreed.

The Secretariat will assist the parties and the facilitators in organizing, meetings, activities, and monitor the evolution of the case.

The facilitation project respects the principles of transparency and inclusiveness and works. Its outcomes can be made public albeit in special circumstances the parties can request full confidentiality over the process.

This project opens new possibilities and chances to find a common ground for the solution of environmental issues. The path is set. Now, it is the time to convince public and private operators of the convenience of dialogue and facilitation in the promotion of sustainability.

References

Camera Arbitrale di Milano. (2016). *La mediazione dei conflitti ambientali: linee guida operative e testimonianze degli esperti*. Available via CAM. Retrieved June 7, 2021, from https://www.camera-arbitrale.it/it/mediazione/mediazione-dei-conflitti-ambientali.php?id=524

Castiglioni, B., & De Marchi, M. (2009). *Di chi è il paesaggio? La partecipazione degli attori nella individuazione*. Coop. Libraria Editrice.

Di Salvatore, L. (2018). Encouraging public entities to settle environmental disputes through mediation in Italy, theoretical analysis, practical evidences, possible solutions. *Ius Publicum, 2*, 1–79.

European Commission for the Efficiency of Justice. (2019). *European handbook for mediation lawmaking*. CEPEJ(2019)9. Available via Mondoadr. Retrieved June 7, 2021, from https://www.mondoadr.it/wp-content/uploads/CEPEJ20199-EN_Handbook.pdf1_.pdf

Giampietro, L. (Ed.). (2018). *Conflitti ambientali: mediazione, transazione*. Wolters Kluwer Italia.

OECD. (2011). United Nations Conference on Straddling Fish Stocks and Highly Migratory Fish Stocks art. 27.

UN Interagency Framework Team for Preventive Action. (2012). *Renewable resources and conflict*. Toolkit and guidance for preventing and managing land and natural resources conflict. Available via UN. Retrieved June 7, 2021, from https://www.un.org/en/land-natural-resources-conflict/pdfs/GN_Renew.pdf

UNDPA & UNEP. (2015). *Natural resources and conflict*. A guide for mediation practitioners. Available via UNEP. Retrieved June 7, 2021, from https://wedocs.unep.org/bitstream/handle/20.500.11822/9294/-Natural_resources_and_conflic.pdf?sequence=2&amp%3BisAllowed=

UNEP. (2009). *From conflict to peacebuilding*. The role of natural resources and the environment. Available via UNEP. Retrieved June 7, 2021, from https://postconflict.unep.ch/publications/pcdmb_policy_01.pdf

UNEP. (2019). *Guidelines for conducting integrated environmental assessments*. Available via UNEP. Retrieved June 7, 2021, from https://www.unep.org/explore-topics/environment-under-review/what-we-do/environmental-assessment/guidelines-conducting

Roberta Regazzoni has a law degree and has been working for the Milan Chamber of Arbitration—a commercial dispute resolution provider—for 20 years now. The focus of her studies and expertise is commercial mediation and dialogue-based dispute resolution processes. She participates in international roundtables focused on alternative/appropriate dispute resolution and gives lectures on the topic.

Index

A
Acidification, 97
Actual climate movements, 13
Adaptation and mitigation strategies, climate change, 6, 98–99
  adaptation options, 294
  AFOLU, 295, 296
  agricultural farming practices, 295
  carbon and nitrogen management, 296
  crop varieties, 295
  determinants, 294
  higher connection through railroads, 295
  limiting food waste and losses, 296
  market integration, 295
  new technologies, 295
  pace consideration, 294
  R&D investments, 295
  twofold objective, 296
Agricultural activities, 321
Agricultural (and nonagricultural) economic outcomes, 307
Agricultural infrastructure, 308
Agricultural productivity determinants, 291
Agricultural yield gap, 294
Agriculture
  climate change in, 6
  GDP growth, 290
  primary source, food and income, 89
Agriculture, forestry, and land used (AFOLU), 295, 296
Albedo, 82–83
Alpine glaciers, 73, 85
Animal-based protein foods, 323
Antarctica, 82–83
Artificial Intelligence, 164
Asymmetric information, 325

B
Backloading, 152
Behavioral intention, 317
Behaviorally-informed and education-based policies, 326
"Behaviorally informed" policies, 317
Blockchain, 164
Border Carbon Adjustment (BCA) mechanism, 153
"Bottom-up" factors, 19
Brisbane Summit, 171
Buenos Aires Summit, 173
Business as usual scenario, 293
Butterfly effect, 49

C
Camp David Summit, 168
Cannes Summit, 170
Carbon Border Tax, 153
Carbon capture, utilization, and storage (CCUS) technologies, 164, 167–168, 174, 194, 265
Carbon dioxide (CO_2) emission, 134, 142, 154, 160, 163, 164, 187, 188, 190, 225, 228, 238, 263, 275, 278
Carbon footprint, 316, 322–324
Carbon-intensive activity, 322
Carbon leakage, 146, 150–154

Carbon-neutral countries
    green structures, 275
    mega projects, mega impact, 277
    smart carbon-neutral cities, 277–278
    sustainable architecture, 274–275
    sustainable land-use planning, 276–277
    sustainable transport, 276
    urban planning, 276
Carbon Offsetting and Reduction Scheme for International Aviation (CORSIA), 150
Carbon pricing, 4
    ETS
        allowance allocation, 151–152
        European Green Deal, 153–154
        international experience, 149
        origins, 150
        oversupply, 152
        price containment, 152
        prices, 152
        reform, 150
        scope and coverage, 150
        tradable permits, 148–149
    price-based vs quantity-based regulation, 154–155
    tax (see Carbon tax)
Carbon tax
    benefits, 145–147
    costs, 145–147
    economic rationale, 144
    European experience, 147–148
    international and institutional level, 143
    optimal structure, 144
    tax base and rates, 144–145
Carbon tax rate, 144–145
Carbon-trading, 101
Centralization vs. decentralization, 236–237
Central Karakoram National Park (CKNP) Glacier Inventory, 74–75
Ceteris paribus, 299
Chamber of Commerce, 371, 372
Charlevoix Summit, 169
China's National Major Science and Technology Projects program, 164
Choice architecture, 327
Chronic hunger and malnutrition, 307
"CIRCLE-2" project, 117
Circular Carbon Economy (CCE) framework, 174
Circular Economy Action Plan (CEAP), 253
Circular economy, EU
    circular economy regulatory measures, 254–256
    circular model, 253
    regulatory trends, 256–257

Circular economy regulatory measures, EU
    EU directives, 254
    landfill directive, 255
    packaging and packaging waste directive (PPW Directive), 255, 256
    plastic bags directive (PB Directive), 256
    single use plastic directive (SUP Directive), 256
Classical science, 45, 47
Clean Energy, 167
Clean Energy Ministerial (CEM), 167
Climate, 6
Climate action tax credit, 146
Climate change, 2, 3, 63
    in agriculture, 6
    and consumer behavior, 7
    economic behavior, 290
    environmental policy, 290
    human progress, 290
    rural revival and coastal areas, 93–95
        exposure, 96–97
        hazards, 95–96
        vulnerability, 98
    and sustainability
        climate skepticism and right-wing populism, 17–21
        decision-making process, 12
        European Investment Bank, 14
        green, 13–14
        protest movement's ideology, evolution of, 15
        science comprehension, 13
        scientific community approval, 17
        scientific data, 12
        students, 16
        United States and European Union, 14
Climate Change Adaptation Program, 218
Climate change impacts
    adaptation and mitigation strategies (see Adaptation and mitigation strategies, climate change)
    agricultural land expansion, 293
    agricultural yields, 294
    average yield gap, 293
    CO_2 concentration, 291
    economic costs estimation (see Economic costs estimation, climate change)
    environmental consequences, 293
    environmental damages, 292
    Fifth Assessment Report of the IPCC, 291
    food security, 292
    food supply loss, 292
    precipitation patterns, 291
    rise in temperatures, 291

Index 377

semiarid and tropical zones, 291
socioeconomic factors, 292
threats agricultural and food production, 292, 293
undernourished people, 293
water availability, 293
weather-related natural disasters, 291
Climate change litigation, 7
  activists and environmental associations, 335
  climate lawsuits, 334
  communities, 342
  environmental code, 340
  environmentalists' claims, 335
  in Europe, 338
  feature, 335
  global warming, 338
  governments, 334
  greenhouse gas emissions, 337
  IPCC, 336
  in Italy, 337
  judicial protection, 342
  and movements, 335
  scientific evaluations and policies, 343
  techniques and instruments, 340
  temperatures, 342
  three-dimensionality, 338
Climate crisis, 28
Climate deniers, 114
Climate emergency, 28
Climate Finance, 130, 138
Climate impacts and mitigation policy heterogeneity
  ceteris paribus, 299
  climate policy, 298
  damage functions, 297
  development policy strategy, 298
  economic damage, 297
  exposure, 297, 298
  fundamental dimensions, 297
  linear temperature damage function, 298
  mitigation policy, 297
  nonlinear relationship, 297, 298
  socioeconomic characteristics, 297
  vulnerability, 297, 298
Climate inertia, language in
  global trend, climate change debate, 30–31
  La Repubblica, metaphors in, 33
    illness metaphor, 36–37
    journey metaphor, 37–38
    war metaphor, 34–36
  metaphors, living in, 31–33
Climate justice principles, 346
Climate Law, 316

Climate skepticism, 13, 17–19
  denial countermovement, 20–21
  right-wing ideology, 19–20
Climate, sustainability, and waste
  economic and waste regulation models, 248–249
  interconnection between, 247–248
  waste by numbers, 246–247
Climate Sustainability Working Group (CSWG), 175
Climate variation, 83
Climatic variables, 303
Coastal areas, rural revival and challenges, 102
  climate change, 93–95
    exposure, 96–97
    hazards, 95–96
    vulnerability, 98
  coastal systems, adaptation of, 98–99
  rural communities, 94, 100–102
Coastal engineering, 102
Coastal systems, adaptation of, 98–99
Cognitive systems, 320
"Command and control" policies, 142
Communication, 28
Complexity
  socio-economic phenomena
    causal explanation, 46
    circular relationship, 49
    classical science, 45, 47
    contextuality, 45
    experimental method, 46
    innovations, 49
    linearity, 46
    negentropy, 48
    non-equilibrium, 48
    plexum, 44
    reductionism, 47
    relationship with knowledge, 46
    stylization of reality, 45
    thermal equilibrium, 48
  and systems, 50–51
Conceptual metaphors, 32
Conference of the Parties (COP), 128, 130, 136, 167
Consensus, 108, 109, 114
Consensus factory, 108
Construction industry, 262
  energy efficiency, 266
  environmental impact
    designing for reuse, 273
    life-cycle requirements, 271–272
    reusing existing buildings, 272–273
    reusing materials, 273–274

Construction industry (*cont.*)
    fights pollution
        bio-carton, 271
        cleaning up, 270
        smog-eating concrete, 271
        smog-eating photo-catalysis, 270–271
    renewable energy projects, 265
    smart energy infrastructure, 265–266
    strategic planning, 264–265
    sustainability, 262–263
    sustainable water use
        construction of water infrastructure, 267–268
        flood protection, 269
        integrated resource management, 267
        reuse, 268
        sustainable water infrastructure, 266–267
        water and energy, 268–269
Construction industry and sustainability, 6
Consumer awareness, climate change, 317–318
Consumer behavior, 7, 326, 327
Consumption activities, 321
Consumption habits, 323
Contextuality, socio-economic phenomena, 45
Copenhagen negotiations, 129
Copernicus, 4, 127
Corporate Average Fuel Economy (CAFE) standards, 206
Country-specific socioeconomic factors, 294
COVID-19, 116, 127, 128
    European Green Deal, 131–133
    technology innovation in energy sector, 164, 165
Credence attributes, 324
Cross-sectional Ricardian studies, 303
Cultural beliefs, 31

D

Damage function estimation methods
    empirical estimation, 301–303
    empirical evidence, 303–306
    formal model, 300–301
    long-differences approach, 299, 300
    panel data models, 299
    Ricardian models, 299
Deauville Summit, 168
Debate, 108, 110, 111, 113, 115
Decision-making process, 319
Demand-side policies, 317
    information and education-based policies, 325–326
    information provision, 325
    nudging-based policies, 326–327
    product labeling, 327
Democracy, 107, 110, 113–115, 121
Denial countermovement, 20–21
Denialism, 115
Development policy strategy, 298
Dichloro-diphenyl-trichloroethane (DDT), 64
Digital elevation models (DEMs), 80
Dimensions, 55, 57
Dispute prevention, 365
"Domain dependence" issue, 318
Double dividend, 146
Dutch government, 343

E

Earth Summit of Rio de Janeiro, 143
Economic costs estimation, climate change
    climate econometric literature, 296
    economic behavior, 297
    economic outcomes, 296
    environmental regulations, 296
    heterogeneity (*see* Climate impacts and mitigation policy heterogeneity)
    modern panel data econometrics, 296
    temperature–productivity relationship, 296
Economy-wide damage function, 304
Ecosystem, 96
Ecosystem services, 100
Efficient energy transition
    electricity decarbonization
        DG through community, 191
        government buildings, 191
        municipalization, 190
        public-owned utilities, 190
        solar and storage microgrids, 192
        vertically-integrated electric utilities, 190
    electrification
        improving transit, 192
        vehicle electrification, 193
    energy efficiency and circularity
        building energy performance standards, 187–188
        energy building codes, 186–187
        energy consumption buildings, 187–188
        low-and moderate-income residents, 187
        managing traffic, 188–189
    hard-to-electrify sectors
        carbon capture, utilization, and storage (CCUS), 194
        hydrogen cities, 193–194

Effort-sharing legislation, 133–134
Electricity decarbonization
　DG through community, 191
　government buildings, 191
　municipalization, 190
　public-owned utilities, 190
　solar and storage microgrids, 192
　vertically-integrated electric utilities, 190
Electrification
　improving transit, 192
　vehicle electrification, 193
Elmau Summit, 169
Emission trading system (ETS), 4, 143, 145, 148, 149, 155
　allowance allocation, 151–152
　European Green Deal, 153–154
　international experience, 149
　origins, 150
　oversupply, 152
　price containment, 152
　prices, 152
　reform, 150
　scope and coverage, 150
　tradable permits, 148–149
Emotions, 31
Empirical estimation, damage functions
　ceteris paribus, 302
　cross-sectional research design, 302
　cross-sectional Ricardian approach, 301
　dependent variable, 302
　desired damage effect, 302
　empirical studies, 302
　hybrid approach, 302
　long-run adaptation, 302
　long-term damages, 301
　panel data approach, 302
　precipitations, 303
　time-invariant unobserved characteristics, 302
　unit of observation, 301
　variable bias, 303
Empirical mindset, 53
Energy, 135
Energy efficiency and circularity
　building energy performance standards, 187–188
　energy building codes, 186–187
　energy consumption buildings, 187–188
　low-and moderate-income residents, 187
　managing traffic, 188–189
Energy sector, 14
Energy storage, 164
Energy Sustainability Working Group, 171
Energy taxation, 147, 148, 154

Energy transition policies
　challenges, 177
　G7/G8 Summits
　　Camp David Summit, 168
　　Charlevoix Summit, 169
　　Deauville Summit, 168
　　Elmau Summit, 169
　　Gleneagles Summit, 167
　　Hokkaido/Toyako Summit, 168
　　Ise Shima Summit, 169
　　L'Aquila Summit, 168
　　Lough Erne's Summit, 168
　　Muskoka Summit, 168
　　St. Petersburg Summit, 168
　　Taormina Summit, 169
　G20 Summit, 167
　　Brisbane Summit, 171
　　Buenos Aires Summit, 173
　　Cannes Summit, 170
　　CCE framework, 174–175
　　CSWG, 175
　　ETWG, 175
　　G20 energy ministerial meeting, 174
　　Istanbul Summit, 171
　　Pittsburgh Summit, 170
Energy Transition Working Group (ETWG), 175
Environmental conflict
　definition, 366
　dynamics, 366
　facilitation, 372
　infrastructure construction projects, 369
　judicial proceeding, 368
　prevention, 372
　public administrations, 370
　scholars, 366
　situations, 368
Environmental economists, 142
Environmental judicial review
　administrative standards, 352
　auto-generate and substitute, 352
　laws and regulations, 352
　sustainable development, 351
　transversal characters
　　administrative action, 358
　　administrative decisions, 357
　　administrative law, 356
　　characterization, 357
　　flexibility, 358
　　power principle, 357
　　subjective characterization, 358
　　trias politica, 357
Environmentally friendly food products, 321
Environmental sustainability, 293

e-Procurement, 237–238
Erosion, 95
EU Adaptation Strategy, 134
EU and US regulatory approaches, 248
EU directives, 254
EU Emissions Trading System (EU ETS), 133
EU-harmonized sustainable food labeling framework, 324
Eurobarometer survey, 321
Euro-Mediterranean Centre, 336
European climate legal proceedings, 338
European Coal and Steel Community, 131
European Convention on Human Rights, 339
European Environmental Agency (EEA), 117, 316
European Environment Information and Observation Network (Eionet), 117
European Free Trade Association States, 150
European Glacier Inventory, 72–74
European Green Deal (EGD), 2, 14, 128, 148
    adaptation to climate change, 134
    COP, 128
    Copenhagen negotiations, 129
    COVID-19 impact, 131–133, 138–139
    Effort-Sharing, 133–134
    energy, 135
    EU ETS, 133
    European Green Deal Call, 137
    Glasgow, 137–138
    Kyoto Protocol, 129
    LULUCF, 134
    next-generation EU and Paris, 131
    Paris, 129–130
    research and innovation, 135–136
    2030 Climate and Energy Framework, 130
European Green Deal Call, 137
European Investment Bank, 14
European research, 116, 136
European Southern regions, 304
European Space Agency (ESA), 72
European Union (EU), 245
    circular economy regulatory measures, 254–256
    circular model, 253
    regulatory trends, 256–257
European Union Allowances (EUAs), 150
European Union Emission Trading System (EU ETS), 145, 147
Eurozone, 131
EU waste framework directive, 257
Evidence-based policy, 110
Evidence-informed policy, 110
Explicit attitudes, 320
Extended producer responsibility, 255

F

FacilitaAmbiente, 372
Facilitation project, 373
Farm-2 Fork Strategy, 324
Fast-food-based eating, 323
Federal Energy Regulatory Commission (FERC), 183
Finance track, 170
Flexible nonlinear damage functions, 308
Flooding, 95, 96
Food and Agriculture Organization (FAO), 322
Food-insecure people, 307
Food intake, 289
Food overconsumption trends, 316
Food production and climate change
    agricultural sector, 321
    choosing sustainable product alternatives, 323–324
    day-to-day consumption decisions, 322
    ecosystem services, 321
    FAO, 322
    food-system activities, 321
    intensive agriculture, 321
    overconsumption trends, 322
    plant-based food products, 322
    sustainability issues, 321
    sustainable diet, 322
    unsustainable (m)eat, 322–323
Food production and consumption, 316
Food purchasing, 321
Food-related behaviors, 317
Formal model of climate change impact
    conceptual differences, 300
    conventional varieties, 300
    crop varieties, 300
    production functions, 300
    quadratic production technology, 300
    temperature shifts, 300
Forni Glacier, 65–68, 73
Freedom of information, 120
Future generations, 131, 248

G

G20 Climate and Energy Action Plan for Growth, 173
G20 Energy Efficiency Leading Programme (EELP), 172
GEO-Heritage, 85
German reunification, 131
GHG mitigation, 307
G20 Karuizawa Innovation Action Plan on Energy Transitions, 173
Glacier-darkening phase, 85

Index 381

Glaciers, 63, 292
  albedo variations, 82–83
  Antarctica, 82–83
  Forni Glaciers, 65–68
  hazards, 78–81
  ice of, 63
  inventories, 69
    European Glacier Inventory, 72–74
    Karakoram Anomaly, 74–76
    New Italian Glacier Inventory, 69–72
  melting, 66, 75, 85
  remote-sensing, 67–68, 82–83
  slow process causes, 64
  substances, 65
  techniques, 67
  temperature increase, 64
  unmanned aerial vehicles, 79–81
Glasgow, 137
Gleneagles Plan of Action on Climate Change, 167
Global agricultural system, 306
Global Energy Architecture, 172
Global Environment for Sustainable Growth, 173–174
Global warming, 63, 84, 306
Goals, 116–119
Go green, 2, 13
Great Recession, 152
Green, 13–14
Green and social procurement, 221–223
Greenhouse effect, 32
Greenhouse gas emissions (GHG), 129, 130, 135, 142, 160, 247, 316
Green hydrogen, 163, 193
Green recovery, 132, 133, 170, 175
Green stimuli, 165
Group of Seven/Eight (G7/G8) Summit, 167
  Camp David Summit, 168
  Charlevoix Summit, 169
  Deauville Summit, 168
  Elmau Summit, 169
  Gleneagles Summit, 167
  Hokkaido/Toyako Summit, 168
  Ise Shima Summit, 169
  L'Aquila Summit, 168
  Lough Erne's Summit, 168
  Muskoka Summit, 168
  St. Petersburg Summit, 168
  Taormina Summit, 169
Group of Twenty (G20) Summit, 167
  Brisbane Summit, 171
  Buenos Aires Summit, 173
  Cannes Summit, 170
  CCE framework, 174–175
  CSWG, 175
  ETWG, 175
  G20 energy ministerial meeting, 174
  Istanbul Summit, 171
  Pittsburgh Summit, 170
  G20 Toolkit, 172

H
Hazards, 95–96
Heritage, 85
Heterogeneous marginal effects, 308
Hokkaido/Toyako Summit, 168
Horizon Europe, 164
Housing-and mobility-related consumption, 316
Huge model uncertainty, 304
Human activities, 315
Human welfare, 289

I
Ideological bias, 17, 18, 20, 21
IEA International Low Carbon Energy Technology Platform, 168
IECC 2018 code, 182
Illness metaphor, 33, 36–37
Implicit attitudes, 320
Indigenous communities, 339
Industrial emission directive, 257
Information and education-based policies
  asymmetric information, 325
  behavioral changes, 325
  labeling, 325
  labels and logos, 326
  policy instrument, 326
  policy interventions, 325
  specific knowledge, 325
  System I and System II reasoning system, 326
Institutional engagement, 316
Intended Nationally Determined Contribution (INDC), 130
Intensive agriculture, 321
Intention–behavior gap
  definition, 318
  double reasoning system, 319–320
  endogenous and exogenous factors, 318
  exogenous factors, 320–321
  time preferences, 319
Intergovernmental Panel on Climate Change (IPCC), 3, 19, 28, 32, 38, 94, 95, 98, 111, 112, 115, 128, 131, 143, 160, 247, 291, 315, 336

International Civil Aviation Organization (ICAO), 150
International Energy Agency (IEA), 160, 166
International Energy Conservation Code (IECC), 182
Internationally Transferred Mitigation Outcomes (ITMOs), 156
International Maritime Organization (IMO), 169
International Partnership on Energy Efficiency Cooperation (IPEEC), 168
International Renewable Energy Agency (IRENA), 166
Internet of things, 164
Ipsos survey, 318
Ise Shima Summit, 169
Istanbul Summit, 171
Italian Glaciers, 73
Italian Glaciological Committee, 73

J
Journey metaphor, 33, 37–38
Judicial control, 113
Judicial decisions
  climate agenda, 346
  climate change, 343
  environmental policies, 345
  institutional awareness, 344
  mediation, 344
  policy change, 344
  preceding, 345
  sustainability theories, 345
Judicial protection, 341
Judicial review
  administrative decisions, 350
  characteristics, 350
  climate change, 349
  environmental claims
    administrative action, 354
    administrative actions, 356
    administrative denial, 353
    authority, 353
    definition, 353
    in environmental matters, 355
    human activity, 353
    proceedings, 356
  on environmental matters
    interpretation of the law, 358–359
    secondary judgment, 360–362
  flexibility, 358
  legal framework, 358
  public matters, 350
  public powers, 350
  public–private dynamic, 350
  on sustainable development, 362–363
Justification, 55

K
Karakoram Anomaly, 74–76, 84
Khurdopin Glacier, 76
Knowledge, complexity and
  causal explanation, 46
  circular relationship, 49
  classical science, 45, 47
  contextuality, 45
  experimental method, 46
  innovations, 49
  linearity, 46
  negentropy, 48
  non-equilibrium, 48
  plexum, 44
  reductionism, 47
  relationship with knowledge, 46
  stylization of reality, 45
  thermal equilibrium, 48
Kyoto Protocol (KP), 129

L
Labeling, 325
Labels, 325
Landfill directive, 255
Land use, land-use change, and forestry (LULUCF), 134
Language, climate inertia of
  global trend, climate change debate, 30–31
  information and knowledge, 28, 31
  La Repubblica, metaphors in, 33
    illness metaphor, 36–37
    journey metaphor, 37–38
    war metaphor, 34–36
  metaphors, living in, 31–33
L'Aquila Summit, 168
La Repubblica, 33
  illness metaphor, 36–37
  journey metaphor, 37–38
  war metaphor, 34–36
Latin American countries, 289
Life-cycle costing (LCC), 220–221
Linearity, 46
Linguistic, 31, 32
Livestock production, 322
Long-difference research design, 302
Long-differences approach, 299, 300

Index 383

Lough Erne's summit, 168
Low-carbon challenges, 307

M
Major Economies Forum (MEF), 168
Market consultation, 230–231
Market Stability Reserve (MSR), 152
Massachusetts v. Environmental Protection Agency, 359
Mass media, 120
Meat consumption, 323
Meat production, 323
Media, 28, 29
Mediation
  damage mitigation tool, 365
  dialogue, 367
  environmental disputes, 365
  in environmental disputes, 7
  and facilitation, 371
  feature, 367
  mediator, 367
  phases, 367
  settlements, 368
  supporting dialogue, 369
Metaphors
  in environmental issues, 31–33
  in *La Repubblica*, 33
    illness metaphor, 36–37
    journey metaphor, 37–38
    war metaphor, 34–36
Milan Chamber of Arbitration (CAM), 371, 372
Mission Innovation (MI), 166
Mitigation and adaptation policies, 4, 6, 116
Modeling uncertainty, 305
Moderate Resolution Imaging Spectroradiometer (MODIS) sensors, 82
Multiannual R&D funding program, 164
Multi-indicator approach, 57
Muskoka Summit, 168

N
NASA, 4, 82, 127
National energy and climate plans (NECPs), 135, 166
National Key R&D Projects, 164
Nationally determined contributions (NDCs), 156
Natural capital, 100
Negentropy, 48
Neoclassical economic theory, 319

Net-zero emissions, 183
"New global" movement, 15
New Italian Glacier Inventory, 69–72
Next-Generation EU, 129, 131, 133
1998 Italian Carbon Tax project, 147
Nitrous oxide (N_2O) emissions, 150
Nominal definition, 53
Nonlinear weather variables, 303
Not In My Back Yard (NIMBY), 366
Nudges
  aim, 326
  behaviorally informed, 326
  characteristics, 326
  definition, 326
  reminders, warnings, salient disclosures, 327
  responses, 326
  sustainable choices, 327

O
Occupy, 15
Ocean acidification, 292
Operational definitions, 53
Operationalization, 55, 56
Orthophotos, 80
Ortles-Cevedale glaciers, 74
Out-of-sample prediction, 304

P
Packaging and packaging waste directive (PPW Directive), 255, 256
Pakistan-Italian Debt for development Swap Agreement (PIDSA), 74
Panel data models, 299, 305
Paris Climate Agreement, 18, 31, 36, 128, 129, 131, 132, 137, 138, 156, 166, 169, 173, 262, 307, 316, 342
Perfluorocarbon (PFC) emissions, 150
Pittsburgh Summit, 170
Plant-based diets, 322
Plastic bags directive (PB Directive), 256
Platonic hyperuranium, 46
Polarization, 20, 22
Policy implications, 306
Policy makers
  congestion pricing, 204–205
  inequitable outcomes, 203–204
  land use leaders, 203
  leaders in jurisdictions, 203
  mileage/kilometer fees, 204–205
  transit infrastructure, 203
Political determination, 108, 121

Politics, 307
Politics and science
　awareness, 119–120
　climate deniers, 114
　complexity, 112–116
　convergence, 107–109
　data, 112
　democracy, 107, 110, 113–115, 121
　denialism, 115
　evidence-based policy, 110
　evidence-informed policy, 110
　friction, 110
　global warming and pollution, 110
　identification and implementation of goals
　　"CIRCLE-2" project, 117
　　COVID-19, 116
　　EEA, 117
　　Eionet, 117
　　European Union, 116–117
　　mitigation and adaptation policies, 118
　　treaty, 118
　IPCC, 111
　judicial control, 113
　knowledge, 119–120
　lack of consequentiality, 115
　measurement of phenomena, 119
　similarities, 113–114
　technical proposals, 112
　tension, 110
Polluter pays principle, 254
Pollution, 64
Population-growth dynamics, 292
Posidonia meadows' stocks, 101
Potential distributional effects, 308
Power, 108
Princeton University, 108
Province of Sondrio, 86
Public interventions, 325
Public policies, 4, 6, 117, 120, 121, 162, 333, 335, 336, 343
Public–private dynamic, 350
Public procurement
　active and strategic tool, 219
　circular economy, 221
　command and control, 235
　economical level, 218
　e-Procurement, 237–238
　EU Green New Deal Communication, 219
　gender qualities, 232–234
　green and social procurement, 221–223
　life-cycle costing (LCC), 220–221
　market consultation, 230–231
　minorities, 234–235
　program, and monitoring system, 238
　purchasing authorities
　　design evaluation, 229–230
　　evaluation of quality, 227–229
　　fixed cost, 229
　　lowest price criterion, 226–227
　purchasing authority, 223
　small-and medium-size
　　enterprises, 231–232
　supply chain, 220
　transition enhancement, 224–225
　young professionals, 234
Purchasing authorities
　design evaluation, 229–230
　evaluation of quality, 227–229
　fixed cost, 229
　lowest price criterion, 226–227

Q
Quadratic reaction function, 304

R
Random error, 52
RCRA's cooperative framework, 251
R&D investments, 294
Reductionism, 47
Regional Greenhouse Gas Initiative (RGGI), 149, 156
Regional transmission organization (RTO), 183
Remote-sensing, 67–68, 82–83
Renewable energy, 163
Representative concentration pathway (RCP), 304
Research and innovation, 135–136
Resource Conservation and Recovery Act (RCRA), 250, 251
Resource-efficient competitive economy, 316
Response function, 306
Ricardian models, 296, 299, 303
Right-wing populism, 19–20
Rural communities, 94, 100–102
Rural revival and coastal areas
　challenges, 102
　climate change, 93–95
　　exposure, 96–97
　　hazards, 95–96
　　vulnerability, 98
　coastal systems, adaptation of, 98–99
　rural communities, 94
　　coastal natural capital and opportunities for, 100–102

S

Santer, Ben, 115
Scholarship, 1
"Science for all," 120
Scientific research, 55, 107, 108, 117, 119
Scientists' social responsibility, 109
Sea ice, 64
Sea-level rise, 95
17 National Laboratory system, 164
Sherpa track, 170
Shingchukpi Glacier, 75
Short-run reaction function, 306
Single indicator approach, 57
Single market, 131
Single use plastic directive (SUP Directive), 256
Social loafing, 318
Social science, measurement, 52–54
Socio-economic phenomena, 43
    complexity and knowledge
        causal explanation, 46
        circular relationship, 49
        classical science, 45, 47
        contextuality, 45
        experimental method, 46
        innovations, 49
        linearity, 46
        negentropy, 48
        non-equilibrium, 48
        plexum, 44
        reductionism, 47
        relationship with knowledge, 46
        stylization of reality, 45
        thermal equilibrium, 48
    complexity and systems, 50–51
    measurement, 54–57
    social science, measurement in, 52–54
Soft power of numbers, 57
Solar energy adoption, 321
Solid Waste Disposal Act (SWDA), 251
South Asia regions, 292
St. Petersburg Summit, 168
Strategic litigation, 333, 368
Sub-Saharan Africa (SSA), 292, 306
Sustainability, 6
    climate skepticism and right-wing populism, 17–21
    decision-making process, 12
    European Investment Bank, 14
    green, 13–14
    protest movement's ideology, evolution of, 15
    science comprehension, 13
    scientific community approval, 17
    scientific data, 12
    students, 16
    United States and European Union, 14
Sustainability-involved consumers, 321, 327
Sustainability labeling, 324
Sustainability labels, 324, 327
Sustainability-oriented behavior, 319
Sustainability-related decisions, 319
Sustainability-related literature, 318
Sustainable behaviors, 317
Sustainable construction enablers
    digital technologies, 282–283
    education and training, 283
    financial instruments, 280
    lean thinking, 283
    regulations, 278–279
    sustainability standards, 279–280
    sustainable procurement, 283–284
    voluntary certification, 279–280
Sustainable development, 13
Sustainable development goal (SDG), 290
Sustainable diet, 322
Sustainable food production and consumption, 316
Sustainable global economic growth, 307
Sustainable product alternatives, 323–324
Sustainable product policy legislative initiative, 257
Sustainable transportation, 5
    addressing climate change, 199–200
    battery electricity vehicles, 208–209
    hydrogen, 207–208
    improve fuel economy, 205–206
    low-carbon liquid biofuels, 207
    policy makers
        congestion pricing, 204–205
        inequitable outcomes, 203–204
        land use leaders, 203
        leaders in jurisdictions, 203
        mileage/kilometer fees, 204–205
        transit infrastructure, 203
    reduce carbon content, 205–206
    transportation emissions, 201–202
    zero-emission fuels and transportation
        electricity rate reform, 213–214
        incentives, 211–212
        mandates, 210–211
        regulatory reform, 213–214
        subsidies, 212–213
Systematic error, 53

T

Taormina Summit, 169
Tax base, 144

Technology Collaboration Programme (TCP), 166
Technology innovation, 5
　distributed generation, 185–186
　emerging technologies, 184
　energy storage, 185
　smart grids, 184–185
Technology innovation in energy sector
　Covid-19 pandemic, 164, 165
　energy transition policies (*see* Energy transition policies)
　governments role, 161–163
　importance, 159–161
　international cooperation on technology innovation, 165–167
Terminology and data, 246
"The Future of Hydrogen," 174
Thermal equilibrium, 48
"3E + S" approach, 173
Time preferences, 319
Top-down social strategy, 19
Tourism, 95
Traditional linear economy, US
　linear model, 249–250
　state and local waste regulation, 251–252
　US Resource Conservation and Recovery Act, 250–251
　waste laws, 250
　waste management trends, 252
2020 CEAP outlines, 256
2030 Climate and Energy Framework, 130
2021 Global Risks Report, 128

U

United Nations (UN), 289
United Nations Economic Commission for Europe (UNECE), 370
United Nations Framework Convention on Climate Change (UNFCCC), 129, 130, 143, 156, 167
United Nations Framework Convention on Climate Change's Paris Agreement in 2015, 307
United States (US), 245
　traditional linear economy
　　linear model, 249–250
　　state and local waste regulation, 251–252
　　US Resource Conservation and Recovery Act, 250–251
　　waste laws, 250
　　waste management trends, 252
Unmanned aerial vehicles (UAVs), 79–81
Unsustainable (m)eat, 322–323
Urbanization, 289
U.S. Advanced Research Projects Agency–Energy (ARPA-E) program, 164–165
US Environmental Protection Agency (US EPA), 247, 257
US EPA policy, 252
US Resource Conservation and Recovery Act, 250–251
US *vs.* EU regulatory approaches, 257–259

V

Variable sectoral labor productivity, 306
Voluntary Action Plan on Renewable Energy, 172
Voluntary green claims, 324
Vulnerability, 98, 297

W

War metaphors, 33–36
Waste framework directive, 254
Waste hierarchy, 254
Waste management projects, 269–270
Weather variables damage function, 304
"Western" diet, 322
Westernized meat, 323
World Bank, 15, 132, 156, 167, 218, 370
World Economic Forum, 128
World Trade Organization (WTO), 147
Worse-scenario RCP, 304, 305

Z

Zero-emission, 170
Zero-emission fuels and transportation, 5
　electricity rate reform, 213–214
　incentives, 211–212
　mandates, 210–211
　regulatory reform, 213–214
　subsidies, 212–213

CPSIA information can be obtained
at www.ICGtesting.com
Printed in the USA
BVHW052301230223
659102BV00005B/149

Lakewood Memorial

Book One of The Memorial Trilogy

Robert R Best

Copyright © 2009 by Robert R Best. All rights reserved.

Second Edition, 2012

ISBN-10: 1479352349
ISBN-13: 978-1479352340

Without limiting the rights under copyright reserved above, no part of this publication may be reproduced, stored, or introduced into a retrieval system, or transmitted in any form, or by any means (electronic, mechanical, photocopying, recording or otherwise) without the prior written permission of the copyright owner, except in the case of brief quotations embodied in critical articles and reviews.

This book is a work of fiction. People, places, events and situations are the product of the author's imagination. Any resemblance to actual persons, living or dead, or historical events, is purely coincidence.

Cover painting by Cara Crocker, who needs a website because her art is cool and the world needs to know.

Title font used is "Zombified," by Chad Savage: www.sinisterfonts.com.